The Particle Century

THE PARTICLE CENTURY

EDITED BY GORDON FRASER

Routledge
Taylor & Francis Group

LONDON AND NEW YORK

First published 1998 by Institute of Physics Publishing

2 Park Square, Milton Park, Abingdon, Oxfordshire OX14 4RN
52 Vanderbilt Avenue, New York, NY 10017

Routledge is an imprint of the Taylor & Francis Group, an informa business

First issued in paperback 2019

British Library Cataloguing-in-Publication Data

A catalogue record for this book is available from the British Library.

ISBN 978-0-7503-0543-3 (hbk)
ISBN 978-0-367-40070-5 (pbk)

Library of Congress Cataloging-in-Publication Data are available

Typeset in LaTeX

Contents

CONTENTS

INTRODUCTION

The discovery of the electron in 1897 heralded a new conception of the small-scale structure of matter in terms of subatomic particles. Unravelling the details of this microscopic world in the twentieth century—the 'Particle Century'—has revolutionized our picture of the Universe and led to a totally new understanding of the oldest question of all: how was the Universe created?

Towards the end of the nineteenth century, physicists believed that our surroundings could be described as an assembly of more or less familiar objects (particles, fluids, etc) moving about in space and in time in accordance with definite laws of force, described by force 'fields'. From these ideas it was easy to build up a mental picture. We now know that Nature is much more subtle and sophisticated. To understand the Universe now requires a special effort: preconceptions from years of everyday experience have to be discarded and unfamiliar new ideas adopted instead. The reward is a new depth of understanding. We are able to comprehend what happens far out in space and what happened far back in time, and at the same time to appreciate the inner workings of a minute subnuclear world equally remote from our own experience.

At the dawn of the twentieth century, new discoveries quickly began to erode the neat picture inherited from nineteenth-century physics. Soon after the discovery of the electron, the advent of quantum theory suggested that Nature behaves in a different, and very unfamiliar, way at this ultra-microscopic level. In the first few decades of the twentieth century, the logical foundation of basic physics underwent a major overhaul. Quantum mechanics caused the philosophy of the microworld to be completely reappraised. Meanwhile relativity showed that understanding the large-scale structure of the Universe also required radical thinking. Everyday ideas cannot be applied to physics on very large and very small scales without introducing what appear to be paradoxes and irrelevancies.

These advances went hand in hand with fresh knowledge of first the subatomic and then the subnuclear domains. In the 1920s, the atom was known to be composed of electrons quantum mechanically 'orbiting' around a tiny central nucleus, itself composed of protons. Subsequently, this nucleus was revealed to contain neutrons as well as protons.

While the atom could be understood as a miniature planetary system held together by electromagnetic force, what held the nuclear protons and neutrons together, overcoming the tremendous electrical repulsion between positively charged protons so close to each other? This strong nuclear force was attributed to a heavy carrier particle, postulated by Hideki Yukawa in the 1930s. If this new subnuclear particle could be found, physicists believed, then nuclear forces would be understood.

To probe the atomic nucleus, to 'see' its constituent particles and to study how they interact needed fast-moving projectiles capable of penetrating nuclear matter. At first this was done using radioactive debris, but eventually physicists exhausted all experimental possibilities using the particles emitted in natural radioactivity. Further insights needed artificial means of accelerating particles to probe deeper into the nucleus. To provide these fast-moving particles requires high energies, and 'high-energy physics' soon became a synonym for this branch of science.

Fortunately, Nature has its own source of high-energy particles: cosmic rays. For many years these extraterrestrial particles were the main source of such nuclear probes and provided a steady stream of discoveries. Owen Lock in his chapter 'Cosmic rain' describes this romantic period when physicists frequently had to transport their detectors to the tops of mountains.

Then a new era opened. The nuclear and microwave physics advances of the Second World War showed physicists the value of major collaborations, while the same collaborations had also shown the politicians how useful physics could be. Major plans were drawn up to build large new particle accelerators at existing laboratories, and to create new laboratories for even larger accelerators. This increase in government support for basic physics is described in a historical overview by Catherine Westfall and John Krige, who cover the USA and western Europe respectively. (An editorial addition summarizes what happened in parallel in the Soviet Union and Japan.)

However, this considerable post-war investment needed several years before it bore its first fruits, and in the intervening period in the late 1940s cosmic rays continued to provide the discoveries. The first was the π meson or pion. Excitement mounted as physicists speculated that here was the carrier particle predicted by Yukawa as the key to nuclear forces. Even while the new large particle accelerators were being built, cosmic-ray discoveries found new subnuclear states whose existence was totally unexpected. Once the new particle accelerators were in action in the 1950s, they brought a rich harvest of exotic new particles, covered by Nicholas Samios in 'First accelerator fruits'. Initially physicists were excited to find such a rich field. New particles and resonances, it seemed, were everywhere but, with so many new subnuclear states turning up, new questions had to be asked. Why was the subnuclear world so complex? What was behind all these new particles?

In the same way as the periodic table of elements in the nineteenth century had suggested underlying atomic regularities, the new experimental bonanza required some kind of basic pattern, or symmetry, to described the exotic subnuclear tapestry. This quest and its outcome is vividly described by Yuval Ne'eman, co-discoverer of SU(3) symmetry, in 'The three-quark picture'.

The multitude of subnuclear particles could be explained in terms of just three basic elements, the famous quarks of Murray Gell-Mann. The experimentally observed particles could be interpreted as the different ways in which quarks could be put together.

But were these quarks just a mathematical facet of symmetry, or were they real particles, a still lower level in the subdivision of Nature? In their chapter on 'structure', Jerome Friedman and Henry Kendall recall how in the late 1960s and early 1970s a series of milestone experiments established that quarks do exist deep inside protons and neutrons. (For their work, Friedman and Kendall, together with Richard Taylor, were awarded the 1990 Nobel Physics Prize.)

With the establishment of this quark picture, attention turned to the inter-quark force. While quarks lurk deep inside subnuclear particles, under ordinary conditions they cannot exist by themselves. Quarks are locked together in subnuclear particles, and the individual quarks cannot be broken off. What makes quarks stick so tightly together and behave in this way?

Instead of acting between subnuclear particles and being carried by pions, as physicists had supposed in the days of Yukawa, the primary force is now understood to be due to gluons shuttling backwards and forwards between quarks. The gluon carrier of the inter-quark force plays a role analogous to that of the photon in electromagnetism, but the inter-quark force is very different. Paradoxically, the more effort is expended in prising a quark away from its partners, the tighter the retaining force becomes. Probing the behaviour of particles which cannot exist in their own right and are permanently confined calls for considerable experimental ingenuity. Closely confined sprays ('jets') of particles reflect the inner quark and gluon mechanisms. These continuing efforts to understand the quark force and its gluon carriers are described by Sau Lan Wu in 'Quark glue'.

However, this quark binding of the strong nuclear force is not the whole story. In the early years of the twentieth century, it had become clear that some nuclei were unstable. Beta decay released electrons, particles not normally associated with the nucleus. Even the neutron could decay this way. What was behind this special kind of nuclear instability? Dominated under laboratory conditions by the strong nuclear force and even by electromagnetism, the feebleness of this beta decay force is deceptive. The emergence of this 'weak force' as a major player in Nature's complex subnuclear scenario is covered in the chapter by physicist and science writer Christine Sutton. Ahmed Ali goes on to describe how quarks are not stable and can transform into each other under the action of the weak force. This aspect of the weak force is only now beginning to be explored in depth. In striving to make sense of these weak interactions, in the 1930s and 1940s, theorists pointed out curious but compelling parallels between these phenomena and the familiar theory of electromagnetism. These analogies hinted that the two forces, at first sight so different, might have some common origins. Just as electromagnetism is carried by the photon, bold theorists postulated that weak interactions might be mediated by carrier particles. However, while the photon is massless and can be transmitted across long distances, the weak-interaction carriers would have to be heavy to ensure that the weak force has its characteristic limited range.

The standard tool of theoretical particle physics is field theory. As Yuval Ne'eman describes in his chapter, a length of pipe is invariant under rotations about the pipe's axis of symmetry. Such a pipelike symmetry, where the rotation is the same for all points along the pipe, is said to be 'global'. However, a soft rubber pipe can be rotated by different amounts at different points along its length. The symmetry is now 'local'; the pipe still looks the same, but this apparent symmetry has distorted the material of the pipe, and this distortion produces stresses in the pipe. Such local effects, called 'gauge symmetries' as they can be applied according to any arbitrary gauge, give rise to forces.

In his elegant chapter 'Gauge theory', Martinus Veltman explains how these theories are constructed, and in particular how the requirement of 'renormalization' (that they are mathematically well behaved and capable of providing consistent results) leads to incisive explanations of how Nature works. Nature, it appears, wants to be renormalizable. The realization in 1971 that new gauge theories could be made renormalizable was a milestone in understanding. At about the same time, physicists saw how these theories could explain the apparent paradox that quarks, while permanently confined inside nuclear particles, can nevertheless approximate to free particles when probed by very-high-energy beams—the concept of 'asymptotic freedom'.

One such outcome is the 'neutral current' of the weak interactions, in which quarks can interact, as in weak interactions, but without switching electric charges. This unfamiliar new manifestation of subnuclear forces, seen for the first time at CERN in 1973, is covered by Christine Sutton. Another outcome of gauge theory is the need for a fourth quark, now called 'charm'. This was discovered in 1974, as described by Roy Schwitters. The neutral current had been predicted, but many physicists were sceptical that it existed. The charm quark too had been thought of, but few people had taken any notice. The new particles discovered in 1974 came as a surprise and the general acceptance of their interpretation in terms of charm did not happen overnight.

The discovery of the neutral current emphasized the strong parallels between the weak and electromagnetic forces. Using new field theoretic ideas—the now famous 'Higgs mechanism'—the separate effects of electromagnetism and the weak nuclear force had been combined in a single 'electroweak' picture. The emergence of this idea had been the work of a host of theoreticians, spearheaded by Sheldon Glashow, Abdus Salam and Steven Weinberg. Slowly a whole new physics scenario had been pieced together. Its crowing glory, the 1983 discovery by Carlo Rubbia of the W and Z particles, the carriers of the weak force, is one of the great sagas of twentieth-century physics. It is related here by Carlo Rubbia, who shared the 1984 Nobel Prize for his achievement.

In 1977 a fifth quark was discovered and, to preserve the symmetry of the overall picture, a sixth was required. There are some phenomena which can only occur if there are at least six types of quark, arranged as three pairs. The remarkable coherence of this six-quark picture is covered in Ali's chapter, while the search for the sixth, namely the 'top' quark (the only quark to be systematically hunted) is the subject of Mel Shochet's contribution.

At the close of the twentieth century, the dominant picture in particle physics is the 'Standard Model', using six quarks grouped pairwise into three 'generations': 'up' and 'down'; 'strange' and 'charm'; 'beauty' and 'top'. Each of these quark pairs is twinned with a doublet of particles, 'leptons', which do not feel the quark force: respectively the electron and its neutrino, the muon and its neutrino, and the tau and its neutrino. All 12 Standard Model particles feel the electroweak force.

Experiments have shown that there are no more such quarks and leptons. All of Nature seems to be encompassed by the symmetric six-quark six-lepton scenario, with a strong inter-quark force carried by gluons, the electromagnetic force carried by the photon, and the related weak interaction carried by the W and Z. These carrier particles were unknown at the dawn of the twentieth century, and their discovery symbolizes the advance of physics understanding. Several authors make a point of placing their story in the context of the Standard Model, even though the developments that they describe preceded the post-1979 era when the Standard Model finally came into its own. With the Standard Model now the major focus of contemporary particle physics, Guido Altarelli explains how the Standard Model is probed and how accurate measurements are made. Altarelli's chapter displays the cutting edge of today's physics, showing how the emphasis has changed, at least temporarily, from searching for new particles to systematic precision studies. However, the Standard Model cannot be the ultimate picture. It has too many parameters that cannot be predicted and can only be measured by experiment. There is no explanation of why there are three, and only three, generations of particles and no understanding of why the relative strengths of the three underlying forces, strong, electromagnetic and weak, are so different. After a final appraisal of the Standard Model, Graham Ross tries to look beyond it to see whether there is a more fundamental mechanism, perhaps synthesizing more forces together and providing a more incisive explanation of why things are the way they are. If there is, it will probably require a reappraisal of current thinking, in the same way that quantum mechanics and relativity revolutionized physics early in the twentieth century.

Underlying the physics accomplishments of the twentieth century are new laboratory techniques, which, as well as exploiting new technologies, in some cases have themselves catalysed new technological breakthroughs. A vivid example is Wilhelm Röntgen's 1896 discovery of x-rays, which, as well as opening up the study of the atom, also changed the face of clinical examination and medical diagnosis. The particle accelerators which supplanted cosmic rays and brought intense beams of subnuclear particles to the laboratory are covered by E J N Wilson, while David Saxon deals with the detectors, the 'eyes' which record the results of scattering experiments.

As well as the synthesis of forces and the elucidation of the underlying quark stratum of Nature, one of the major accomplishments of twentieth-century physics is the realization that the Universe began in a 'Big Bang', and everything that we see around us is the outcome of what has often been described as the ultimate particle physics experiment. The large-scale results of this 'experiment' are studied by astronomers and astrophysicists, while particle physics looks at the other end of the distance spectrum and tries to understand the forces unleashed in the Big Bang.

John Ellis highlights this symbiosis in his chapter on 'Particle physics and cosmology'. As he says, 'Researchers working in the fields of particle physics and cosmology use arguments from each other's fields as naturally as they draw breath'. Just as the Standard Model has become the catechism of particle physics, so the particle physics scenario of the Big Bang has become standard cosmology dogma. Also covering this particle physics picture of the Big Bang is Qaisar Shafi, who traces its foundation in Einsteinian general relativity and the discovery earlier this century by US astronomer Edwin Hubble that the Universe is expanding—the twentieth-century equivalent of the Copernican revolution. Shafi also introduces the crucial concept of inflation, the mechanism which catapulted a microscopic quantum oddity into a self-sufficient Universe and set the scene for the Hubble expansion. Both Ellis and Shafi point out the enigma of 'dark matter', that the Universe appears to behave gravitationally as though it contains more matter than we can actually see.

Classical optical astronomy, blanketed by the Earth's atmosphere, has been compared to bird spotting from the bottom of a swimming pool. However the development of detectors monitoring other parts of the electromagnetic spectrum, less hindered by the atmosphere, together with the mounting of telescopes and other detectors into space vehicles, has enabled astronomers to study the Universe in new depth and detail. Andy Fabian summarizes the results.

Among its many other accomplishments, twentieth-century physics has made us aware just how small the human scale is in the grand order of things. We look out towards the mighty distances of outer space, and we look in towards the tiny subdimensions of the building blocks of matter, but this new understanding is no excuse for complacency. Many major questions remain unresolved. Why are there only three families of quarks and leptons? Why do they have their particular properties? The exact nature of the neutrino is still a puzzle, and the Higgs mechanism at the root of electroweak symmetry breaking is a complete mystery. Perhaps most intriguing of all is the 'dark matter' question, i.e. the part of the Universe that is directly observable seems to be only a fraction of the whole. Where and what is all this missing matter?

Physics has made incredible advances in the twentieth century. At the end of one great new adventure in understanding the Universe around us, we could be on the threshold of a new one, with breakthroughs no less dramatic than Planck's introduction of the quantum concept and Einsteinian relativity in the first few years of the twentieth century.

For the future, big new experiments are being prepared and ambitious facilities planned to answer the big questions. Whatever the next particle century may bring in the way of discovery, it will certainly bring increased international collaboration as nations pool their resources to match the increasing demands of this complex research.

This collection of essays is not meant to be a textbook and deliberately does not aim to cover all aspects of particle physics. There are some gaps. There are also some overlaps. The contributions highlight different aspects of the science, explaining how these developments have come about and what they mean—a sort of scientific travelogue in which editorial intervention has been kept to a minimum to preserve as far as possible the flavour of the original. Many technical terms which recur throughout are defined in an initial glossary. Mathematics has been limited.

Each contribution has a short list of books for further reading. For a general overview, Abraham Pais' *Inward Bound: of Matter and Forces in the Physical World* (Oxford: Oxford University Press) is a masterly account of the development of subatomic physics from the late nineteenth century to the mid-1980s, providing a valuable historical introduction and background material.

The Editor would like to thank Robin Rees for suggesting the idea for this book, Al Troyano for his painstaking work and continual attention to detail, and all the contributors for their diligence and enthusiasm.

Gordon Fraser
July 1998

ABOUT THE EDITOR

Born in Glasgow in 1943, Gordon Fraser received his physics education from the Imperial College of Science and Technology, London, obtaining a first-class degree in 1964 and a doctorate in the theory of elementary particles in 1967. After several postdoctoral research positions, he moved into technical journalism. After a spell as Information and Press Officer at the UK Rutherford Laboratory, he switched to

CERN, the European Laboratory for Particle Physics in Geneva, where for the past eighteen years he has edited *CERN Courier*, the international journal of high-energy physics. He is also a Contributing Editor of *Beamline*, published by the Stanford Linear Accelerator Center, and is a visiting lecturer in science communication at several UK universities.

With Egil Lillestøl and Inge Sellevåg, he is the author of *The Search for Infinity* (London: Mitchell Beazley, 1994; New York: Facts on File, 1995; London: Philip's, 1998), which has been translated into nine languages. He is also the author of *The Quark Machines* (Bristol: Institute of Physics Publishing, 1997) and is currently writing a book on 'Antimatter'.

GLOSSARY

Many commonly encountered technical terms used throughout this book are defined here, rather than being laboriously re-defined in each chapter. Words in *italics* refer to other entries in the glossary.

Abelian   Mathematical operators are said to be Abelian if they commute: $AB = BA$.

antiparticle   The antimatter counterpart of a particle.

antiquark   The antimatter counterpart of a *quark*.

asymptotic freedom   The remarkable property of certain *non-Abelian* gauge theories, such as *quantum chromodynamics*, in which the mutual force between particles becomes weaker as the particles approach each other. If the particles are close enough, the force becomes so weak that the particles, to a first approximation, behave as though they have no mutual interaction and can be considered as being free.

baryon   Any *fermion* subnuclear particle which feels the *strong force*. Contains three *quarks*.

beauty b   A quark *flavour*.

BeV   See GeV. 10^9 eV.

boson   Any subnuclear particle carrying integer *spin*.

CERN   The European Particle Physics Laboratory in Geneva, a consortium of 19 European Member States.

charged current   The electrically charged component of the *weak force*, carried by a *W boson*.

charm c   A quark *flavour*.

chirality   Having distinguishable properties under rotation: right handed and left handed.

collider   A machine, usually a *storage ring* or rings, for colliding particle beams. A linear collider uses two *linear accelerators*.

colour   Charge-like quantity carried by *quarks* and *gluons*, coming into play in *quantum chromodynamics*, the mechanism ultimately responsible for the *strong interaction*. Colour can have one of three values, conveniently labelled as red, green and blue.

critical density   The density at which the mutual gravitational pull of the constituent masses within a Universe would eventually just stop its Big Bang expansion; above the critical density, the Universe will eventually contract under this gravitational pull. In these cases, the Universe is said to be 'closed'; below the critical density, the Universe will continue to expand for ever and is said to be 'open'.

cross section   The reaction rate per unit incident flux of particles per target particle. Usually expressed in fractions (micro, nano, etc) of a barn (10^{-28} m^2).

dark matter   The mysterious invisible content of the Universe inferred by observed gravitational effects.

DESY   The German particle physics laboratory (Deutsches Elektronen-Synchrotron) in Hamburg.

down d   A light *quark* carrying electric charge $-\frac{1}{3}$ that of the electron.

electron e⁻   The lightest electrically charged *lepton*. A *lepton flavour*.

electronvolt (eV)   The energy needed to move one *electron* across 1 V. 1 eV = 1.602 177 33 × 10^{-19} J.

fermion   Any particle carrying half-integer *spin*.

fine-structure constant   A dimensionless quantity, $\alpha = e^2/2hc$, where h is Planck's constant, e is the charge of the *electron* and c is the velocity of light. Its numerical value is close to 1/137. Used to measure the strength of electromagnetic effects on particles.

flavour There are six flavours, or types, of quark: *up*, *down*, *strange*, *charm*, *beauty* and *top*. The three types of *lepton*, namely *electron*, *muon* and *tau*, are also referred to as flavours. See *Standard Model*.

gauge invariance, gauge theory A field theory which is not tied to a particular reference frame, or 'gauge', and is invariant under certain changes of reference frame.

gauge particle A messenger particle (a boson of spin 1) which carries the interaction described by a *gauge theory*.

GeV 10^9 eV. Formerly known as 1 BeV.

gluon The *gauge particle* of the *colour* force between *quarks*. Carries *colour*.

hadron Any strongly interacting subnuclear particle.

helicity The component of a particle's *spin* in the direction of the particle's motion.

HERA The Hadron Electron Ring Accelerator electron–proton collider of 6.3 km circumference at *DESY*.

hyperon Any *hadron* heavier than a *nucleon*.

infrared slavery Because the *quantum chromodynamics* force between *quarks* does not decrease as the distance between them increases, *quarks* cannot easily be separated and remain 'confined' (bound to each other) in *hadrons*.

isotopic spin (isospin) A quantity, analogous to ordinary *spin*, whose components are used to label particles with different electrical charges but nevertheless related, such as the *proton* and the *neutron*.

Kaon (K meson) A *meson* carrying the flavour *strange*.

LEP The Large Electron–Positron collider of 27 km circumference at CERN.

lepton Any weakly interacting subnuclear *fermion*. There are three distinguishable varieties, or *flavours*: *electron*, *muon* and *tau*. They are thought to be indivisible.

LHC The Large Hadron Collider, a proton collider being constructed at *CERN* in the 27 km tunnel originally used for *LEP*.

linear accelerator (linac) A machine which accelerates particles in a straight line.

Megaparsec 3.26 million light-years (1 parsec = 3.26 light-years).

meson Any *boson* lighter than a *proton* and heavier than an *electron*. Usually restricted to *hadrons*, like the *pion*. Normally composed of a *quark* bound to an *antiquark*. However, the *muon*, a *lepton*, is anomalously known as the μ meson, reflecting initial confusion between this particle and the *pion*.

MeV 10^6 eV.

muon μ A heavy *lepton*, often described as a heavy version of the *electron*. A lepton *flavour*.

neutral current The electrically neutral component of the *weak force*, carried by a *Z boson*. Discovered at *CERN* in 1973.

neutrino ν Any electrically neutral *lepton*. There are three distinct varieties, or *flavours*: the *electron* neutrino, *muon* neutrino and *tau* neutrino.

neutron n Electrically neutral component of most nuclei.

non-Abelian Mathematical operators are said to be non-Abelian if they do not commute: $AB \neq BA$.

nucleon A generic term for *protons* and *neutrons*.

parity Left–right space symmetry. *Weak interactions* have a 'handedness' and are not parity symmetric.

photon The gauge boson which carries the electromagnetic force.

pion (π meson) The lightest *hadron*.

Planck mass The Planck mass M_P is the mass of a particle whose gravitational coupling becomes appreciable (unity): $M_P = (hc/K)^{1/2} = 10^{19}$ GeV, where h is Planck's constant, c is the velocity of light and K is the Newtonian gravitational constant. The gravitational force is so weak that it can usually be neglected in particle physics.

polarization The *spin* component in a particular direction. Polarized particles have their spins aligned in the same direction.

positron e^+ The *antiparticle* of the *electron*.

proton p Electrically charged component of nuclei.

quantum chromodynamics (QCD)   The *gauge theory* of *colour* forces between *quarks*. Being *non-Abelian*, it has the remarkable properties of *asymptotic freedom* and *infrared slavery*.

quantum electrodynamics (QED)   The *gauge theory* of electromagnetism.

quark   The (supposedly) ultimate component of *hadrons*. Quarks are *fermions* and are electrically charged, carrying charges $\frac{2}{3}$ or $-\frac{1}{3}$ that of the *electron*.

quark confinement   The inability of *quarks* to appear as free particles. *Quarks*, victims of *infrared slavery*, are imprisoned inside *hadrons*.

radiative corrections   Secondary quantum *gauge theory* effects in which additional gauge bosons come into play.

renormalization   The avoidance of infinities in *gauge theory* calculations by carefully defining certain basic quantities, such as charge and mass. A renormalizable theory can thereby be freed of such infinities, permitting reliable calculations to be made and making unique predictions.

SLAC   The Stanford Linear Accelerator Center, California, built around a 2 mile *linear accelerator* for *electrons*.

SLC   The Stanford Linear Collider, an electron–positron *linear accelerator* at SLAC.

SPEAR   A small electron–positron *collider* built at SLAC in the early 1970s.

spin   A quantum number carried by particles, usefully (but erroneously) compared with intrinsic angular momentum. Expressed in integer or half-integer multiples of the Bohr magneton $eh/2\pi m_e c$, where h is Planck's constant, c is the velocity of light and e and m_e are the charge and mass respectively of the *electron*.

SPS   The Super Proton Synchrotron of 7 km circumference at *CERN*.

Standard Model   The minimalist description of particle physics in terms of six *quarks* and six *leptons*, and involving all forces, except gravity. The six *quarks*, arranged in three pairs (*up* and *down*, *strange* and *charm*, *beauty* and *top*), are twinned with three pairs of *leptons*, the electrically charged *electron*, *muon* and *tau* respectively, each with its associated *neutrino*.

Quarks		Leptons	
Up	Down	Electron	Electron neutrino
Strange	Charm	Muon	Muon neutrino
Beauty	Top	Tau	Tau neutrino

Only the *quarks* feel the *colour* force of *quantum chromodynamics*. All *Standard Model* particles feel the *weak force*.

storage ring   A *synchrotron*-like device in which particles are accumulated and circulate freely but are not extracted.

strange s   A quark *flavour*.

strong interaction, or strong force   Originally understood as the force holding *nucleons* together in nuclei, but now understood in terms of the forces between particles carrying *colour*, and described by *quantum chromodynamics*.

synchrotron   A circular machine for accelerating particles to very high energy.

tau τ   A heavy *lepton*, much heavier than the *muon*. A lepton *flavour*.

TeV 10^{12} eV.

top t   A quark *flavour*. The top quark is the heaviest of the six *quarks*.

unitary matrix   A square matrix which is the inverse of its Hermitian conjugate.

up u   A light quark carrying electric charge $\frac{2}{3}$ that of the *electron*.

W boson   The gauge boson of the electrically charged component of the *weak interaction*.

weak interaction, or weak force   A universal force which permutes quark *flavours* and makes *leptons* interact, while respecting lepton *flavour*. Once better known as nuclear β decay.

Z boson   The gauge boson of the electrically neutral component of the *weak interaction*.

NOBEL PRIZES IN PHYSICS

In the following list of Nobel Prize awards, contributions directly relevant to particle physics and cosmology are in bold type. Adapted from Allday J 1998 *Quarks, Leptons and the Big Bang* (Bristol: Institute of Physics Publishing).

YEAR	NAME OF WINNER	CITATION
1901	**Wilhelm Conrad Röntgen**	**Discovery of X-rays**
1902	**Hendrick Antoon Lorentz** **Pieter Zeeman**	**Effect of magnetic fields on light emitted from atoms**
1903	**Antoine Henri Becquerel** **Pierre Curie** **Marie Sklodowska-Curie**	**Discovery of radioactivity** **Research into radioactivity**
1904	Lord John William Strutt Rayleigh	Discovery of argon and measurements of densities of gases
1905	**Philipp Eduard Anton von Lenard**	**Work on cathode rays**
1906	**Joseph John Thomson**	**Conduction of electricity by gases**
1907	**Albert Abraham Michelson**	**Optical precision instruments and experiments carried out with them**
1908	Gabriel Lippmann	For his method of producing colours photographically using interference
1909	Carl Ferdinand Braun Guglielmo Marconi	Development of wireless telegraphy
1910	Johannes Diderik van der Waals	Equations of state for gases and liquids
1911	Wilhelm Wien	Laws governing the radiation of heat
1912	Nils Gustaf Dalen	Invention of automatic regulators for lighthouse and buoy lamps
1913	Heike Kamerlingh Onnes	Properties of matter at low temperatures leading to the discovery of liquid helium
1914	**Max von Laue**	**Discovery of X-ray diffraction by crystals**
1915	**William Henry Bragg** **William Lawrence Bragg**	**Analysis of crystal structure using X-rays**

1917	**Charles Glover Barkla**	**Discovery of characteristic X-rays from elements**
1918	**Max Planck**	**Discovery of energy quanta**
1919	**Johannes Stark**	**Discovery of Doppler effect in canal rays and the splitting of spectral lines in magnetic fields**
1920	Charles-Edouard Guillaume	Services to precision measurements in physics by his discovery of anomalies in nickel steel alloys
1921	**Albert Einstein**	**Services to theoretical physics especially explanation of photoelectric effect**
1922	**Niels Bohr**	**Structure of atoms and radiation from them**
1923	**Robert Andrew Millikan**	**Measurement of charge on electron and work on photoelectric effect**
1924	**Karl Manne Georg Siegbahn**	**X-ray spectroscopy**
1925	**James Franck** **Gustav Hertz**	**Experimental investigation of energy levels within atoms**
1926	Jean Baptiste Perrin	Work on discontinuous nature of matter especially discovery of sedimentation equilibrium
1927	**Arthur Holly Compton** **Charles Thomson Rees Wilson**	**Discovery of Compton effect** **Invention of the cloud chamber**
1928	Owen Willans Richardson	Work on thermionic phenomena
1929	**Prince Louis-Victor de Broglie**	**Discovery of the wave nature of electrons**
1930	Sir Chandrasekhara Venkata Raman	Work on scattering of light
1932	**Werner Heisenberg**	**Creation of quantum mechanics**
1933	**Paul Adrien Maurice Dirac** **Erwin Schrödinger**	**Discovery of new productive forms of atomic theory (quantum mechanics)**
1935	**James Chadwick**	**Discovery of the neutron**
1936	**Carl David Anderson** **Victor Franz Hess**	**Discovery of the positron** **Discovery of cosmic rays**
1937	**Clinton Joseph Davisson** **George Paget Thomson**	**Discovery of electron diffraction by crystals**
1938	**Enrico Fermi**	**Discovery of nuclear reactions brought about by slow neutrons (fission)**
1939	**Ernest Orlando Lawrence**	**Invention of cyclotron**

1943	Otto Stern	Discovery of the magnetic moment of the proton
1944	Isidor Isaac Rabi	Resonance recording of magnetic properties of nuclei
1945	Wolfgang Pauli	Discovery of the exclusion principle
1946	Percy Williams Bridgman	Invention of apparatus to produce extremely high pressures and discoveries made with it
1947	Edward Victor Appleton	Investigation of the physics of the upper atmosphere
1948	Patrick Maynard Stuart Blackett	Development of Wilson cloud chamber and discoveries made with it
1949	Hideki Yukawa	Prediction of mesons
1950	Cecil Frank Powell	Photographic method for recording particle tracks and discoveries made with it
1951	John Douglas Cockcroft Ernest Thomas Sinton Walton	Transforming atomic nuclei with artificially accelerated atomic particles
1952	Felix Bloch Edward Mills Purcell	Development of new methods for nuclear magnetic precision measurements and discoveries made with this technique
1953	Frits Zernike	Invention of the phase contrast microscope
1954	Max Born Walter Bothe	Fundamental research in quantum mechanics The coincidence method and discoveries made with it
1955	Willis Eugene Lamb Polykarp Kusch	Discoveries concerning the fine structure of the hydrogen spectrum Precision measurement of the magnetic moment of the electron
1956	John Bardeen Walter Houser Brattain William Shockley	Discovery of transistor effect
1957	Tsung Dao Lee Chen Ning Yang	Prediction of parity non-conservation
1958	Pavel Aleksejevic Čerenkov Il'ja Michajlovic Frank Igor' Evan'evic Tamm	Discovery and interpretation of Čerenkov effect
1959	Owen Chamberlain Emilio Gino Segrè	Discovery of the antiproton
1960	Donald Arthur Glaser	Invention of the bubble chamber

1961	**Robert Hofstader**	**Studies in the electron scattering of atomic nuclei**
	Rudolf Ludwig Mössbauer	**Research into the resonant absorption of γ rays**
1962	Lev Davidovic Landau	Theory of liquid helium
1963	J Hans Jensen	Theory of atomic nucleus
	Maria Goeppert Meyer	Discovery of nuclear shell structure
	Eugene P Wigner	
1964	Nikolai G Basov	Quantum electronics and masers/lasers
	Alexander M Prochrov	
	Charles H Townes	
1965	**Richard Feynman**	**Development of quantum electrodynamics**
	Julian Schwinger	
	Sin-itiro Tomonaga	
1966	Alfred Kastler	Discovery of optical methods for studying Hertzian resonance in atoms
1967	**Hans Albrecht Bethe**	**Contributions to the theory of energy production in stars**
1968	**Luis W Alvarez**	**Discovery of resonance particles and development of bubble chamber techniques**
1969	**Murray Gell-Mann**	**Discoveries concerning the classification of elementary particles**
1970	Hannes Alfvén	Discoveries in magnetohydrodynamics
	Louis Néel	Discoveries in antiferromagnetism and ferrimagnetism
1971	Dennis Gabor	Invention of holography
1972	John Bardeen	Jointly developed theory of superconductivity
	Leon N Cooper	
	J Robert Schrieffer	
1973	Leo Esaki	Discovery of tunnelling in semiconductors
	Ivar Giaever	Discovery of tunnelling in superconductors
	Brian D Josephson	Theory of super-current tunnelling
1974	**Antony Hewish**	**Discovery of pulsars**
	Sir Martin Ryle	**Pioneering work in radioastronomy**
1975	Aage Bohr	Discovery of the connection between collective motion
	Ben Mottelson	and particle motion in atomic nuclei
	James Rainwater	
1976	**Burton Richter**	**Independent discovery of J/ψ particle**
	Samuel Chao Chung Ting	

1977	Philip Warren Anderson Nevill Francis Mott John Hasbrouck Van Vleck	Theory of magnetic and disordered systems
1978	Peter L Kapitza	Basic inventions and discoveries in low temperature physics
1978	**Arno Penzias** **Robert Woodrow Wilson**	**Discovery of the cosmic microwave background radiation**
1979	**Sheldon Lee Glashow** **Abdus Salam** **Steven Weinberg**	**Unification of electromagnetic and weak forces**
1980	**James Cronin** **Val Fitch**	**Discovery of K^0 CP violation**
1981	Nicolaas Bloembergen Arthur L Schalow	Contributions to the development of laser spectroscopy
	Kai M Siegbahn	Contribution to the development of high resolution electron spectroscopy
1982	**Kenneth G Wilson**	**Theory of critical phenomena in phase transitions**
1983	**Subrahmanyan Chandrasekhar**	**Theoretical studies in the structure and evolution of stars**
	William Fowler	**Theoretical and experimental studies of nucleosynthesis of elements inside stars**
1984	**Carlo Rubbia** **Simon van der Meer**	**Discovery of W and Z particles**
1985	Klaus von Klitzing	Discovery of quantum Hall effect
1986	Ernst Ruska Gerd Binnig Heinrich Rohrer	Design of electron microscope Design of the scanning tunnelling microscope
1987	Georg Bednorz K A Müller	Discovery of high temperature superconductivity
1988	**Leon M Lederman** **Melvin Schwartz** **Jack Steinberger**	**Neutrino beam method and demonstration of two kinds of neutrino**
1989	Normal Ramsey **Hans Dehmelt** **Wolfgang Paul**	Separated field method of studying atomic transitions **Development of ion traps**
1990	**Jerome Friedman** **Henry Kendall** **Richard Taylor**	**Pioneer research into deep inelastic scattering**

1991	Pierre-Gilles de Gennes	Mathematics of molecular behaviour in liquids near to solidification
1992	**Georges Charpak**	**Invention of multiwire proportional chamber**
1993	**Russell Hulse** **Joseph Taylor**	**Discovery of a binary pulsar and research into gravitational effects**
1994	Bertram Brockhouse Clifford Schull	Development of neutron spectroscopy development of neutron diffraction techniques
1995	**Martin Perl** **Frederick Reines**	**Discovery of tau-meson** **Discovery of electron-neutrino**
1996	David M Lee Douglas D Osheroff Robert C Richardson	Discovery of superfluid helium-3
1997	Steven Chu Claude Cohen-Tannoudji William D Phillips	Development of methods to cool and trap atoms with laser light

1   THE PATH OF POST-WAR PHYSICS

Catherine Westfall and John Krige

Editor's Introduction: The advance of twentieth-century physics is inextricably entwined with global factors such as war, economics and politics. In this chapter, Catherine Westfall and John Krige chart the post-Second World War story. As they confine their attention to, respectively, the USA and Europe, some remarks on what happened in Japan and the former USSR are appended.

THE UNITED STATES AND POST-WAR PHYSICS
Physics and the Cold War, 1946–58

The astonishing destructiveness of atomic bombs transformed international affairs. The new possibility that nuclear war would lead to the extinction of civilization made actual war too risky. As a result, the two major opposing postwar powers, the USA and the USSR, and their allies, entered into a 'Cold War', avoiding confrontation by amassing an increasingly large and sophisticated arsenal of atomic weapons as a show of force.

Wartime research affected post-war society in a number of ways. The development of nuclear reactors laid the groundwork for the civilian nuclear power industry, as well as improved reactors for plutonium production and research. The US federal investment in the atomic bomb project had other, more unexpected dividends for research. For example, the use of computing machines acted as a catalyst for the post-war computer revolution. In addition, the development of pure isotopes led to the crucial discovery in the 1950s by Emmanuel Maxwell and Bernard Sein of the 'isotope effect' in superconductors. This discovery set John Bardeen on the path to the development of the microscopic theory of superconductivity. Radar development also had considerable impact on post-war life, leading to better probes for studying condensed-matter physics, such as magnetic resonance imaging, and gave rise to such diverse inventions as blind landing systems, atomic clocks, world-wide communication and navigation systems, microwave ovens, and improved linear accelerators.

Perhaps post-war research was most affected by the alliance which formed during the war, binding the physics community with the US government and industry. Hope and fears concerning nuclear energy and the success of wartime projects led to enormous funding increases for research and development. The advent of the Korean War further convinced government leaders of the utility of physical research and strengthened ties between the federal government and the scientific community. During the Second World War, federal funding for research and development multiplied tenfold, from less than US$50 000 000 to US$500 000 000. Afterwards, the increase continued with a high of US$1 000 000 000 in 1950, and more than US$3 000 000 000 by 1956. Three separate government agencies were formed to support physical science research: the Office of Naval Research (ONR), the National Science Foundation (NSF) and the Atomic Energy Commission (AEC).

Although before the Second World War scientists had, in general, been wary of political control accompanying federal funding, this attitude changed drastically after the war. In his book *Alvarez: Adventures of a Physicist*, Nobel Prize winner Luis Alvarez, who served in both the radar and the atomic bomb projects, relates: 'we had gone away as boys... and came back as men. We had initiated large technical projects and carried them to completion as directors of teams of scientists and technicians.' Warmed by wartime successes, many scientists grasped at the opportunities for large-scale projects made possible by government support.

The wartime alliance led to major institutional changes in US physics. One particularly important development was the establishment of the national laboratory system, which reinforced the wartime model of hierarchically organized multidisciplinary government-supported research. When the AEC was formed in 1947, it inherited from the atomic bomb project a system of laboratories: Los Alamos, Oak Ridge, Hanford, and Argonne (which had been part of the Chicago Metallurgical Laboratory). These laboratories all had in common ample federal funding, a research organization built on the coordination of multidisciplinary research groups working collaboratively, and a gradually fading distinction between the experimental scientist and the engineer. Lawrence's Radiation Laboratory was included in the system of laboratories, as well as a laboratory formed just after the war, Brookhaven National Laboratory on Long Island. Throughout the 1950s and 1960s, these laboratories would sponsor the nation's largest accelerator projects. The wartime model of research also found wider use in other large-scale federally funded projects, e.g. space, microwaves, and lasers.

As Europe and Japan struggled to recover from the war, US physicists increasingly dominated physical research, as indicated by the increasing number of Nobel Prizes awarded to Americans. In this climate of accomplishment, the vigorous and productive American research enterprise became an important model for laboratories worldwide at a time when advances in communication and transportation made science more international than ever before.

Federal largess and domination of research were not the only legacies of the war, however. Robert Seidel, in an essay entitled 'The postwar political economy of high energy physics', argues that even the development of basic research instruments, such as accelerators, was tied to national security considerations and that, by accepting federal funding for these and other projects, the physics community contracted three new obligations: to provide manpower, to offer military advice as needed, and to be ready to turn expensive equipment 'into instruments of war'.

In the post-war era, attitudes about physicists and their work also changed. Although most citizens tended to see the development of the atomic bomb as an heroic act in the immediate post-war period, the stage had been set for a more sinister view, which drew on long-standing fears of nuclear power and the stereotype of the mad scientist. Physicists also grew increasingly concerned and sometimes even guilty about the dangers that accompanied the development of atomic weaponry. Individuals, such as Leo Szilard and Hans Bethe, and groups, such as the Federation of American Scientists, lobbied for international cooperation, promoted public debate on such issues as nuclear power and strategic weapons and encouraged physicists to play an active role in public education and science policy. In addition to guilt, individual physicists, like other citizens, suffered from the excesses of the anti-Communist movement associated with its most famous adherent, Joseph McCarthy. For example, in 1953, the AEC suspended the security clearance of J R Oppenheimer for suspected Communist sympathies.

In the atomic age, research itself was transformed. For example, due to war-time experiences, as Peter Galison observes, experimentalists continued to work more closely with engineers. In addition, as Sylvan S Schweber has noted, theorists continued their focus on experimental results and tied their work more closely to observable phenomena. More abstract topics not amenable to experimental study tended to be ignored. This tendency was then reinforced by the rapid development of expensive and powerful new experimental tools. Paul Forman also notes that federal support, especially from the military, 'effectively rotated the orientation of academic physics toward techniques and applications'.

US physics and the space age, 1958–80

In October 1957, the Soviets launched Sputnik I, the world's first artificial satellite. In December, America's attempt to launch a satellite failed. Worry that the USA was lagging behind its rival in technological development, and thereby military strength, spurred a massive effort to establish the USA as first in all areas of science and technology. At the same time, the desire to ease tensions between the USA and the USSR facilitated collaborative efforts between the two nations in research with high prestige but remote applications. For example, in 1959, AEC chairman John McCone and his Soviet counterpart Vasily S Emelynaov signed an information exchange agreement encouraging collaboration in high-energy physics. The gains for the physics community were bolstered in 1960 with the election of President John F Kennedy, who gave physicists unprecedented visibility and influence in public affairs. For example, Kennedy appointed Jerome Wiesner, who had served at both the Massachusetts Institute of Technology's Radiation Laboratory and at Los Alamos, to act as his advisor, a post created in response to the Sputnik crisis, and to head a President's Science Advisory Committee (PSAC). He also appointed Nobel Laureate Glenn Seaborg, who had worked at Ernest Lawrence's Berkeley Laboratory, as AEC chairman; for the first time the AEC was headed by a research scientist. Although the physics community would enjoy unprecedented prosperity in the next ten years because of the favourable, post-Sputnik environment, the alliance that made this prosperity possible was attacked at the highest government level, before Kennedy even took office. In his January 1961 farewell speech, President Dwight Eisenhower warned: 'In the councils of government, we must guard against the acquisition of unwarranted influence... by the military–industrial complex.... We must also be alert to the... danger that public policy could itself become the captive of a scientific–technological elite.'

Throughout the mid-1960s, the post-Second World War trend of steadily increasing funding and growth continued. From 1957 to 1961, American research and development expenditures more than doubled to reach US$9 billion annually, and, by 1965, the figure rose to almost US$15 billion annually. Physical research in both academia and industry received its share of the bounty. Since basic research experienced a two-and-a-half-fold increase from 1959 to 1965, even fields remote from space or military applications flourished. For example, high-energy physicists received two new laboratories, the US$100 million Stanford Linear Accelerator Center (SLAC) and the US$250 million Fermi National Laboratory (Fermilab), which were added to the AEC laboratory system in the 1960s. At these and other laboratories, research was increasingly conducted by large teams, giving rise to what Alvin Weinberg called 'big science'. University physics also boomed; from 1959 to 1967 the number of schools offering PhDs in physics nearly doubled. The base of financial support for university physics research was also solidified because, as in other parts of the world, the number of academic posts in the field doubled every 10–15 years from the end of the war through the 1970s owing to the growth in higher education and curriculum changes

which required engineers and others to take physics courses. Applied physics flourished as well, receiving support from government laboratories funded by the Department of Defense (DOD), the far-flung centres of the National Aeronautics and Space Administration (NASA), and at the National Bureau of Standards. By this time, aerospace, electrical and instrument industries, in particular, depended heavily on physics research.

As the scepticism in Eisenhower's 1961 remarks signalled, however, the relationship between leaders in government, industry and science was headed for change. In fact, from the mid-1960s to the mid-1970s, a number of factors prompted a 'crisis', which transformed this alliance. By the mid-1960s, public complaint about the highly technological war in Vietnam, the development of civilian nuclear power, and environmental pollution prompted politicians to debate the social value of science. In this critical atmosphere, scepticism rose about the role of scientists in policy making.

Political differences caused further divisions between leaders of the scientific community and top government officials. For example, eminent scientists, including members of PSAC, vehemently opposed President Richard Nixon's plans to develop antiballistic missiles and a supersonic transport. This reaction annoyed Nixon, who was already irritated because many scientists opposed the Vietnam War. Although a 1970 House Subcommittee headed by Emilio Daddario and a White House Task Force recognized the escalating divisiveness between the scientific community and government and advocated increasing the number and influence of science advisory officers, Nixon abolished the PSAC and the attendant Office of Science and Technology and gave the role of science advisor to H Guyford Stever, who simultaneously directed the NSF. The executive branch science advisory system established in the wake of Sputnik was not reinstated until 1977.

Changes in Washington complicated the administration of government-sponsored projects. Prompted by the concerns for promoting new non-nuclear energy sources and for separating nuclear development from nuclear safety, President Gerald Ford, in 1974, abolished the AEC, which had supported the nation's largest accelerators since its formation in 1946. The AEC's research and development function was transferred to the newly formed Energy Research and Development Administration (ERDA), which brought together, for the first time, major research and development programs for all types of energy. In 1977, ERDA was reorganized into a cabinet-level Department of Energy (DOE). William Wallenmeyer, former Director of the Division of High Energy Physics, explained that as the funding agency got larger, accelerator laboratories had to compete for funding with a wider range of programs. Also, with size came greater bureaucracy and less scientific and technical understanding at the higher levels of the agency. As a result, the numerous DOE laboratories—which supported a wide range of applied and basic research—had to adhere to more regulations and to produce more paperwork to account for their activities. In 1977, in the midst of the transition from ERDA to DOE, the 30-year-old Joint Committee on Atomic Energy (JCAE) was disbanded. Thereafter, budget items were considered by established Congressional committees instead of by the JCAE, which had included members well versed in science and technology who were willing to champion deserving projects through the Congressional funding process.

These changes coincided with an economy weakened by the expense of the Vietnam War and President Lyndon Johnson's Great Society programs. In the resulting unfavourable environment, physics funding languished. Although the field had enjoyed an almost exponential increase in funding in the USA from 1945 to 1967, funding reached a plateau in 1968, when both the DOD and NASA drastically cut support. This plateau continued with decreases in the mid-1970s and early 1980s.

Physicists devised a number of strategies to cope with funding difficulties. For example, at SLAC and Fermilab, which were squeezed in the vice of tight budgets and drastic increases in the size of instruments and teams, leaders such as Robert Wilson and Wolfgang 'Pief' Panofsky devised creative financing schemes and methods for more frugal accelerator construction. The DOE laboratory system was increasingly forced by fiscal stringency to reduce the scope of research, to cut older facilities and to keep operating funding at a much lower than expected level. As had been the case in the 1930s, individual physicists coped with declining employment prospects by remaining as long as possible in academia, often by accepting one post-doctoral

position after another. Nonetheless, not everyone found a job; from 1964 to 1974, for example, more than 6000 physicists left the field.

The shifting alliance: physics in the USA, 1980–90s

Although it is impossible, owing to lack of information, to write with certainty in 1997 about the development of the physics community in the last two decades of the century, panel discussions and other informal forums suggest the issues confronting physicists as they approach the end of the century. Of primary concern is the continued deterioration in the relationship between government, industry and the scientific community.

As the 1980s wore on, the economy remained sluggish, and numerous environmental problems, such as poorly managed radioactive waste disposal programs and global warming, gained increasing attention. In this environment, public wariness about the evils of 'the military–industrial complex' rose as did questions about the objectivity and morality of scientists. At the same time, government leaders became even more sceptical about the value of big science and technology and more insistent that scientists account for money and research progress. One symptom of government scepticism and insistence upon accountability was the advent of 'tiger teams' at DOE laboratories in the early 1990s. During tiger team visits, groups of outside experts carefully checked that all regulations had been strictly followed. Non-compliance was simply not allowed; in some cases, laboratories were temporarily shut down until corrections were made. Pressure for compliance continued even though the US economy improved as the decade neared its end. Even though Brookhaven National Laboratory successfully weathered the tiger team era, the laboratory came under fire over the timely implementation of environmental heath and safety measures, and as a result, in 1997, the laboratory's contracting agency, the Associate Universities Incorporated, was dismissed by DOE.

Many physicists worried about the consequences of changing government attitudes toward science. For example, former SLAC director Pief Panofsky warned: 'We are seeing a shift from the partnership between government and science to "acquisition" of science by government', an approach non-conducive to creative problem solving and the advancement of scientific knowledge. 'Nothing short of restoring a spirit of mutual trust and confidence between government and the scientific community can reverse' the trend and reconcile the partners so that they can continue to accomplish mutually beneficial goals. For many physicists, a distressing sign of the federal government's lack of confidence in basic research came in 1993, when Congress cancelled funding for the partially completed Superconducting Supercollider (SSC), which would have been the next major DOE laboratory devoted to high-energy physics.

The dissolution of the USSR, which ended the Cold War, provided another complication in the troubled relationship between science, government and industry. At this juncture, the many facilities devoted to weapon research—both government and industrial—were forced to search for ways to redirect expertise and resources as the demand for weapon development sharply declined. Those working in laboratories focusing on basic research were also affected by the end of the Cold War, which altered the dynamics of international cooperation and competition. Building large basic research projects was no longer a crucial demonstration of technological and, thereby, military strength, and international scientific cooperation was no longer a means for avoiding nuclear war. Researchers working on both basic and applied projects had to adjust to the rising importance of competition for economic advantage which rose as national security concerns receded and the global economy widened.

As US physicists look to the future of their discipline at the end of the century, a number of issues are raised. George Heilmeier, president of Bellcore, noted in 1992 that, since Congress was becoming more concerned about the economic value of research, scientists needed to 'explain very clearly, with no jargon, what it is [they're] trying to do... how it's done today and what the limitations of current practice are, articulating what is new and what is the opportunity, what's new in our approach and why do we think it can succeed and, assuming we are successful, what difference does it make.' Nobel Laureate Leon Lederman added that scientists needed to communicate with the general public as well as Congress. 'It's not only communicating with the public to sell science, it's

contributing to the science literacy in this nation.' When discussing the future role of DOE laboratories in 1993, Hermann Grunder, director of the DOE-sponsored Thomas Jefferson National Accelerator Facility, summarized: 'We must identify and articulate what we have to contribute to education, to the economic well-being of our nation, and to the national culture. With change comes loss, but also opportunity. After all we *must* take advantage of the largest research and development organization known to man.'

A host of other issues concern the physics community. Has the success of studying smaller and smaller building blocks of nature led to a shift towards the study of 'emergent phenomena, the study of the properties of complexes whose "elementary" constituents and their interactions are known', as Silvan Schweber suggests? Does the cancellation of the SSC signal federal unwillingness to support big science further? Will the cancellation of the SSC encourage greater interdisciplinary mixing between the particle and nuclear physics communities? Can big science survive through greater reliance on collaboration with both European and Japanese physicists? Given the current intellectual and funding environment, should greater priority and status be given to smaller projects, or to projects more likely to lead to practical applications, e.g. in condensed-matter physics research? Perhaps the greatest challenge facing the US physics community is to adjust to an era of stable funding and changing opportunities after a century of extraordinary success and explosive unrestricted growth and development.

BIG PHYSICS IN EUROPE

When thinking about the post-war development of big physics in Europe, it is important not simply to export the American 'model' across the Atlantic. Certainly in the first two or three decades after the Second World War, Europeans had to learn to do physics, particularly particle physics, in the American style, *competition oblige*. However, this should not blind us to the fact that physics in Europe has been conducted in a very different socio-political context, within different constraints, and with different priorities. This brief survey can do little more than quickly identify some of these specificities, illustrating them with a few examples.

Big physics in (western) Europe, meaning basic research carried out with heavy equipment, has been conducted on two levels: the national and the European. At the national level in Europe, as elsewhere, individual governments, in consultation with their scientists, decide upon the apparatus that they want and they pay for it. Thus, in the 1950s and 1960s the 'big four', namely Britain, France, Italy and Germany, all invested in reasonably powerful national high-energy accelerators, for example. The UK commissioned a high-intensity 7 GeV proton synchrotron named Nimrod in 1963. The year before, the government had agreed to supplement this with a 4 GeV electron synchrotron (NINA). France's 3 GeV proton synchrotron (Saturne) began operations in 1957. Italy, for her part, chose to build electron–positron colliding beam machines. The first storage ring, Ada, was put into operation at Frascati near Rome in 1961, being soon extended to become Adone. As for Germany, in 1964, its 6 GeV electron synchrotron (Deutsches Elektronen-Synchrotron (DESY)) accelerated its first beam to full energy.

In addition to these activities at the national level, big physics is also done at the European level, in which a number of governments, as few as two or as many as a dozen, decide to pool resources and to build a piece of equipment in common. CERN, the European Laboratory for Particle Physics as it is now called, and which was established officially in 1954, was the first of such ventures. It has been followed by a number of similar organizations. In the early 1960s most of the member states of CERN decided to establish a 'sister' European Space Research Organization (ESRO), dedicated initially to scientific research alone. The main initiative for setting up ESRO came from Pierre Auger and Edoardo Amaldi, physicists and CERN pioneers. Indeed, physicists interested in space have always played and important role in setting the broadly based European Synchroton Radiation Facility, built on the same site in Grenoble. Fusion research has also been 'Europeanized'. A Tokamak, labelled the Joint European Torus (JET), was officially inaugurated at Culham near Oxford, UK, in 1984. It is the most successful project undertaken within the framework of the European Economic Community's Euratom program, which was initially, and idealistically, intended to mount a common fission effort among the six signatories of the Treaty of Rome in 1957.

The difficulties confronted by Euratom, whose goal of creating a common atomic energy program was partly sabotaged by the centrifugal pull of national interests, are instructive. For example, they are indicative of the types of area in which European collaboration is possible. Put rather too simply, it has to be basic research, with no apparent commercial or military spin-off. Thus, it is precisely in areas such as high-energy physics and fusion, where the applications, if any, are in the remote future, that European collaboration can be a viable option. To make the same point another way, as soon as governments are persuaded that a field is of strategic importance and is potentially ripe for exploitation, they prefer to keep control themselves, to develop a national capability whose development they can monitor at first hand, and whose benefits do not have to be shared with partners. This is one reason why Euratom fell so short of expectations. Coming into being in the late 1950s and early 1960s, when Britain and France already had major atomic energy programs of their own, there was little hope that either would be willing to share know-how and materials perceived to be of such great national importance.

This tension between following the national or the collaborative roads is ever present in European big science. The weight that is given to one or the other varies enormously over time and between countries. For smaller European states (the Scandinavian countries, the Low Countries and the central 'neutral' countries such as Austria and Switzerland), or for larger but less-developed countries (such as Spain and Portugal until recently, and now the eastern European states), collaboration is the cheapest and safest way of guaranteeing access to the prestige, skills and high-technology industrial contracts which a European venture ensures them. The situation is more ambiguous for the larger countries, who are always tempted to maintain major national programs in parallel with the joint ventures. That temptation, it should be said at once, cannot outlast the cold logic of economic rationality. When CERN was set up, the British, who were the leading accelerator builders in Europe at the time, saw little interest in joining. They only did so, and that with considerable reluctance, when the physics establishment realized that the laboratory's plans for building a 30 GeV strong-focusing proton synchrotron were so well advanced that they either had to join it or be reduced to marginality in the field. Many UK physicists remained highly sceptical about what was happening in Geneva and pushed hard and successfully for a supplementary national program.

Only in the 1970s was it finally accepted that Britain simply could not afford both. Nimrod was closed down in 1978 as the CERN 300 GeV Super Proton Synchrotron came on stream, and the machine was converted into a spallation neutron source. France and Italy had reached these conclusions a decade before. In the early 1960s, there was a lively and divisive debate in the former over whether the country should have its own 60 GeV accelerator, a debate which ended in a victory for those who felt that the European collaborative project should have overriding priority. Essentially the same choice was made in Italy. Indeed, since the late 1970s, only Germany, the richest of the European nations, has been able to maintain a national high-energy physics laboratory, DESY, in parallel with its relatively heavy commitment to CERN. The international success of this laboratory, the quality of its accelerators and its detectors, and the physics done with them, have made DESY a jewel in the crown of the German national physics program. Indeed in national eyes it is now seen as something of a competitor to CERN, rather than as a complement, and has led the Ministry concerned recently to have severe doubts about the costs of its CERN involvement.

It would be wrong to infer from the above that financial considerations dominate the thinking of governments and scientists when choosing to collaborate at the European level. Rather, in the most general sense, they collaborate because they believe it is in their interests to do so. Within that framework there is an enormous diversity of motives, which change over time, between groups, and even between individuals in the same group. Cost sharing is, of course, one such motive, but it is not necessarily the only or even the main motive. When CERN was established, for example, financial considerations were important for some key decision makers in Britain, but far less so in France. Here, at least initially, the establishment of a European laboratory dovetailed with the general policy in the Department of Foreign Affairs to favour supranational initiatives, a policy which found expression in the launching of the European Coal and Steel Community in 1950, and

the (abortive) attempt to build a European army. Germany's case was different again. Some of her scientists, at least, saw CERN membership as a way back into international respectability, a way of burying the past, along with the doubts about their role under the Nazi regime during the war. In similar vein, physicists building large electronic detectors in the 1980s and 1990s do not set up multinational multi-institutional collaborations just to share costs, nor, one should add, do they do so because they are fired by the ideal in 'internationalism'. It is rather because they need the specialized skills, the access to computing time, and the man and woman power that is required to build and operate modern giant detectors, and to exploit fruitfully the data that pour from them.

One key difference between the post-war expansion in American and European laboratories is that, in the European case, very low weight was given to military considerations in building a laboratory such as CERN. CERN, put differently, was not seen in Europe as playing a role in the technological Cold War. Of course, at the time most people associated nuclear physics with bombs. When the laboratory was set up, some scientists and government officials certainly played this card; they did so on an *ad hoc* basis and more out of convention than out of conviction. Indeed, those 'product champions', scientists, science administrators and politicians, who launched CERN went out of their way to build an image of the laboratory which insisted on the distinction between basic and applied research, and which locked CERN firmly in the former. This need to demilitarize and to depoliticize the laboratory's role was, in fact, a necessary condition for the project winning support in national state apparatuses. For, as we have already said, if it was defined otherwise, if it was seen by governments as doing work of strategic importance, the major countries at least would have insisted on keeping the research strictly under national control. Also, they would certainly not have been willing to join with nations such as Germany, with whom they had recently been at war, and Yugoslavia, which had one foot squarely in the eastern Communist camp.

It is only to be expected that, without the stimulus of the technological Cold War, funds for big science in Europe have not have flowed as smoothly or as copiously as has sometimes been the case in the USA. The general trend has, however, been the same, at least in high-energy physics: relative richness until the mid-1960s, followed by a tightening of the belt and a reassessment of the cost-effectiveness of the organization in the late 1960s and early 1970s, and the related demand that new projects be built within constant annual budgets. Within that broad picture, there has been the gradual phasing out of most national programs, and considerable vacillation in governments before they commit themselves to climbing further up the spiral to ever higher energies.

Another difference from the USA is to be noted here. To launch a major European scientific project it is often necessary to have many governments, and usually, at least, those in the 'big four', individually agree that it is worth doing. This can and does take years. As seen from across the Atlantic the process may appear to be enormously cumbersome and 'inefficient'. Yet it has distinct advantages for the scientists and for industry. First, at a political level, the participating states are locked into collaborative projects by a kind of international treaty, which is extremely difficult to break. Once having engaged themselves they cannot, therefore, easily withdraw. Second, and related to this, major programs such as the construction of a new accelerator are adopted along with an agreed construction schedule and budget envelope. The project is thus not subject to the annual budgetary uncertainties which have plagued similar schemes in the USA, and which have led to the waste and demoralization in the physics community ensuing on the cancellation of the SSC.

In conclusion, a few comments need to be made about the physicists themselves, and the European high-energy physics community at CERN in particular. One striking feature is its initial lack of experience, compared with competitors across the Atlantic, in building and running a major high-energy physics laboratory. The physicists who found themselves with access to the most powerful accelerator in the world in 1960 were physicists who were ill prepared for the demand of big science. They had not had the benefit of working on machines such as the Brookhaven Cosmotron and the Berkeley Bevatron. They had to learn how to manage an experimental program in which different pieces of heavy equipment, accelerators, beams, detectors and data-handling devices needed to be simultaneously available and properly interfaced. They had to overcome

prejudice, often deeply ingrained, against spending months and years building the instruments that they needed to do their research; they had to become applied physicists in the American sense of the term. They had to learn how to deal with engineers and accelerator builders, whom they left initially to manage the laboratory, and who used their power to impose machines on the physics community, such as the Intersecting Storage Rings—which initially it did not want. They had to establish mechanisms for dealing with outside users long before their American colleagues were 'forced' to do so, this being demanded from the outset by the international character of the laboratory at which they worked. Above all they had to conquer their inferiority complex vis-à-vis their US colleagues, a complex which was repeatedly fuelled as they watched one Nobel Prize after the other elude their grasp between the early 1960s and the late 1970s. It took 30 years and the emergence of an entire new generation of men and women, many of whom spent considerable periods in US laboratories, to mould European physicists into the shape demanded for success by the world leaders. The Nobel Prizes for Physics awarded to CERN scientists in 1984 (Carlo Rubbia and Simon van der Meer) and again in 1992 (Georges Charpak) are indicative of the extent to which that transformation has occurred.

JAPAN

Science does not belong to any particular nation, and the advance of science is an international effort. However, from time to time progress is strongly influenced by developments in a particular country. Such was the case with subnuclear physics in Japan, following Hideki Yukawa's 1935 meson hypothesis of nuclear forces, and the prediction of a strongly interacting meson.

These compelling arguments were a new departure in physics thinking, and this hypothesis and the search for the Yukawa meson dominated physics for many years. However, it led to some confusion (see the chapter by Lock), and to a grossly exaggerated view of the importance of the π meson, or pion. Only towards the end of the twentieth century has the Yukawa edifice been seen in its true light. Except for special cases at a low energy, pions cannot be viewed as the carriers of the nuclear force.

Yukawa developed a prestigious school, as did Sin-itiro Tomonaga in Tokyo. Many of their students went on to become influential post-war figures, both in Japan and in the USA.

Important advances were made in Japan in the 1940s, particularly in underlying theory and in cosmic-ray studies but, because of communications and language difficulties, these advances were not appreciated, or in some instances not even known, in the West (see the chapter by Lock). Tomonaga's remarkable work on quantum electrodynamics, in almost total isolation, paralleled that of Schwinger and Feynman in the USA, as recognized by his share of the 1965 Nobel Prize.

The continuing dynamism of Japanese physics was also reflected in the independent work in the early 1950s which led to the idea of the strangeness quantum number (see the chapter by Samios) and by the work towards a composite picture of hadrons (see the chapter by Ne'eman).

With the establishment of major new research centres at the national KEK Laboratory, Japan begins the twenty-first century as a major contemporary player on the world scene, with a flourishing national effort, attracting researchers from overseas, and collaborates in major experiments at USA and European accelerator centres.

RUSSIA AND THE FORMER SOVIET UNION

The harsh conditions of Russia breed genius in many fields of human endeavour, and science is no exception.

Immediately after the Second World War, US responsibility for nuclear matters passed from the military to a new Atomic Energy Commission. The Soviet Union, yet to produce its first nuclear bomb, kept its scientific effort hidden from view under a pall of secrecy, and it was only with the advent of perestroika and the break-up of the Soviet Union that the story of these years finally emerged.

The Soviet nuclear research program abruptly changed gear following the outcome of the Anglo-American wartime effort centred at Los Alamos. Anxious to follow suit, Stalin appointed no less a figure than KGB chief Lavrenti Beria to manage the program, and no effort was spared.

Endowed with such legendary talents as Niko-lai Bogolyubov, Pyotr Kapitsa, Yuli Khariton, Igor Kurchatov, Lev Landau, Isaak Pomeranchuk, Andrei Sakharov, Igor Tamm, Vladimir Veksler and Yakov Zel-dovich, the Soviet effort not only caught up but went on to rival the US program.

This immense Soviet effort needed a ready supply of trained scientists and the ability to handle particle beams. New research centres came into being, and a new laboratory at Dubna, north of Moscow, built world-class accelerators, culminating in the 10 GeV synchro-phasotron which, when it began operation in 1957, was the highest energy machine in the world. This high energy honour was quickly snatched away by CERN, and briefly regained when the 70 GeV synchrotron at Serpukhov, south of Moscow, was commissioned in 1967.

In particle physics, Russian influence has been most marked in underlying theory and in accelerator and detector development.

In their quest to push back the frontiers of knowledge, scientists have surmounted other barriers too. Even in the depths of the Cold War, the scientific message somehow managed to get transmitted, with free exchange of new results as long as the research added to basic knowledge and understanding. In the 1960s, a politically significant collaboration effort began between CERN and the USSR. CERN specialists and European equipment helped build the new 70 GeV machine at Serpukhov, in return for which European scientists were able to participate in the experimental program at the new machine. This led to joint experiments conducted at both at CERN and at Serpukhov, and to major Soviet investment in experiments in Europe and the USA.

Despite the difficulties of the Cold War, this cooperation flourished, and was influential in establishing contact and mutual trust between East and West so that, when the Iron Curtain finally fell, full-scale scientific cooperation could get off to a flying start.

MAJOR SOURCES
Brown L, Dresden M and Hoddeson L (eds) 1989 *Pions to Quarks: Particle Physics in the 1950s* (Cambridge: Cambridge University Press)
Hoddeson L, Brown L, Riordan M and Dresden M (eds) 1996 *The Rise of the Standard Model: A History of Particle Physics from 1964 to 1979* (Cambridge: Cambridge University Press)

Forman P 1984 *Hist. Stud. Phys. Biol. Sci.* **18** 149–69

Kevles D 1978 *The Physicists: The History of a Scientific Community in Modern America* (New York: Alfred A Knopf)

Hermann A, Krige J, Mersits U and Pestre D 1987 *History of CERN I: Volume I—Launching the European Organization for Nuclear Research* (Amsterdam: Elsevier Science)

Physics Survey Committee 1972 *Physics in Perspective* vol I (Washington DC: National Academy of Sciences)

Physics Survey Committee 1986 *Physics Through the 1990s: An Overview* (Washington DC: Smithsonian Institution Press)

Stine J K 1986 *A History of Science Policy in the United States, 1940–1985* Committee on Science and Technology, House of Representatives, 99th Congress, Second Session (Washington, DC: US Government Printing Office)

Trenn T 1986 *America's Golden Bough: The Science Advisory Interwist* (Cambridge, MA: Oelgeschlager, Gunn, and Hain)

SUGGESTED FURTHER READING

Surveys of the history of large-scale research and particle physics that include the period after the Second World War:

Galison P (ed) 1992 *Big Science: The Growth of Large-Scale Research* (Stanford, CA: Stanford University Press)

Schweber S 1994 *QED and the Men Who Made it* (Princeton, NJ: Princeton University Press)

For an interesting look at the history of experimentation in particle physics:

Galison P 1997 *Image and Logic: A Material Culture of Microphysics* (Chicago, IL: Chicago University Press)

For more information on the Atomic Energy Commision and post-war national laboratories:

Furman N S 1990 *Sandia National Laboratories: The Postwar Decade* (Albuquerque, NM: University of New Mexico Press)

Hewlett R G and Anderson O E 1962 *The New World, 1939/1946: A History of the United States Energy Commission* vol 1 (University Park, PA: Pennsylvania State University Press)

Hewlett R G and Holl J M 1989 *Atoms for Peace and War: 1953–1961* (Berkeley, CA: University of California Press)

Riordan M 1967 *The Hunting of the Quark: A True Story of Modern Physics* (New York: Simon & Schuster)

For CERN:

Hermann A, Krige J, Mersits U and Pestre D 1987 *History of CERN I: Volume I—Launching the European Organization for Nuclear Research*, 1990 *Volume II—Building and Running the Laboratory* (Amsterdam: North-Holland)

Krige J (ed) 1996 *History of CERN, Volume III* (Amsterdam: North-Holland)

Fraser G 1997 *The Quark Machines* (Bristol: Institute of Physics Publishing)

ABOUT THE AUTHORS

Catherine Westfall was born in 1952 in Loma Linda, California, USA. She received her BA from the University of California, Riverside, in 1974 and her PhD in American Studies from Michigan State University in 1988. She has spent most of her career writing contract histories on the founding, building and science of accelerators for a variety of US national laboratories, including Fermi National Accelerator Laboratory, Los Alamos National Laboratory, the E O Lawrence Berkeley National Laboratory and the Thomas Jefferson National Accelerator Facility. She is an Adjunct Assistant Professor in the Lyman Briggs School at Michigan State University and has taught history of science and technology courses both at Lyman Briggs and in the Physics Department. Major publications include: *Critical Assembly: A Technical History of Los Alamos During the Oppenheimer Years, 1943–1945* (with Lillian Hoddeson, Paul Henriksen and Roger Meade), Cambridge University Press, 1993; 'Thinking small in big science: the founding of Fermilab, 1960–1972', *Technology and Culture* (with Lillian Hoddeson), 37, July 1996; and 'Panel session: science policy and the social structure of big laboratories', in *The Rise of the Standard Model: Particle Physics in the 1960s and 1970s*, Lillian Hoddeson, Laurie Brown, Michael Riordan and Max Dresden, eds, (Cambridge University Press, 1997).

Born in Capetown, South Africa, in 1941, John Krige has a doctorate in physical chemistry from the University of Pretoria and a doctorate in philosophy from the University of Sussex, UK. He has researched and published extensively on the history of European scientific and technological collaboration, having contributed to the three volumes of the *History of CERN* (Hermann A, Krige J, Mersits U and

Pestre D 1987 *History of CERN, Volume I—Launching the European Organization for Nuclear Research, 1990; Volume II—Building and Running the Laboratory*; Krige J (ed) 1996 *History of CERN, Volume III* (Amsterdam: North Holland)) and has led the project to write the history of the European Space Agency (partially summarized in Krige J and Russo A 1994 *Europe in Space, 1960–1973* ESA SP-1172). He is also one of the editors of *Science in the Twentieth Century* (Harwood Academic Press, 1997). He is a Directeur de Recherche Associé in the French Centre National de la Recherche Scientifique and is Director of the Centre de Recherche en Histoire des Sciences et des Techniques at the Cité des Sciences et de l'Industrie, la Villette, Paris.

2 COSMIC RAIN

W O Lock

Editor's Introduction: Cosmic-ray specialist W O Lock painstakingly traces how evidence for new particles was disentangled from the sometimes flimsy evidence gathered during the early part of the twentieth century from the thin rain of high-energy extraterrestrial particles which manages to penetrate the blanket of the Earth's atmosphere. Unlike today's accelerator laboratories, where physicists equipped with sophisticated detectors have intense beams of particle on tap, the cosmic-ray specialists often had to take crude detectors to remote mountain tops or to fly them in balloons, where the research conditions were far from optimal. Nevertheless dramatic discoveries were made from this romantic science, and it was on this apparently flimsy foundation that the particle accelerators of the latter half of the century were able to build their new research monuments. For the dawn of the new century, new satellite-borne cosmic ray experiments will probe extraterrestrial particles beyond the screen of the atmosphere.

'Coming out of space and incident on the high atmosphere, there is a thin rain of charged particles known as the primary cosmic radiation.'   C F Powell, Nobel Prize Lecture 1950

EXTRATERRESTRIAL RADIATION: THE 'COSMIC RAIN'

In the twilight years of the nineteenth century, three major and unexpected discoveries opened the way for the nuclear and particle physics of the twentieth century. X-rays were discovered by Wilhem Röntgen in 1895, natural radioactivity by Henri Becquerel in 1896 and the first 'elementary particle', the electron, by Joseph John (J J) Thomson in 1897. These discoveries attracted great interest although the detectors of the time were very simple, such as the electroscope. In its basic form this consists of two suspended gold leaves which, given an electric charge, stay apart due to mutual repulsion. When charged particles such as those emitted from radioactive substances pass through matter, they can strip electrons from atoms. This process is known as 'ionization' and the electrons so removed and the remaining charged atoms are known as ions. A trail of ions acts as a conductor of electricity. Thus a charged electroscope will gradually lose its charge when exposed to a radioactive source due to the ionized paths created by the incident particles.

A surprising discovery was that the electroscope continued to lose charge no matter how well it was shielded or distanced from any radioactive sources. To explain this residual conductivity, as early as 1901 the inventor (in 1912) of the expansion cloud chamber, Charles (C T R) Wilson, postulated an extraterrestrial radiation falling upon the earth. It was not until 1912 that this was confirmed by an Austrian physicist, Victor Hess, in balloon flights up to an altitude of 5300 m, heroically observing the rate of loss of charge of his electroscopes. Even with this rudimentary apparatus he found that the ionization increased with increasing altitude to values many times that at sea level, suggesting that radiation was incident on the high atmosphere. His observations were confirmed after the First World War and the American physicist Robert Millikan coined the term 'cosmic rays' in 1926.

It required many years before the nature and composition of cosmic rays were well understood, and even now their origin is still not entirely clear. Most of the primary radiation reaching the top of the atmosphere consists of high-energy protons together with a small amount (less than 10%) of helium and heavier nuclei. It originates most probably within our Galaxy in the form of supernovae and similar explosions. The Sun provides a small percentage of low-energy electrons, protons and neutrinos, while there are also a few ultra-high-energy electrons and γ-rays, possibly of extra-galactic origin.

The various components of this incoming primary radiation collide with the nuclei of atmospheric nitrogen and oxygen and in the ensuing nuclear disintegrations give rise to a whole range of secondary particles which in turn cause further collisions lower down in the atmosphere. The heavier nuclei in particular are rapidly absorbed or split into smaller nuclei.

Much of our initial knowledge of elementary particles stemmed from classic cosmic-ray studies from 1930 to around 1955. These striking advances, using simple apparatus, motivated the building of the high-energy particle accelerators which have enabled collisions to be studied under controlled conditions.

ANALYSING THE COSMIC RADIATION IN THE ATMOSPHERE

Experimental techniques

The main obstacle in unravelling the nature and composition of the cosmic radiation was the need to be able to see, record and identify individual charged particles. In the late 1920s, two types of detector, developed just before the First World War, were adapted and improved specifically for cosmic ray studies. The first was the *Geiger–Müller counter* and the second the *Wilson cloud chamber*. A third, the *photographic emulsion technique*, became of considerable importance following the Second World War. All these detectors depend on the consequences of the ionization caused by charged particles. Ionization makes the incident particle lose energy. This energy loss decreases with increasing particle energy until it reaches a minimum of ionization, before rising slightly (about 10%) to a plateau value for relativistic particles. The *Geiger–Müller counter*, in its simplest form, consists of a thin wire running down the centre of a cylindrical metal foil. With a high voltage between the central wire and the metal foil and the gas at a low pressure, the creation of a single ion in the gas by a charged particle initiates an electronic spark which can trigger a counter.

In the *Wilson cloud chamber*, a charged particle passing through a supersaturated vapour generates ions, which act as nuclei on which molecules of the vapour tend to condense, forming a vapour trail of droplets. The supersaturation is achieved by rapidly expanding the vapour in a piston with a glass window through which photographs may be taken. The trails, called tracks, show the trajectory of the incident particle, as footprints in snow show the path of a walker. In general, the slower a particle, the more ionization it causes, and the increasing droplet density of the resulting tracks shows the direction of motion.

The *photographic emulsion technique* achieves its high sensitivity by a higher concentration of silver bromide grains and a much greater thickness than that employed for normal photographic work. A charged particle passing through the emulsion leaves a latent image which on development shows up as a series of silver grains along its path. Under a high-power microscope, measurements of the grain density and of the multiple scattering of the particle give information on its mass. Further, as for the cloud chamber, the slower the particle, the greater is the density of silver grains, giving a pointer to the direction of motion.

The products commercially available from Ilford Ltd from 1937 were 'Half Tone' emulsions which could only register particles giving ionizations greater than 25 times the minimum value, i.e. slow protons and α-particles. However, even with these emulsions Cecil Powell at Bristol had demonstrated that the technique could be used as a precision tool to study the scattering of protons of a few MeV. Detecting particles of energy greater than a few MeV needed an increased concentration of silver bromide. Ilford's first attempts were halted by the war. Patrick Blackett then played a decisive role through his influence with the Ministry of Supply of the 1945 UK Labour Government. He was largely responsible for the setting up of two panels: one to plan accelerator building for the UK (which he chaired) and one to encourage the development of sensitive emulsions (chaired by Joseph Rotblat, who went on to win the 1996 Nobel Peace Prize for his Pugwash work). For the emulsions, the Ministry placed contracts with both Kodak and Ilford, and with Cooke, Troughton and Simms of York for precision microscopes.

As early as 1946, Ilford was able to supply 'Nuclear Research Emulsions' which could detect particles of six times minimum ionization. These were used to discover the π meson in the following year, gaining Powell the

Nobel Prize in 1950 (see page 18). In the autumn of 1948, Kodak was able to supply emulsions which could detect particles even at minimum ionization, the so-called 'electron-sensitive' emulsions. Ilford achieved the same goal a few months later and rapidly became the leader in the field.

The composition of the cosmic radiation

From 1928 to 1933, several experiments used Geiger counters placed one above the other and separated by thicknesses of lead absorber. The passage of a cosmic ray was only recorded when both counters discharged simultaneously: a 'coincidence'. In particular, in 1933 Bruno Rossi in Florence, using three counters, recorded the number of threefold coincidences as a function of absorber thickness. He found that the number of particles traversing a given thickness of lead decreased fairly rapidly in the first 10 cm followed by a much slower decrease with increasing absorber thickness, such that one half of the particles emerging from 10 cm of lead were then able to penetrate 1 m of lead. Later workers, using both counter arrays and multiplate cloud chambers, observed not only single particles but also groups of penetrating particles capable of traversing great thicknesses of lead.

To investigate further, Rossi placed three counters in a triangular configuration so that a single particle travelling in a straight line could not traverse them all. With the counters enclosed in a lead box with walls a few centimetres thick he observed a large number of threefold coincidences. Removing the lid of the box caused a marked decrease in coincidences. He concluded that most of the coincidences observed with the box closed were due to groups of easily absorbed particles arising from cosmic-ray interactions in the top of the box.

Similar results were obtained by Pierre Auger and Louis Leprince Ringuet in France, who in 1935 pointed out that these results could only be understood in terms of three components of the cosmic radiation: the primary radiation from outer space; a soft component easily absorbed in a few centimetres of lead, giving easily absorbed showers of particles; a hard penetrating component able to pierce great thicknesses of lead without appreciable energy loss or degenerating into showers.

The composition of the soft component, although not its origin, became clear after the cloud chamber experiments of the early 1930s which led to the discovery of the positron (see page 15) and of electron–positron pair production. A fast photon may be converted into an electron–positron pair and then, if this fast electron and/or positron is electromagnetically deflected by a nucleus, a photon is radiated, possibly giving more pairs. Thus the electrons and photons rapidly lose energy and at the same time multiply rapidly before being brought to rest by normal ionization. Following initial work by Auger, the theory of these 'cascade' showers was elaborated in the years 1935–37. At this time it was assumed that high-energy electrons in the primary cosmic radiation were the origin of the cascade of showers in the atmosphere.

With the discovery of charged π mesons in 1947, followed by that of the neutral π meson in 1950 (see page 19), the whole of the cosmic-ray cascade in the atmosphere at last became clear. In the collisions of the primary protons and heavier nuclei with atmospheric nitrogen and oxygen nuclei, both charged and neutral π mesons are produced together with protons and neutrons from the colliding particles. The charged pions interact with the air nuclei or decay in flight to the weakly interacting muons and neutrinos and form the major part of the hard component. The neutral pions decay rapidly to photon pairs which in turn materialize as the electron–photon cascade of the soft component.

THE 'POSITRON'
Prediction

Following the development of quantum mechanics from 1926 to 1932, Paul Dirac went on to develop a version of quantum mechanics consistent with special relativity. His equation for a single electron had solutions of negative energy and in 1930 he proposed that the states of negative energy are normally all full (the 'sea'), with one electron in each state. An unoccupied negative energy state, a 'hole', would appear as something with a positive energy and positive charge, since to make it disappear, i.e. to fill it up, one would have to add an electron with negative energy. Dirac at first proposed that these holes were protons (being unwilling to predict a new particle) but soon became convinced that there should be symmetry between holes and electrons.

The holes should be positive electrons (positrons) having the same mass as electrons but opposite charge. Dirac further postulated that a positive energy electron could drop into an unoccupied state of negative energy, in which case an electron and a positron would disappear simultaneously, their energy being emitted as radiation. The inverse process is the creation of an electron and a positron from the conversion of a γ-ray in the electromagnetic field of a nucleus, called pair production. The positive electron or positron is said to be the antiparticle of the electron. The antiproton, the antiparticle of the proton, was discovered in 1955 at a new large particle accelerator at Berkeley (see the chapter by Samios).

Discovery of the positron and of electron–positron pair production

In 1930, the veteran American physicist, Millikan, asked Carl Anderson, a young graduate student at the California Institute of Technology (Caltech) in Pasadena, to build a large cloud chamber inside a large magnet for cosmic-ray studies. This would enable the momentum and charge of particles to be determined which, together with ionization measurements, would indicate the particles' mass. Anderson also placed a 6 mm lead plate across the middle of the chamber. In traversing it, particles would lose energy and the change in curvature (which depends on momentum) of the tracks in the magnetic field above and below the plate would reveal the direction of motion of the particles.

In 1932 Anderson took more than 1000 chamber pictures with the chamber in Pasadena (altitude 220 m). In some cases he was fortunate enough to see a particle or particles crossing the chamber at the moment of expansion and leaving measurable tracks. Fifteen photographs showed a positive particle crossing the lead plate in the middle of the chamber, none of which could be ascribed to particles heavier than $20m_e$, where m_e is the electron mass. Anderson concluded that he had seen positive particles of unit charge and much lighter than the proton, which he called positrons.

While Anderson was constructing his chamber, a series of events brought together in Cambridge two outstanding experimental physicists: Blackett, who had been working on cloud chambers in Rutherford's laboratory since the early 1920s, and a young Italian physicist from the laboratory of Rossi, Giuseppe Occhialini, an expert on the new coincidence techniques.

According to Rossi, Occhialini had already thought of pooling his expertise with Blackett's to build a cloud chamber which would be triggered by a coincidence arrangement of counters signalling the arrival of one or more particles. In this way the cloud chamber could be sure of capturing tracks. Occhialini arrived in Cambridge towards the end of 1931 and by the following summer such a counter-controlled cloud chamber was operational inside a high magnetic field. In the autumn of 1932, Blackett and Occhialini took some 700 photographs, 80% of which showed single cosmic-ray particles and the rest more complex showers. Both the single particles and the shower particles were positively and negatively charged particles, in equal numbers, and the shower particles tended to produce further showers.

Having discussed Dirac's new theory with him, Blackett and Occhialini were able not only to confirm Anderson's conclusions but also to interpret the pair events as the creation of electron–positron pairs, as predicted by Dirac, who received the 1933 Nobel Prize. Anderson had published his positron results in 1932, while Blackett and Occhialini delayed publishing until 1933 in order to be quite sure of their interpretation. It was, therefore, Anderson who was awarded the 1936 Nobel Prize for his epic discovery; in due course Blackett obtained the Nobel Prize in 1948.

'It seems to me that in some ways it is regrettable that we had a theory of the positive electron before the beginning of the experiments. Blackett did everything possible not to be influenced by the theory, but the way of anticipating results must to some extent, be influenced by the theory. I should have liked it better if the theory had arrived after the experimental facts had been established.'

Lord Rutherford, Solvay Conference, 1933

THE NEED FOR 'MESONS'

The prediction of nuclear mesons and the subsequent unravelling of experimental puzzles is one of the sagas of twentieth-century physics. Following the discovery of the proton and neutron, the next question was to ask what held these nuclear particles together, overcoming the repulsion between such closely packed protons, which is immense, 10^8 that of the electric attraction of atomic electrons. In 1932, Werner Heisenberg suggested that protons and neutrons were held together by the exchange of electrons.

A young Japanese physicist, Hideki Yukawa, in 1935 realized that the range of any quantum force is inversely proportional to the mass of the exchanged particle. In electromagnetism the exchanged particle is the photon of mass zero; so the range of the force is infinite. Yukawa knew that nuclear forces are very short range and postulated that they arose from the exchange of a massive particle, several hundred times the mass of the electron, and later called a 'meson'.

The discovery of the mesotron or muon

In contrast with the clear-cut discovery of the positron, it took many years to establish that the penetrating radiation consisted mainly of hitherto unknown particles which were heavier than the electron and lighter than the proton.

Anderson and his student Seth Neddermeyer had been puzzled by this penetrating radiation. It could not consist of protons as the energy of the electron secondaries produced was too high, particles of similar masses tending to share their energy equally in collisions. In addition the penetrating particles had positive and negative charges; so, if they included protons, one also would have to invoke the existence of negative such particles. On the other hand they could not be electrons as they did not produce electron–positron showers.

In order to exploit the greater cosmic-ray intensity at high altitudes, in 1935 Anderson moved his chamber to Pike's Peak in Colorado at an altitude of 4300 m. There, Neddermeyer and he observed nuclear disintegrations where ionization and momentum measurements suggested that the secondary particles were heavier than the electron but lighter than the proton.

To confirm this hypothesis they placed a 1 cm platinum plate in the chamber and measured the energy loss of particles in it. Convinced that the majority of the single particles were of intermediate mass, they published their results in May 1937. They suggested the term 'mesoton' for

the new particle (from the Greek 'meso' for intermediate) which, much to their annoyance, was changed to 'mesotron' by Millikan. It was not until the 1947 International Cosmic Ray Conference in Cracow that the term 'meson' was formally adopted for any particle of mass between that of the electron and that of the proton.

In mid-1938 Anderson and Neddermeyer obtained a striking picture (figure 2.1) of a positive mesotron crossing a Geiger counter in the middle of the chamber and finally stopping in the gas. Momentum and range showed the mass to be about $240m_e$. Later measurements by other workers gave a mass of the order of $200m_e$.

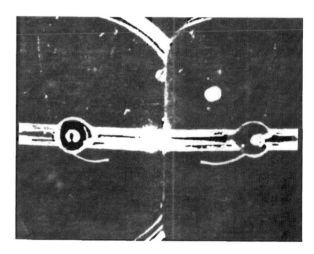

Figure 2.1. 1938 cloud chamber picture of a positively charged 'mesoton' by Neddermeyer and Anderson. The particle traverses a Geiger–Müller counter in the cloud chamber and stops in the chamber gas.

The lifetime of the mesotron

None of the mesotron pioneers was aware of Yukawa's theory, which had been published in English, but in a Japanese journal. However, the possible identification of his massive quanta with the observed meson was pointed out by the theoreticians J Robert Oppenheimer (who had received a reprint from Yukawa) and Robert Serber in the USA and later by Yukawa himself.

Yukawa had also predicted that, in order to explain β decay, his meson should decay to an electron and a neutrino.

Although there was indirect evidence that the mesotron was unstable, it was not until 1940 that the first direct visual evidence came from a high-pressure cloud chamber photograph taken by Evan James ('E J') Williams and George Roberts in Aberystwyth. This showed a mesotron coming to rest and emitting what appeared to be a positron. In the following years, ingenious counter arrangements studied the emission of particles from mesotrons brought to rest in various absorbers. That the mesotrons were at rest was determined by requiring a signal from counters placed above the absorber together with the absence of a simultaneous signal from counters below the absorber, an 'anticoincidence'. Using coincidence and anticoincidence techniques, it was possible to measure the time between a mesotron stopping and the subsequent emission of a decay electron. A typical experiment by Rossi (by then in the USA) gave a lifetime of $2.26 \pm 0.07 \times 10^{-6}$ s, very close to the present-day value of 2.197×10^{-6} s.

Thus most experimenters assumed that the mesotron was indeed the Yukawa meson, although the lifetime was a hundred times longer than that expected (10^{-8} s) and it did not appear to interact with matter as a particle responsible for nuclear effects should. In fact in 1940 it had been pointed out by Sin-itiro Tomonaga and Toshima Araki that, because of the electric charge of the nucleus, the competition between nuclear capture and spontaneous decay should occur at different rates for mesons of different charges. They calculated the capture probability for mesons in different materials, showing that the probability is negligible for the positive mesons (which are repelled by positively charged nuclei and are compelled to wander around in matter until they decay), whereas for negative mesons capture is always the more likely outcome.

In 1942 two young Italian physicists in war-time Rome, Marcello Conversi and Oreste Piccioni (not knowing of the work of Rossi and others), decided to make a direct precision measurement of the mesotron lifetime, and then to carry out a series of experiments to test the ideas of Tomonaga and Araki. With the first experiment nearly ready in July 1943, Rome was bombed by the Americans for the first time and nearly 80 bombs fell on the university campus. Conversi and Piccioni therefore decided to move their equipment to a semiunderground classroom near the Vatican, which they hoped would be better protected!

They were not able to move back into the Physics Institute on the other side of the city until Rome was liberated on 4 June 1944. They were joined by Ettore Pancini, who had been a leader in the resistance in Northern Italy. The team used magnetic lenses, developed by Rossi in 1931, to distinguish between positive and negative mesotrons. Their striking result, published in 1947, confirmed the ideas of Tomonaga and Araki, demonstrating that in light absorbers, such as carbon, a negative mesotron almost always decayed instead of being captured by the positive nucleus. This was in complete contrast with the expectations that negative Yukawa mesons slowed down in any material should be captured by nuclei before the decay lifetime of 2.2×10^{-6} s. In fact interaction between the mesotron and the nucleus was too weak by many orders of magnitude! The cosmic-ray mesotron was not the Yukawa nuclear force meson.

DISCOVERY OF THE PION AND THE IDENTIFICATION OF THE MESOTRON AS THE MUON

Unknown to most people, a possible solution to this dilemma had already been suggested in Japan by Yasutaka Tanikawa and by Shoichi Sakata and Takeshi Inoue in 1942–43, but because of the war their ideas were not published in English until 1946 and 1947 and only reached the USA at the end of 1947. They proposed that a Yukawa meson, strongly interacting with matter, decayed into the penetrating cosmic-ray mesotron. In early June 1947 a small conference of American theoreticians on Shelter Island (off Long Island) discussed the difficulties of identifying the mesotron with the Yukawa meson, and an independent two-meson hypothesis was put forward by Robert Marshak, which he published together with Hans Bethe. Unknown to anyone at the meeting such a decay had already been observed shortly before the conference by Powell and his co-workers in Bristol, again without any knowledge of the Japanese ideas.

Towards the end of the war, Blackett had invited his erstwhile collaborator Occhialini, then in Brazil, to join the British team working with the Americans on the atomic bomb. Occhialini arrived in the UK in mid-1945, only to learn that, as a foreign national, he could no longer work on the project. Instead, he joined Powell in Bristol, becoming the driving force behind the development of the new emulsion technique. He was joined by one of his research students, Cesar Lattes, towards the end of 1946.

As described earlier, early in 1946 Ilford were able to supply improved 'Nuclear Research Emulsions' and Occhialini exposed several dozen plates at the Pic du Midi in the French Pyrenees at 2867 m. Their examination by trained microscope observers revealed, as Powell put it: '...a whole new world. It was as if, suddenly, we had broken into a walled orchard, where protected trees flourished and all kinds of exotic fruits had ripened in great profusion.'

At the same time, some of the new emulsions had been exposed at 9100 m in an RAF aeroplane by Don Perkins of the emulsion group established at Imperial College, London, by Sir George Thomson. Perkins observed a clear example of what appeared to be the nuclear capture of a meson in the emulsion and producing a nuclear disintegration. Measurements of the multiple scattering as a function of residual range indicated a mass between $100m_e$ and $300m_e$. Perkins' observations, published in late January 1947, were confirmed by Occhialini and Powell, who published details of six such events only two weeks later.

Yet more exotic fruits followed. By the summer of 1947 Powell and his co-workers observed two clear examples of a meson coming to rest in the emulsion and giving a second meson (figure 2.2). In ignorance of the two-meson theory of the Japanese, and of Marshak, but knowing of the findings of Conversi *et al*, Lattes, Muirhead, Occhialini and Powell wrote in their May 1947 paper: 'Since our observations indicate a new mode of decay of mesons, it is possible that they may contribute to a solution of these difficulties.'

Clearly more evidence was needed to justify such a radical conclusion. For some time no more meson chains were found in the Pic du Midi plates and it was decided to make more exposures at much higher altitudes. Lattes proposed going to Mount Chacaltaya in the Bolivian Andes, near the capital La Paz, where there was a meteorological station at 5500 m. Arthur Tyndall, the Head of the Bristol Physics Department, recommended that Lattes should fly by a British Overseas Airway Corporation (BOAC) plane to Rio de Janeiro. Lattes preferred to take the Brazilian airline

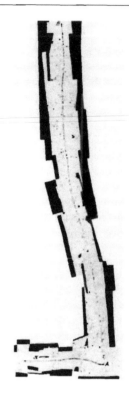

Figure 2.2. First observations of the decay chain of the π meson by Powell's group in Bristol in 1947. The pion enters the emulsion at the bottom left and decays at the point labelled A, producing a μ meson, or muon which stops at point B, at the top.

Varig, which had a new plane, the Super Constellation, thereby avoiding a disaster when the British plane crashed in Dakar and all on board were killed.

Examination of the plates from Bolivia quickly yielded ten more two-meson decays in which the secondary particle came to rest in the emulsion. The constant range in all cases of the secondary meson led Lattes, Occhialini and Powell to postulate a two-body decay of the primary meson, which they called π or pion, to a secondary meson, μ or muon, and one neutral particle. The process was termed $\pi\mu$ decay. Measurements on 20 events gave the pion and muon masses as $260 \pm 30m_e$ and $205 \pm 20m_e$ respectively while the lifetime of the pion was estimated to be about 10^{-8} s. Present-day values are $273.31m_e$ and $206.76m_e$ respectively and 2.6×10^{-8} s.

The number of mesons coming to rest in the emulsion and causing a disintegration was found to be approximately equal to the number of pions decaying to muons. It was, therefore, assumed that the latter represented the decay of π^+ mesons and the former the nuclear capture of π^- mesons. Clearly the π mesons were those postulated by Yukawa. This led to the conclusion that most of the mesons observed at sea level are μ mesons (the weakly interacting mesotrons) arising from the decay in flight of π mesons, created in nuclear disintegrations higher up in the atmosphere.

The advent, in the autumn of 1948, of emulsions sensitive to relativistic particles at minimum ionization enabled Peter Fowler (the grandson of Rutherford) and co-workers at Bristol to demonstrate in 1950 that the majority of the particles in the showers from nuclear disintegrations observed in the new emulsions exposed during high-altitude balloon flights were π mesons. The emulsions also revealed the complete decay sequence

$$\pi^+ \rightarrow \mu^+ + \nu \quad \text{and} \quad \mu^+ \rightarrow e^+ + \nu + \nu \text{ (antineutrino)}$$

which enabled the energy spectrum of the decay positrons to be determined.

Discovery of the neutral pion

The probable existence of a neutral companion to the charged Yukawa particle had been suggested by Nicholas Kemmer in 1938 on the grounds that nuclear forces should be independent of nuclear charge, and predicted that it should decay into two γ-rays with a lifetime of about 10^{-16} s. In 1947–48 it was suggested by Mituo Taketani in Japan, and by Rossi and by Oppenheimer and their co-workers in the USA, that the decay of such a meson would explain the cascade showers of the soft component. Decisive evidence for a π^0 meson came from an accelerator experiment by Jack Steinberger, 'Pief' Panofsky and Jack Steller at Berkeley in 1950.

At the same time, a group at Bristol studied the energy spectrum of electron pairs observed in the electron-sensitive emulsions exposed in balloon flights (figure 2.3). Assuming that the pairs originated from γ-rays from π^0 decay, the corresponding photon spectrum gave the π^0 mass as $295 \pm 20 m_e$ (the present-day value is $261.41 m_e$). An analysis of the geometry of the electron pairs materializing near parent nuclear disintegrations gave a π^0 lifetime less than 5×10^{-14} s (the present-day value is 0.8×10^{-16} s).

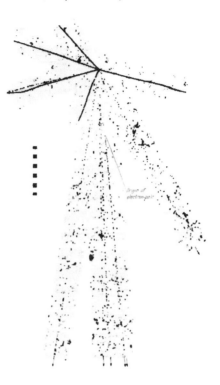

Figure 2.3. Emulsion tracks of the electron–positron pairs produced by γ-rays (photons) from the decay of a neutral π meson (π^0) produced in the interaction 'star' at the top.

THE SITUATION IN 1950

Thus with the pion identified, at least temporarily, as the nuclear meson of the Yukawa theory (although it decayed to a muon and not to an electron), in 1950 the picture of elementary particles appeared to be simple and clear. However, there were two major questions. Firstly there appeared to be no reason for the existence of the muon, which behaved as a heavy electron. In the words of Murray Gell-Mann and E P Rosenbaum in 1957: 'The muon was the unwelcome baby on the doorstep, signifying the end of days of innocence.' Secondly, evidence had accumulated for the existence of particles of mass between that of the pion and that of the proton and even beyond, later termed 'strange particles'. Once more cosmic-ray studies would provide the beginnings of answers to these two problems before the advent of the large particle accelerators soon to come into operation.

An important factor was the complementarity of the cloud chamber and emulsion techniques, where the steady increase in the number of research groups and advances in the technical sophistication of the detectors led to significant progress. For example, a typical cloud chamber arrangement, particularly effective in the discovery and study of neutral particles, might consist of two chambers one above the other. Emulsions were particularly effective for the study of charged particles, while the simplicity and relative cheapness of this technique led to many new groups being established. Technical advances made it possible to construct large stacks of stripped emulsions, each typically 600 μm thick. At the same time collaborations grew up between the newly formed groups in different countries, sharing the cost of the large emulsion stacks and the analysis of the events found. These collaborations later formed the model for the large-scale collaborative experiments at the high-energy accelerators.

HEAVY MESONS

'The particles described at this Conference are not entirely fictitious and every analogy with the particles really existing in nature is not purely coincidental.' From the *Proceedings of the Bagneres de Bigorre Cosmic Ray Conference (July 1953)*, entirely devoted to 'strange' particles.

The first suggestion of mesons heavier than the pion came from the cloud chamber observation in 1943 by Leprince Ringuet and L'Heritier of a positive particle colliding with an electron. Analysis indicated, but could not prove, a primary mass of $990 \pm 120m_e$. Decisive evidence came in 1946–47 from two cloud chamber pictures, taken at Manchester (where Blackett had moved in 1937) by George Rochester and Clifford Butler (figure 2.4). One showed a neutral particle decaying to two charged particles and the second the deflection by 19° of a single charged particle, suggesting its decay into a lighter particle. In both cases (later termed V^0 and V^+ particles respectively) a primary

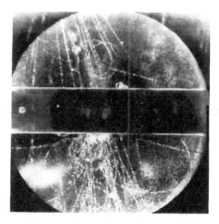

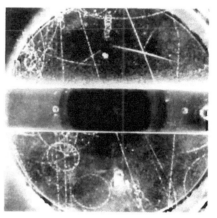

Figure 2.4. The classic 'V particles' seen by Rochester and Butler in Manchester in 1947: top, the V^0 decaying on the right, just below the central lead plate; bottom, the V^+ decaying at the top right of the picture.

mass of about $1000m_e$ was indicated. In late 1948 an event found by Powell and co-workers in the new electron-sensitive emulsions, exposed on the Jungfraujoch (altitude 3500 m), showed a primary particle (initially called τ) of mass about $965m_e$ stopping in the emulsion and decaying to three charged π mesons (figure 2.5), one of which produced a small nuclear disintegration.

Charged K particles

In 1950, Cormac O'Ceallaigh at Bristol found two events which he interpreted as the decay at rest of a single charged

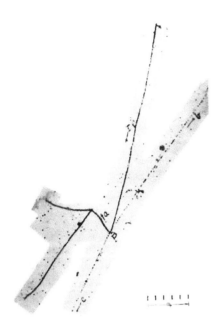

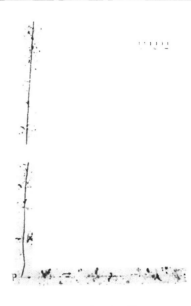

Figure 2.6. An early example of a K meson decay seen by O'Ceallaigh at Bristol in 1951.

Figure 2.5. Early sighting (1951) of a τ meson in emulsion at Bristol. It stops at the point P.

particle, heavier than the pion. In one case the secondary particle was a muon of twice the range of the muon from pion decay, identified by its decay to an electron. In the second event, the secondary particle was either a fast pion or a muon, which left the emulsion (figure 2.6). More events of both kinds were observed in the following months and in 1952 'Goku' Menon, and O'Ceallaigh proposed that there were two different decay modes of the heavy meson (later termed K), one to a muon and two neutral particles, because the secondary muons had a spread of energy, and the other to a pion and one neutral particle, since the pions had a unique energy.

This hypothesis was later confirmed by a series of multiplate cloud chamber studies, both at the Pic du Midi by the Ecole Polytechnique (Paris) and Manchester groups, and in the USA. In particular, a 1954 picture taken by the Princeton group showed the decay producing $\pi^+ + \pi^0$ in which the π^0 decayed via the rare (1 in 29 000!) channel directly to two electron pairs. The Ecole Polytechnique group also obtained decisive evidence for a decay to a muon and a single neutral particle. In the following years decay

modes to an electron, a π^0 and a neutrino, and also to a π^+ and two π^0 particles (called τ') were discovered. The similarity of the mass values and of the lifetimes for the different decays strongly suggested that they all originated from a single K particle of about $970m_e$ and lifetime about 10^{-8} s. At this time, 1954, the parent particle of the two pion decay was termed θ^+ and the three-charged-pion mode was still being termed τ.

However, there was a powerful theoretical argument against this hypothesis. Richard Dalitz of Birmingham University showed that the three-π system had negative parity whereas that of the two-π system was positive. Parity, the inversion or reflection of the space coordinates, is an important underlying symmetry in quantum theory. If nature does not distinguish between left and right, then the parity of the K particle must be the same as the overall parity of its decay products. Since the τ and the θ^+ secondary particles had opposite parities, the initial view was that they could not arise from the same parent particle.

This eventually led to the realization that the τ and the θ^+ were in fact the same particle, and parity is not conserved, a major turning point in weak interaction physics (see the chapter by Sutton).

NEUTRAL V PARTICLES, HYPERONS AND THE 'CASCADE' PARTICLE

Neutral V particles

Following the discovery of V particles, groups moved their chambers to the mountains in order to obtain more events. In 1950 Anderson's group (working on White Mountain at 3200 m) published details of 34 V events of both kinds and confirmed the earlier findings of Rochester and Butler. They were able to establish that the secondaries of the V^0 events included at least one meson, π or μ, and at least a strongly interacting particle, π meson or proton. The Manchester group, now at the Pic du Midi, were able to show quantitatively the existence of two distinct V^0 decay modes, namely a V_1^0 or 'superproton', subsequently called a Λ^0, decaying to a proton and a π^- meson, and a meson decaying to a $\pi^+ + \pi^-$. The latter was later identified as a decay mode of the neutral K meson and for a time was termed a θ^0. The same parity considerations hold for the comparison of this two-π decay with the three-π τ decay mode outlined above.

Hyperons

The discovery of the Λ^0 was the first indication of the existence of what came to be known as 'hyperons', particles heavier than a neutron and lighter than a deuteron. Another type of hyperon was found in 1953 by the Milan and Genoa emulsion groups in a photograph of a particle which came to rest in the emulsion, with mass $2500 \pm 600 m_e$ and which decayed to a proton, which also came to rest. The primary particle was termed Σ with the decay scheme $\Sigma^+ \rightarrow p + \pi^0$, confirmed by a cloud chamber event observed by Anderson's group and by observations by the Milan group of the alternative decay $\Sigma^+ \rightarrow \pi^- + n$. The corresponding negative particle, Σ^-, was not discovered until the end of 1953 by an accelerator experiment at Brookhaven.

Yet another new particle was discovered by the Manchester group in 1952. They observed a picture of a V^- decay to a V^0 (Λ^0) and a meson, the Λ being identified by its decay to $p + \pi^-$. Shortly afterwards Anderson's group observed three more unambiguous events of this nature, confirming the existence of what became known as a 'cascade' particle, $\Xi^- \rightarrow \Lambda^0 + \pi^-$.

THE ACHIEVEMENTS OF COSMIC-RAY STUDIES 1945–55

The years after the end of the war witnessed a steady increase in the number of research groups working in the cosmic-ray field as well as in the technical sophistication of the detectors. Not only was the size of cloud chambers increased but a typical arrangement might consist of two chambers one above the other, the top chamber in a high magnetic field for momentum measurements and the bottom chamber a multiplate chamber to observe the subsequent collisions or shower development. For example such an arrangement was installed on the Pic du Midi by the group of Leprince Ringuet and operated from 1950 onwards. New groups were formed, particularly in the USA, who concentrated mainly on large multiplate chambers.

The simplicity and relative cheapness of the emulsion technique also led to many new groups being established, especially in Italy. Technical advances made it possible to construct large stacks of stripped emulsions, each 600 μm thick. After exposure in balloon flights the stack could be disassembled and the pellicles mounted on glass and processed in the usual way. An x-ray grid printed on each emulsion made it possible to follow particle tracks through the stack from one pellicle to another.

At the same time collaborations grew up between the newly formed emulsion groups in the different countries, including several in eastern Europe, to share both in the cost of the large emulsion stacks and in the subsequent searching of the emulsions and in the analysis of the events found. For example, the so-called G-stack collaboration, which flew a 63 kg block of stripped emulsions by balloon over northern Italy in 1954, was initially composed of groups in Bristol, Copenhagen, Genoa, Milan and Padua, who later attracted more groups. The first of the resulting publications carried 36 names, very different from the small collaborations of two or three of the 1930s, but still a long way from the large scientist communities, more than a thousand strong, now planning experiments for the Large Hadron Collider (LHC) under construction at CERN.

An important factor in the unravelling of the complexities in the field of the new particles discovered from 1946 onwards was the complementary nature of the cloud chamber and emulsion techniques. The former was particularly effective in the discovery and study of neutral particles and the latter of charged particles.

For some 20 years, cosmic-ray studies provided the first glimpses of the world of subnuclear particles. With the advent of large particle accelerators in the early 1950s, cosmic-ray studies shifted to the background. Despite the great advances in latter years using successive generations of high-energy machines, ironically the highest energies known to Man are still those of cosmic rays, attaining collision energies of 10^{20} eV (10^8 TeV) and higher. On mountain tops, aboard balloons and high-flying aircraft, and mounted on space missions, cosmic-ray experiments will continue to provide a unique viewpoint of physics at the outer limits.

ACKNOWLEDGMENT

The figures in this chapter are reproduced from *40 Years of Particle Physics* edited by B Foster and P H Fowler 1988 (Bristol: Adam Hilger).

FURTHER READING

For the general reader

Close F, Marten M and Sutton C 1987 *The Particle Explosion* (Oxford: Oxford University Press)

Crease R P and Mann C C 1986 *The Second Creation* (New York: Macmillan)

Fraser G, Lillestøl E and Sellevag I 1998 *The Search for Infinity* (London: Philip's)

Friedlander M W 1989 *Cosmic Rays* (Cambridge, MA: Harvard University Press)

Lederman L with Teresi D 1993 *The God Particle* (New York: Bantam, Doubleday, Dell)

Ne'eman Y and Kirsh Y 1996 *The Particle Hunters* 2nd edn (Cambridge: Cambridge University Press)

Weinberg S 1993 *The Discovery of Subatomic Particles* (Harmondsworth: Penguin)

For the physicist

Brown L M and Hoddeson L (ed) 1983 *The Birth of Particle Physics* (Cambridge: Cambridge University Press)

Brown L M, Dresden M and Hoddeson L (ed) 1989 *Pions to Quarks* (Cambridge: Cambridge University Press)

Brown L M and Rechenberg H (1996) *The Origin of the Concept of Nuclear Forces* (Bristol: Institute of Physics Publishing)

Cahn R N and Goldhaber G 1991 *The Experimental Foundations of Particle Physics* (Cambridge: Cambridge University Press)

Ezhela V V et al 1996 *Particle Physics One Hundred Years of Discoveries (An Annotated Chronological Bibliography)* (New York: American Institute of Physics)

Foster B and Fowler P H (ed) 1988 *40 Years of Particle Physics* (Bristol: Adam Hilger)

Hendry J (ed) 1984 *Cambridge Physics in the Thirties* (Bristol: Adam Hilger)

Pais A 1986 *Inward Bound* (Oxford: Oxford University Press)

Rossi B 1990 *Moments in the Life of a Physicist* (Cambridge: Cambridge University Press)

Sekido Y and Elliot H 1985 *Early History of Cosmic Ray Studies* (Dordrecht: Reidel)

ABOUT THE AUTHOR

W O ('Owen') Lock first studied cosmic rays, obtaining his PhD in 1952 at Bristol, in the research group led by Cecil Powell (Nobel Prize in Physics 1950). He was subsequently a Lecturer in Physics at Birmingham (UK), carrying out experiments using nuclear emulsions at the 1 GeV proton synchrotron.

He joined CERN in 1959 and in 1960 become Joint Group Leader of the emulsion group, working at the new 28 GeV proton synchrotron. In 1965 he moved to the Personnel Division, where he became Deputy Division Leader and then Head of Education Services from 1969–77.

From 1978–88 he was Assistant to successive CERN Directors-General, being particularly involved in the development of relations with the then Soviet Union and later with the People's Republic of China. From 1989 until his retirement in 1992 he was Adviser to the Director-General with particular responsibility for relations with Central and Eastern Europe.

He is the author or co-author of two postgraduate texts on high-energy physics and of some 50 research papers and review articles on elementary particle physics and on the nuclear emulsion technique, as well as numerous reports and articles on international collaboration in science. Together with E H S Burhop and M G K Menon, he edited *Selected Papers of Cecil Frank Powell* 1972 (Amsterdam: North-Holland). He was Secretary of the International Committee for Future Accelerators 1978–92 and a member of the International Council of Scientific Union's Special Committee on Science in Central and Eastern Europe and the Former Soviet Union from 1994–96.

3   FIRST ACCELERATOR FRUITS

Nicholas P Samios

Editor's Introduction: Nicholas P Samios played a major role in particle research at US particle accelerators in the 1950s and 1960s, including the historic 1964 discovery of the Ω^- which confirmed the SU(3) symmetry picture and introduced the idea of quarks. Here he surveys the seemingly unending avalanche of new particle discoveries from those pioneer accelerators. Their discovery potential was boosted by the invention of a powerful new research tool: the bubble chamber. This harvest of particles soon led many physicists to believe that such a complex scenario of supposedly fundamental states hid a deeper level of structure.

The 1950s and 1960s heralded the beginning of a new era for particle physics. As noted earlier (see the chapter by Lock), a great deal of experimental work had been conducted utilizing cosmic rays, particles that descend upon the Earth's atmosphere from outer space. However, this type of work was limited in that it depended on Nature itself both for the low frequency of interesting events as well as the inconvenience and difficulties encountered in performing such experiments, often at high altitudes. The technical, social and economic developments during and after the Second World War also had a profound affect in changing the scale as well as the sophistication of doing science.

The history of the development of the atomic bomb is well known, with the unprecedented concentration of manpower and resources needed to carry out this project. What is less appreciated is the equally important and crucial technical development of radar, sonar, electronics and the establishment of an experienced infrastructure of talented people. As such, at the end of the war, the large pool of available talent, the recognition of governments of their duty in funding research and the desire of scientists to return to academia and basic research, opened the way for a renaissance in particle physics in the 1950s and 1960s (see the chapter by Westfall and Kriege).

E O Lawrence had invented the cyclotron in 1930 for which he received the 1939 Nobel Prize. It was the machine of choice for going to higher energies. It was a rather simple accelerator in concept in that it required a magnet, a good vacuum chamber and a varying electric field (see the chapter by Wilson). In fact, the maximum energy of the projectile, either a proton or an electron, was critically dependent upon the size of the magnet; the larger the magnet, the higher is the energy. With the return of the scientists to research, with the advances in technology and with the desire to learn more about the fundamental nature of matter, a plethora of cyclotrons were built in the USA and Europe.

The scope and breadth of these activities can be illustrated by listing a few of the institutions that established such facilities: the University of California (Berkeley), Chicago, Columbia, Harvard and Rochester in the USA, and Amsterdam, Glasgow, Harwell, Liverpool, Uppsala and CERN in Europe, and Moscow in the USSR. Magnet pole diameters ranged from 100 to 184 in and proton energies from 160 to 720 MeV. It was now possible to study the primary proton–proton interaction in the laboratory at reasonably high energies, hundreds of MeV, and of equal importance was the ability to produce π mesons in great profusion. As such, it was possible in a rather short time to demonstrate that these π mesons, or pions, came in three varieties, i.e. positive π^+, negative π^- and neutral π^0, with similar masses.

This multiplet phenomenon, one particle coming in several states, was not new. The existence of the proton and neutron is another such example, in this instance one particle with two states: one positive, the proton, and the other neutral, the neutron. There are therefore doublets (nucleon) and triplets (pion) that exist as particles and the question arises as to whether there are other combinations of particles: quartets, sextuplets, etc. We shall see that the answer is yes and that the emerging patterns will be crucial in deciphering the underlying structure of matter.

It soon became evident that still higher-energy machines would be extremely useful. Although, as noted earlier, cosmic-ray activities were limited in the rate and ease of obtaining data, some new and interesting phenomena were being observed. These involved the existence of so-called strange particles, particles that decayed into multiple pions or a pion and a nucleon, but with relatively long lifetimes, of the order of 10^{-10} s instead of an expected lifetime of 10^{-23} s, quite a difference. However, these new particles were much heavier than a pion, and some even heavier than a nucleon, and could not be artificially produced in existing cyclotrons. Much higher energies were needed, much higher than a few hundred MeV, in the GeV range, requiring a different type of accelerator.

Such devices were developed but involved a conceptual change. Cyclotrons and synchrocyclotrons were limited in energy by the size of their single magnet. In order to attain higher energies, the geometry had to be changed to that of a synchrotron, with a doughnut-shaped ring, where the size could be much larger, limited essentially by the circumference of the ring and the stability of the beam (see the chapter by Wilson). In this way the Cosmotron, a 3 GeV machine, was built at Brookhaven National Laboratory in 1952 and the 7 GeV Bevatron at Berkeley in 1954. The energies of those machines were sufficiently high to produce strange particles in controlled conditions and in quantity.

This resulted in many noteworthy findings. In particular, the issue of an assortment of a variety of observed strange particles called θ, τ, K_{e3}, $K_{\mu3}$ and τ^1 was resolved by experimentally establishing that they all had the same mass and essentially the same lifetime. Therefore they were all different manifestations of one particle: the K meson. This was a great triumph. It was during these times that Fermi commented, 'Young man, if I could remember the names of these particles, I would have been a botanist.' As a result of this consolidation, he would not have had to resort to such drastic measures. Furthermore, the investigation of different decay modes of the K meson, an episode which

Table 3.1. The list of known particles in the late 1950s.

	Charge	Mass (MeV)	Spin	Strangeness	Parity
Baryons					
p	+1	939	$\frac{1}{2}$	0	+
n	0	940	$\frac{1}{2}$	0	+
Λ^0	0	1115	$\frac{1}{2}$	−1	+
Σ^+	+1	1189	$\frac{1}{2}$	−1	+
Σ^-	−1	1197	$\frac{1}{2}$	−1	+
Σ^0	0	1193	$\frac{1}{2}$	−1	+
Ξ^-	−1	1321	$\frac{1}{2}$	−2	?
Ξ^0	0	1315	$\frac{1}{2}$	−2	?
Mesons					
π^+	+1	140	0	0	—
π^-	−1	140	0	0	—
π^0	0	135	0	0	—
K^+	+1	494	0	+1	—
K^0	0	498	0	+1	—
K^0	0	498	0	−1	—
K^-	−1	494	0	−1	—

became known as the τ–θ puzzle (see the chapter by Lock), had another dramatic consequence, the questioning and ultimately the overthrow of parity, showing that nature is not left–right symmetric in these decays.

However, beyond the K mesons, other strange particles were studied, namely the lambda particle Λ^0, which decayed into a proton and pion ($\Lambda^0 \rightarrow p\pi^-$), and the sigma particles Σ^- and Σ^-, which also decayed into a nucleon and a pion ($\Sigma \rightarrow N\pi$), and a neutral Σ^0 was discovered which decayed into a Λ^0 and a photon. The astonishing feature which emerged was that these particles were always produced in association and not singly; one could not produce a single Λ^0 or a single kaon, but always required a Λ and a K or a Σ and a K. This had the effect of introducing a new label, or quantum number, for each particle and further required that this number be conserved in all their production interactions (involving the strong nuclear force) and violated in their decays (involving the weak nuclear force). By the late 1950s, we had an assortment of particles, labelled by their mass, charge, multiplicity, strangeness and spin, listed in table 3.1.

In addition, the discovery of the antiproton at Berkeley in 1955 essentially doubled the number of known particles. In fact the energy of the Bevatron had been chosen to be able to produce the antiproton if it existed. One then expected the antineutron, antilambda, etc, which were indeed found. But was this the whole story or just the tip of the iceberg? If so, what did it all mean? How many more particles would there be, and how many of them are fundamental? The answer came in the next 20 years, but both higher-energy machines and more advanced experimental techniques were required.

It is more the rule than the exception that great progress is made with the advent of a new idea. One can make steady progress by improving on a given technique or design, such as occurred with the cyclotrons; however, such changes reach a natural limit and further progress needs innovation, and this is where open inquiry in science really shows its worth. As noted earlier, cyclotrons encompassed energies from a few MeV to hundreds of MeV. To surpass 1000 MeV (1 GeV) needed the doughnut geometry of the synchrotron.

For higher energies, ultimately going from 30 GeV to 1000 GeV, a new idea emerged, that of strong alternate focusing of magnetic fields. In the previous doughnut machines such as the Cosmotron and Bevatron, magnetic fields were used to maintain the charged-particle trajectories in the vacuum chamber of the circular doughnut; however, the beams, rather than being held tightly together, were weakly focused and required rather large vacuum chambers. For the Cosmotron (3 GeV) it was approximately $1\frac{1}{2}$ ft wide while the Bevatron (6 GeV) was 3 ft in width, requiring rather large magnets, even for a doughnut configuration. The new idea was to change the magnetic field configuration from uniformity and weak focusing to strong magnetic fields alternately focusing and defocusing, with a resultant net focusing of the beam (see the chapter by Wilson). A direct consequence was the need for a much smaller vacuum chamber; one could envisage an order-of-magnitude jump to 30 GeV with a chamber 6 in across and with commensurately smaller magnets.

Several accelerators were quickly built on this principle: the first being the Cornell electron synchrotron (3 GeV) quickly followed by the Brookhaven Alternating Gradient Synchrotron (AGS) (33 GeV) and the CERN Proton Synchrotron (PS) (28 GeV). This design principle enabled experimentalists to explore the multi-GeV mass region which was essential for deciphering the next stage of complexity in the search for the basic structure of matter, the spectroscopy of particles.

The importance and significance of this energy region is illustrated in figure 3.1 where the rate of proton–proton interactions (the cross section) is plotted as a function of energy. One notes the slow decrease of the rate at very low energies, which then levels off and rises as the energy reaches the 1 GeV range, giving rise to a bump (indicating the emergence of a new phenomenon, namely the production of new particles) and then levelling off with no further significant structure at energies at present attainable. We may have to go to hundreds of GeV or even TeV for newer physics. However, for this discussion it is the bump ranging over several GeV that provided a wealth of new particles and the excitement and vigorous activity that culminated in a simplification in our understanding of the basic constituents of matter.

Equally significant progress was made in experimental techniques. With photographic emulsions, charged tracks were imaged by the silver grains which when developed

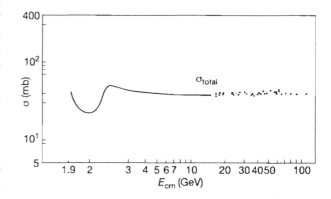

Figure 3.1. The proton–proton interaction rate as a function of energy. After an initial drop, the rate increases significantly. When physicists first explored this region, they found many new particles.

allowed the determination of geometrical and some dynamical quantities (angles and velocities) using specially equipped microscopes. This technique was limited and the processing of its data extremely slow. However, it provided very good spatial resolution and therefore continued to be utilized but to a much lesser extent.

Counters had been developed containing various gases that emitted light when traversed by charged particles. These were coupled to phototubes which detected the emitted light which was then processed by electronic circuits and then catalogued. New materials and electronic circuitry greatly increased the capability of the counter technique. Plastic scintillating material was developed that operated in the nanosecond (10^{-9} s) time domain. Coincidence circuitry now handled microseconds rather than milliseconds, while the phototubes which collected the light and converted it into electron pulses had much greater surface area, higher efficiencies and more sensitivity. Capabilities and data rates increased enormously.

In addition a novel counter, the Čerenkov counter, was developed that measured the velocity of a particle. If one has an independent determination of the momentum of a particle (e.g. by measuring the curvature of its track in a magnetic field) then one can define the mass of the particle by dividing its momentum mv by the velocity v.

Among the more powerful visual techniques was the cloud chamber. This device consisted of a gas in a supersaturated state; although still a gas, the local pressure and temperature favoured a liquid state. Under such unstable conditions, a disturbance by a charged particle creates droplets along its path which can be photographed and the resulting tracks measured.

An enormous advance was made by reversing the above sequence, that of the gas and droplet, in the bubble chamber invented by Donald Glaser. In this case, a liquid is subjected to conditions where its local pressure and temperature would warrant it being a gas. This can be done, for instance, by quickly lowering the pressure on the liquid. Again, one has an unstable situation where the introduction of a charged particle causes bubbles to form along its trajectory. Again, one can introduce charged particles into such a device, have them interact in the liquid and photograph the bubble paths of all the resulting charged particles. The event rate is one thousand times greater in

the bubble chamber than in the cloud chamber owing to the ratio of liquid to gas densities. This, coupled with high spatial resolution (about 0.1 mm), and the superb geometry, make this a most powerful instrument.

From initial volumes of less than a litre, the sizes of bubble chambers increased to 1000 l within a few years to over 10 000 l within 10–15 years. These were built by groups led by some of the most inventive and dynamic experimentalists of the day: Luis Alvarez of Berkeley, Ralph Shutt of Brookhaven, and Jack Steinberger of Columbia, and in Europe Bernard Gregory and Charles Peyrou at CERN. As a consequence this device was utilized in the discovery of many particles and had a profound effect on the development of particle physics during the exciting era from the mid-1950s to the early 1970s.

To find and catalogue the various particles, two generic conceptual techniques were utilized, shown in figure 3.2: formation and production experiments. In the formation approach, one varied the energy at which two particular particles were allowed to collide. When the energy is equal to the mass of a resonance (or particle) R, there will be a substantial increase in the rate at which the two particles interact and, as one goes beyond this mass, a corresponding decrease, giving a characteristic 'bump' in the interaction rate at the mass of the resonance. Furthermore, one can decipher the charge and multiplicity as well as the spin and parity of the particle by studying the properties of the bump and varying the character of the colliding particles. There is, however, the limitation that one is restricted to resonances that can be made up of the two incoming interacting particles.

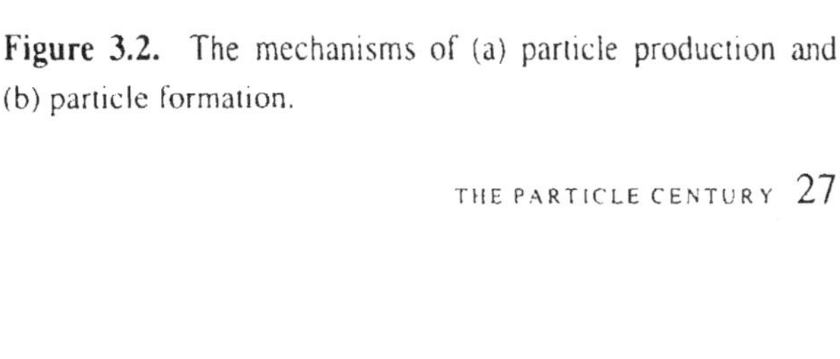

Figure 3.2. The mechanisms of (a) particle production and (b) particle formation.

In the production mode, one instead infers the existence of a resonant state by calculating the effective mass of a combination of the final-state particles. Conceptually what occurs is shown in figure 3.2(a), where a resonance is formed which lives for 10^{-23} s, too short to traverse any observable distance, and then decays. Therefore, for this length of time a resonance has been produced whose effective mass is equal to $m_{ij} = \sqrt{(E_i + E_j)^2 - (P_i + P_j)^2}$ where E_i and P_i are the energy and momentum of particle i. Such a resonance will produce a bump in the effective-mass distribution. Of course sophisticated techniques are used to distinguish real from spurious effects, but the essence of the technique remains the same. Again, by examining details of the decay products of the resonance, one can ascertain its charge, spin and parity. This approach has been especially powerful in finding meson resonances, which decay into multiple pions.

We recall that the ground state of the baryon is the nucleon, composed of the proton and neutron. Possible excited states of the nucleon can be investigated in formation experiments by the interaction of beams of pions π^+ and π^- on hydrogen (protons) or deuterium (protons and neutrons) targets. As such, the counterparts of the neutron could be achieved via π^-p interactions (net charge 0 and nucleon number $+1$) and the proton via π^+n interactions (net charge $+1$ and nucleon number $+1$). One can also create other baryon states of higher multiplicity (not seen before) by the reactions π^+p (net charge $+2$ and baryon number $+1$) and π^-n (net charge -1 and baryon number $+1$). Utilizing the new counter techniques and the accelerators with GeV pion beams, such resonances were found. The results of such endeavours are illustrated in figure 3.3 where the π^+p and π^-p rates (cross sections) are plotted as functions of energy in the centre of mass (the mass of the resonance). One can see several clear bumps.

In fact, the first such resonance observed was in the π^+p system at a mass of 1238 MeV and was completely novel, in that it was an excited baryon but with a charge multiplicity of four, occurring in the $+2$, $+1$, 0 and -1 charge states. It was named the Δ and determined to have a spin–parity of $\frac{3}{2}^+$. In this manner several more states were found, two nucleon doublets, mass 1520 MeV with $\frac{3}{2}^-$ spin–parity and mass 1690 MeV with $\frac{5}{2}^+$ spin–parity, as well as a second Δ at a mass of 1950 MeV

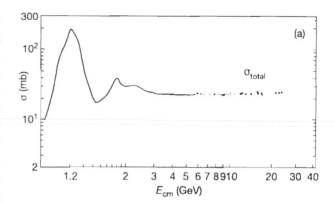

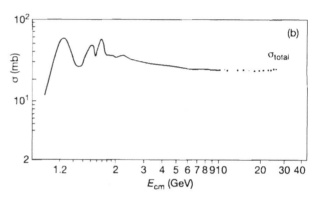

Figure 3.3. The rich structure in (a) π^+p and (b) π^-p reactions.

and a spin–parity of $\frac{7}{2}^+$. Similar but more sophisticated efforts were made using K^+ and K^- beams, again with hydrogen and deuterium targets. This approach was more difficult because kaons were less plentiful than pions, and making kaon beams required more elaborate and meticulous experimental techniques, including extensive use of Čerenkov counters.

Since the K^- has negative strangeness, K^-p interactions can explore excited states of the singlet Λ^0 and/or excited states of the neutral component of the Σ triplet. Analogously K^-n formation experiments would be sensitive to excited Σ^- particles. (Exploring excited Σ^+ particles would need K^0p interactions and therefore require pure K^0 beams, which are very difficult to obtain.) Several excited Λ^0 particles at a mass of 1820 MeV and $\frac{5}{2}^+$ spin–parity, and a mass 1520 with $\frac{3}{2}^-$ spin–parity were found as well as an excited Σ at 1775 MeV with $\frac{5}{2}^-$ spin–parity.

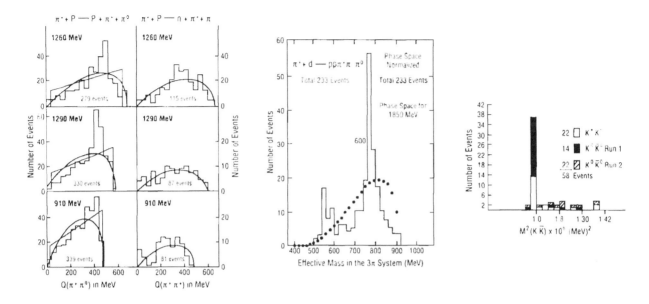

Figure 3.4. Production of meson resonances. Note that resonances appear clearly in $\pi^+\pi^0$ spectra (far left) but not in the $\pi^+\pi^+$ spectra (immediately to the right).

Of particular importance was the lack of any structure or resonances in the K^+p system, resonances which would have signified a positive strangeness baryon.

Experimental groups at many laboratories participated in these activities, the most prominent being those at Berkeley, Brookhaven and CERN. The exploration of more exotic baryon states, namely strangeness minus 2 or even minus 3, required the production reaction technique.

The real strength and power of the production reaction method, however, become even more abundantly apparent in the search for mesonic states, such as ($\pi\pi$, 3π...). (KK, KKπ...) or (Kπ, K$\pi\pi$...) states. These can be explored via reactions such as $\pi^-p \rightarrow (\pi\pi)p$ or $\pi^-p \rightarrow (K\pi...)\Lambda$ or $Kp \rightarrow (KK...)\Lambda$. All are especially amenable to the bubble chamber technique. One can search for excited states of the pions (triplet). Kπ (doublet) and a KK (singlet). Resonances decaying into multiple pions were found, the most noteworthy being the triplet ρ going into two pions and with mass 750 MeV and 1^- spin–parity by Walker and collaborators at the Brookhaven Cosmotron, the ω with mass 785 MeV, also 1^- spin–parity but being a singlet occurring only with one charge, neutral, by Bogdan Maglic

and others of the Alvarez group at Berkeley; and the η, also a singlet decaying into three pions, with net zero charge, a mass of 550 MeV and 0^- spin–parity by Aihud Pevsner and his group. In an analogous manner, several strange Kπ and K$\pi\pi$ resonances were found as well as singlet (KK) states, most notably the φ with a mass of 1020 MeV and 1^- spin–parity by Samios and collaborators at Brookhaven. Several of these production bumps are shown in figure 3.4. It is amazing how many of these mesonic states were found both quickly and accurately, because they were produced profusely and with small background. This was not true for the higher-spin states.

As noted earlier, the baryonic sector is also amenable to exploration via production experiments, especially the negative strangeness sector. As such, the use of kaon beams (K^-) in conjunction with bubble chambers filled with liquid hydrogen was especially productive. The incoming beams had to be made exceptionally clean, with a high kaon purity compared with pions since one was mainly interested in K^-p interactions producing $\Lambda\pi$, $\Sigma\pi$ and $\Xi\pi$ in the final states. This involved the development of sophisticated electromagnetic devices capable of sustaining

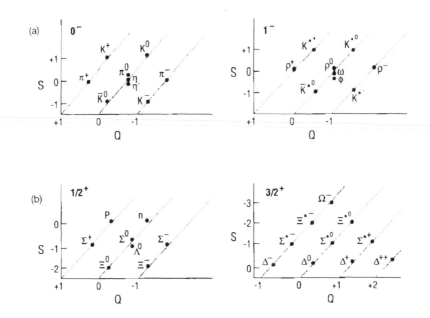

Figure 3.5. Charge–strangeness correlations in (a) mesons and (b) baryons.

high electronic fields, kilovolts, over gaps of several inches, a technical *tour de force*, allowing 10–20 incident K^- particles into the bubble chamber and essentially no pions. A prolific number of resonances were found, the most notable being an excited Σ decaying into $\Lambda\pi$ at a mass of 1385 MeV and $\frac{3}{2}^+$ spin–parity and an excited Ξ at a mass of 1530 MeV, also $\frac{3}{2}^+$ spin–parity.

Examining the spectroscopy of the observed particles, certain emerging patterns led to a classification scheme. This SU(3) symmetry was proposed by Murray Gell-Mann and independently by Yuval Ne'eman and subsequently unveiled an underlying substructure (see the chapter by Ne'eman).

The most pertinent patterns for mesons are shown in figure 3.5(a) and for baryons in figure 3.5(b). The particles seem to form families of eight, or possibly nine in the case of mesons, and a family of eight and ten in the case of baryons. In the meson multiplet of spin parity 0^- and 1^- there are two doublets, a triplet and a singlet with the indicated charges and strangeness. As noted earlier, a ninth member of the 1^- was found, the singlet φ. Subsequently a ninth member of the 0^- multiplet, the singlet η' (960 MeV) was found. The baryon ground state $\frac{1}{2}^+$ is also an octet composed of

two doublets (the N and Ξ) as well as the triplet Σ and singlet Λ. A long-standing debate was whether the Σ–Λ relative parity was the same or different. It took a great deal of time and experimental effort before an affirmative answer of positive parity was derived and until then there were many alternative symmetry schemes (different from SU(3)) with the Σ and Λ in different families.

The most intriguing and possibly simplest of the multiplets is the baryonic $\frac{3}{2}^+$ family of ten anchored by the Δ (1238 MeV) and the Σ (1385 MeV), two of the earliest resonances found. The interesting features are that the mass differences are expected to be equal: $\Sigma - \Lambda = \Xi - \Sigma = \Omega - \Xi$, and that this Ω would be a singlet with strangeness -3, only occurring in the negatively charged state. Furthermore, the expected Ω^- mass would prevent it from decaying into any existing particles conserving strangeness (such as ΞK) and therefore must decay into modes violating strangeness by one unit, namely into $\Xi\pi$ or ΛK final states. This is a weak decay, similar to that of the ground state Λ into πN, and therefore should have a similar lifetime, 10^{-10} s instead of the lifetime of 10^{-23} s of the other members of the family of ten. This astonishing result means that the Ω^- would traverse distances of the

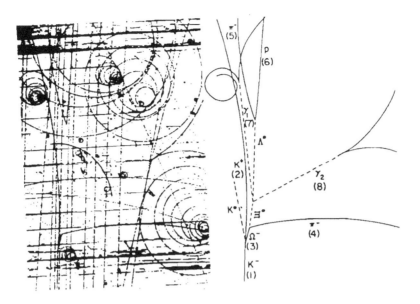

Figure 3.6. One of the most historic photographs in physics: the 1964 bubble chamber discovery of the production of an Ω^- meson and its subsequent decay.

order of centimetres, certainly visible in a bubble chamber.

These considerations emerged at the historic 1962 international conference held in CERN where the present author presented evidence for both the φ (1020 MeV) and Ξ (1530 MeV). Gell-Mann, in a question and comment period after a theoretical talk, pointed out the simplicity and preciseness of the SU(3) prediction for the family of ten spin-$\frac{3}{2}$ baryons, and the unique properties of the Ω^-. This complemented and focused Brookhaven objectives, where further searches for heavier negative strangeness had been planned. A beautiful experiment was performed utilizing a very sophisticated electromagnetic system to produce a pure K^- beam of 5 GeV/c momentum using a 1000 l hydrogen bubble chamber. To produce an Ω^- of strangeness minus 3, the best strategy is to start with as negative a strange baryonic system as possible, in this case K^-p with the expected reaction being $K^-p \rightarrow \Omega^- K^+ K^0$ (which conserves strangeness). The dramatic event (figure 3.6) shows a bubble chamber photograph of a K^- meson interacting with a proton to produce an Ω^- which travels several centimetres before decaying into Ξ^0 and π^- with the Ξ^0 subsequently decaying into a Λ^0 and π^0 each of

which in turn decays visibly in the bubble chamber. The most amazing aspect of this event is the materialization of the two gamma-rays (photons) from the π^0 into electron–positron pairs in the liquid hydrogen. The probability for such an occurrence is one in a thousand!

Since all decay products of the Ω^- were visible, its reconstruction was straightforward and its mass, charge, lifetime and strangeness were determined, all of which fit the pattern of the singlet member of the family of ten particles with spin–parity $\frac{3}{2}^+$. This was the key element in validating the SU(3) symmetry of particle spectra.

It was at least a decade before the spin of the Ω^- was measured to be $\frac{3}{2}$ but this was anticlimactic. In due course, several more meson multiplets of the family of nine were uncovered, namely those with spin–parity 2^+ and 3^-, and several octets among the baryon families with spin–parity $\frac{3}{2}^-$ and $\frac{5}{2}^+$. As noted earlier, no baryon particles were found in the K^+p system with its strangeness of $+1$ and double charge, in the $(\Xi^-\pi^+\pi^+)$ system with strangeness -2 and positive charge, nor in the mesonic K^+K^- strangeness $+2$, doubly charged system which would require SU(3) representations with families larger than eight or ten.

This had major ramifications in that one could mathematically accommodate such a restriction by constructing all these mesons and baryons from three basic units, called quarks by Gell-Mann, using the triplet representation in SU(3). As such, all mesons could be made out of different combinations of a quark and antiquark pair, while baryons would be made up of three quarks.

This was an enormous simplification. Instead of hundreds of elementary particles, π, K, Λ, Σ, etc, there were just three quarks and their antiquarks. These quarks, however, have peculiar properties, such as non-integer charge ($-\frac{1}{3}$, $\frac{2}{3}$ instead of ± 1), and non integer baryon number ($\frac{1}{3}$), but with spin $\frac{1}{2}$. Searches for such non-integral-charge particles have been performed in nearly every conceivable place, in oceans, in Moon rocks, at the highest-energy accelerators, etc, all with negative results. Free quarks have not been found and it is presumed that they do not exist. Quarks only occur bound in hadrons. Furthermore, all these multiplets, with their observed spins and parities, could be derived from a simple model where the quarks are weakly bound when they are close together and their spin–parity is derived from the innate spin–parity of the quarks and their relative angular momenta. However, since quarks never appear as free particles, the force between them must become very strong as they get further apart. The weak binding at close distances became known as 'asymptotic freedom' and the strong binding at large distances as 'infrared slavery'. This experimental observation had to be and ultimately was accommodated in the underlying theory.

These observations of particle spectroscopy were complemented by experiments with high-energy electrons and neutrinos interacting with nucleons (see the chapter by Friedman/Kendall). Hard scattering, *à la* Rutherford, was seen, indicating that the nucleon was not fundamental but was composed of hard centres, called partons. Detailed experiments revealed that the properties of these partons were those of quarks, namely spin $\frac{1}{2}$ and fractional charged entities, and held together by gluons. These studies led to the development of the Standard Model and quantum chromodynamics which to this day describe all experimental observations in terms of the dynamics of quarks and gluons, albeit with many incalculable parameters. To progress further we probably have to go to even higher energies, multi-TeV, and more sophisticated detectors in order to obtain a better understanding of the microscopic physical world.

FURTHER READING

Cahn R N and Goldhaber G 1991 *The Experimental Foundation of Particle Physics* (Cambridge: Cambridge University Press)

Crease R P and Mann C C 1995 *The Second Creation* (New York: Rutgers University Press)

Riordan M 1987 *The Hunting of the Quark* (New York: Simon and Schuster)

ABOUT THE AUTHOR

Dr N P Samios is a distinguished experimental particle physicist as well as an administrator of science. His early career was at Columbia University where he received his AB and PhD, joining the faculty for several years working with Jack Steinberger and Mel Schwartz. In this very productive period this group measured the properties of many new particles—the spin of the Λ and Σ hyperons and the parity of the neutral pion, and discovered more particles such as the neutral Σ hyperon as well as demonstrating parity non-conservation in hyperon decays.

He then joined the scientific staff at Brookhaven National Laboratory where he was active in studying both the spectroscopy of elementary particles as well as their dynamics. He participated in the discovery of the φ (1020 MeV), Ξ^* (1530 MeV), η' (960 MeV) and f' (1250 MeV) resonances as well as the historic Ω^- and the first charmed baryon, all of which laid the groundwork and helped establish the SU(3) symmetry of particle spectra and ultimately of today's Standard Model.

Dr Samios has held many administrative posts including Director of Brookhaven National Laboratory from 1982 to 1997. He has served on numerous national and international advisory committees among which have been the International Committee for Future Accelerators; the US/Japan and US/China Committees on High Energy Physics; Chair, Division of Particles and Fields, and program committees for Fermilab and SLAC.

He has received the E O Lawrence Memorial Award, The New York Academy of Sciences Award in Physical and Mathematical Sciences and the W K H Panofsky

Prize. With Robert Crease he has written several popular articles including 'Cold Fusion Confusion' for the *New York Times Magazine* and 'Managing the Unmanageable' for the *Atlantic Monthly.* Dr Samios is a member of the National Academy of Sciences, and a fellow of the American Academy of Arts and Sciences as well as a corresponding member of the Akademia Athenon of Greece.

4 THE THREE-QUARK PICTURE

Yuval Ne'eman

Editor's Introduction: While the bountiful particle harvest of the 1950s and early 1960s suggested some deeper underlying symmetry, it was not at all obvious what this symmetry had to be. Here, Yuval Ne'eman describes how attention came to centre on the SU(3) Lie algebra. Various SU(3) representations accounted for observed particle multiplets, and missing entries made for spectacular predictions. However, the simplest SU(3) possibility, a triplet, did not correspond to anything seen in experiments. This spin-off prediction of an underlying three-quark picture, in which the observed particles could be explained by fitting the quark building blocks together in the different ways permitted by the SU(3) symmetry, was one of the major physics achievements of the late twentieth century. As well as relating his own role in this search for particle order, the astute and cultivated Ne'eman covers many of the underlying concepts. His explanation of the role of global and local symmetries is especially valuable.

HADRONIC PROLIFERATION IN 1949–62: ISOSPIN AND STRANGENESS

Around 1930, the composition of matter and radiation appeared very simple; atoms were assumed to be made of protons and electrons, and Einstein had shown in 1905 that electromagnetic radiation consists of photons. It was J Chadwick's 1932 discovery of the neutron which put an end to all that simplicity. The strong binding in nuclei, between two protons, two neutrons or a proton and a neutron—a new force appropriately named *strong interaction*—had been measured and found to be short ranged and the same in all three cases, about 1000 times stronger than the electromagnetic repulsion between two protons at that range. Yukawa's suggested carriers of that force, the pions, had eventually been discovered in 1947–50 (see the chapter by Lock)—but a new Pandora's box had opened (see the chapter by Samios). Along with the pions (with masses 135 MeV for π^0 and 139.6 MeV for $\pi^{\pm}$), however, four new and unexpected mesons $K^{\pm}$, K^0 and $\bar{K}^0$ (masses 494 for $K^{\pm}$ and 498 MeV for the K^0 and $\bar{K}^0$) were found (all with 0^- spin-parity assignments) together with six new nucleon-like baryons, the *hyperons* Λ^0 (1116 MeV), Σ^+ (1189 MeV), Σ^0 (1192 MeV), Σ^- (1197 MeV), Ξ^0 (1315 MeV) and Ξ^- (1321 MeV). By 1960, the spins of the first four hyperons had been measured and found to be

$\frac{1}{2}$, like the nucleon's. The number of *hadron* species was increasing fast.

Already in 1952, Fermi had discovered the Δ^{++}, Δ^+, Δ^0 and Δ^- states, at some 1230 MeV. These are very short-lived particles (with a lifetime of around 10^{-23} s), known as *resonances*. By the early 1960s, several such resonances had been discovered: mesons with spin–parity 1^- ('vector' mesons), 1^+, 2^+, decaying into several pions or kaons, and new spin-$\frac{1}{2}$, spin-$\frac{3}{2}$, spin-$\frac{5}{2}$ and spin-$\frac{7}{2}$ baryon resonances. The number of particle species was thus nearing 100.

Right after the 1932 discovery of the neutron, W Heisenberg suggested that protons and neutrons be viewed as two 'states' of one entity. The 1925 discovery of the electron's *spin* had revealed two rotational states of electrons, with the same total amount of angular momentum, $\frac{1}{2}$, but distinguishable in the presence of a magnetic field, through the direction of their intrinsic magnetic moments. Mathematically, spin-'up' and spin-'down' electrons had just been found to span a complex two-dimensional Hilbert space (a space of *states*) upon which the ordinary three-dimensional rotation group can act. Quantum mechanics requires the wavefunctions to be complex and thus allows for the existence of *spinors*, which could not have existed in the classical world.

The strong interaction does not distinguish between

protons and neutrons, and it is only the weaker interaction of electromagnetism which senses the difference (and is responsible for the tiny difference of 0.15% in their masses). What Heisenberg was now suggesting was to apply the same mathematical idea of spinors, in the abstract. Aside from the fact that both protons and neutrons are true spin-$\frac{1}{2}$ spinors, i.e. under rotations in physical space–time we should also regard the two nucleons as the two manifestations of a fictitious spinning motion, *an abstract 'rotation' in the space of states*, a rotation transforming an electrically charged particle into its uncharged partner. This was named *isospin*, since it connects different *isobars* in nuclei, and denoted I, so that the nucleons have $I = \frac{1}{2}$ (the *pseudo*magnetic component I_z being $+\frac{1}{2}$ for the proton and $-\frac{1}{2}$ for the neutron).

The concept was extended to the pions by N Kemmer in 1938. In experiments in which neutrons are scattered off a nucleus, it often occurs that the outgoing particle is a proton ('exchange' reactions). Kemmer showed that three pions, an isospin-1 vector, would fit the bill. *The mathematical statement of the invariance of the strong interactions under isospin 'rotations' was equivalent to 'charge independence' (as it was known in the 1950s), i.e. the strength of the binding between any two nucleons is the same, whatever their electrical charges.*

The new hyperons and kaons are produced in an *associated production*, i.e. two by two, like the animals emerging from Noah's Ark. For example, $p + \pi^- \rightarrow \Lambda^0 + K^0$, or $p + \pi^- \rightarrow \Sigma^- + K^+$, etc. By 1953, T Nakano and K Nishijima in Japan and independently M Gell-Mann in the USA had found the key to this behaviour. A new additive quantum number, *strangeness S*, obeyed by the strong interactions, with the nucleon and pions having $S = 0$, requires the new *strange* particles to come in pairs so as to mutually cancel their strangeness and preserve the initial $S = 0$ state. It is sometimes convenient to replace the strangeness quantum number S by *hypercharge*, which is defined as $Y = B + S$, where B is the 'baryon number', $+1$ for baryons of half-integer spin, 0 for integer spin mesons and -1 for antibaryons. M Gell-Mann and K Nishijima found that the hypercharge (or strangeness) and isospin assignments are constrained by the values of the electric charges Q, namely $Q = I_z + Y/2$. The new particles are all unstable, with the lifetimes indicating, however, that the decays are due to the weak interactions. These have strangeness-changing transitions with $\Delta S/\Delta Q = 1$.

THE SEARCH FOR ORDER (1954–64): STRUCTURAL APPROACHES

This state of affairs, with about 100 different species of hadrons already listed—and yet with no understanding of that list—was clearly intolerable. Theorists were thus drawn by the challenge to find the order behind all that proliferation. Basically, two methods were used: *structural* and *phenomenological*.

The structuralists' point of departure was either a *mechanical model* (one such group, including L de Broglie and D Bohm, reinterpreted isospin as a true physical spin, in a spinning top model with an additional internal spinning motion; the model's weakness was that it predicted unseen transitions in which some isospin might be transferred to the outside as spin) or an *assembling 'game'* of constructing the observed hadrons from an elementary set.

The first such model was suggested by E Fermi and C N Yang in 1949, and extended by S Sakata in Nagoya (Japan) in 1954 to include the *strange* particles. The Fermi–Yang model assumed that the pions are composite particles, bound states of nucleons and antinucleons: $\pi^+ = p\bar{n}$, etc. The various quantum numbers do fit this structure. The strong interaction would also have to be particularly strong: the combined masses of the 'constituents' total about 2 GeV, and yet the compound's mass is only 0.14 GeV, thus pointing to a (negative) binding energy 'swallowing up' some 93% of the input masses (about three orders of magnitude more than nuclear binding energies).

With the discovery of strange particles, S Sakata in Nagoya (and W Thirring and J Wess in Vienna) added the Λ^0 hyperon (with strangeness -1) as a third 'fundamental brick'. The scheme has some elegance and became very popular around 1960. Sakata and several of his collaborators were animated by the philosophical ideas of dialectical materialism, emphasizing the structural aspect as an absolute. They held on to the model for several years after it had been faulted experimentally. It was only after the crystallization of our *quark* model, which amounted to replacing the sakatons by an alternative 'elementary' set, that they finally made their peace with the experimental facts.

At Yukawa's advice, the Sakata school turned in 1959–60 to group theoretical methods in trying to predict hadronic spectra. The group used was SU(3), appropriate to manipulations with three basic complex objects. As we shall see, this happened to coincide with the group that we used in the alternative (phenomenological) approach. At least one of the Japanese physicists working on the Sakata model with SU(3) noticed that there was in SU(3) an alternative assignment for the baryons but discarded the idea, in his dialectical materialist orthodoxy.

With the discovery of the pions in 1947–49, in the wake of the great successes of quantum electrodynamics (QED), it was natural that an attempt should immediately be made to extend the use of relativistic quantum field theory (RQFT) to the strong interaction, in a version in which the pions play a role resembling that of the photon in QED. Although it proves possible to renormalize this theory, the program falls flat as a dynamical theory. This is because the value of the 'charge' coupling a pion to a nucleon is a large number, 15, instead of the $\frac{1}{137}$ of QED, which should not surprise us, since we know that the force is 1000 times stronger. The strong interaction (at this level) does not allow for a perturbative treatment. All this had become clear by 1956.

R P Feynman tried in the late 1950s to adapt the RQFT approach to Einstein's theory of gravity and failed (this is still an unresolved problem). Meanwhile, in 1953, C N Yang and R L Mills had constructed a *model field theory* which had many of the features of Einstein's general relativity.

A piece of pipe is a system which is invariant under rotations around the pipe's axis of symmetry (this is termed an application of the *Lie group* SO(2), the special group of orthogonal transformations in two dimensions; a 'Lie' group, named after the Norwegian mathematician S Lie, is a group of *continuous* transformations, such as translations or rotations, and unlike a mirror reflection). Such a symmetry is termed *global*, i.e. you have to turn the entire pipe as one piece. One may, however, use a pipe made of soft rubber; in that case, the pipe might still look the same *even though we would be rotating it by different amounts at different spots along its axis.* This is a *local* symmetry; its presence introduces stresses in the material of the pipe, namely curvature and torsion, now termed *gauge fields.* Einstein's general relativity indeed uses such ideas, with

the gravitational field representing the curvature of space–time. Between 1919 and 1929, Herman Weyl had provided such a *gauge theory* for electromagnetism, which had been successfully incorporated in QED. In 1953, Yang and Mills constructed a generalization, adapted to any Lie group of transformations. Feynman tried to extend the techniques of RQFT, so successful in QED, to this 'pilot' model (which at that stage was not yet related to a definite interaction). Feynman achieved some progress, but the work of many other theorists was needed before Veltman and 't Hooft completed the formalism of the Yang–Mills field in 1971 and RQFT was vindicated.

THE SEARCH FOR ORDER (1954–64): PATTERN ABSTRACTION FROM THE PHENOMENOLOGY

In contradistinction to all this, there was a phenomenological (exploratory) approach. It consisted in just *charting the observations with no preconceived notions, trying to find clues to the emerging spectral pattern, postponing any guesses as to the dynamics themselves to a later stage.* This has been the traditional approach in other fields: Carl von Linné identified in plants those features which are conserved through genetic transmission and classified the plant kingdom; it took another 100 years before Gregor Mendel's experiments started to reveal the dynamics of genetic transmission, and almost another century before Crick and Watson unravelled the microstructure. In chemistry, Mendeleyev just 'read the pattern'. Once his classification was confirmed by the discovery of the missing elements which it predicted, it could become the basis of the search for structure and dynamics. Hence Rutherford, Bohr and the understanding of atomic structure followed. Kepler's phenomenological charting of regularities in the planetary motions was essential for Newton's construction of the dynamical theory; the *ad hoc* formulae in atomic spectroscopy had to preceed Bohr and quantum mechanics.

The advent of hypercharge brought A Salam and J Polkinghorne to suggest in 1955 embedding isospin in a larger algebraic framework. With isospin corresponding to abstract three-dimensional rotations, they suggested a scheme involving rotations in four abstract dimensions. J Schwinger and M Gell-Mann in 1957 offered a scheme, 'global symmetry', listing all possible isospin and hypercharge conserving interactions between the eight observed stable and metastable baryons (two nucleons, Λ^0, three Σ and two Ξ) and the three π and four K mesons. There are eight such allowed isospin and hypercharge conserving interactions, i.e. eight independent couplings, and the idea was to try to obtain the observed mass spectrum from some simple assumption about these couplings. Indeed, it yielded a formula $2(m_N + m_\Xi) = 3m_\Sigma + m_\Lambda$, accurate to 5%. The theory was, however, overthrown when experiments showed that the assumptions about symmetric pion couplings were wrong. Meanwhile, J Tiomno had shown that this scheme is equivalent to an algebraic assumption of a seven-dimensional rotational isospace, in which the mesons behave as a seven-vector and the baryons as spinors (which have eight components in that dimensionality). This led A Salam and J C Ward in 1960 to suggest models involving eight- and nine-dimensional isospace rotations, etc.

Some comments about the mathematical tools provided by *group theory* and the theory of *group representations* are in order here. In the introduction to the second edition of his *Group Theory and Quantum Mechanics* (1930) Herman Weyl wrote, 'It has been rumoured that the *group pest* is gradually being cut out of physics. This is not true, in so far as the rotation and Lorentz groups are concerned.' Beyond these groups of transformations, G Racah had applied the theory to the understanding of atomic spectra. In 1951, he was at the Institute of Advanced Studies at Princeton and gave a well attended series of lectures 'Group theory and spectroscopy', presenting the entire algebraic 'toolkit'. Most of the future participants in the search for the hadron order were present, M Gell-Mann, A Pais, A Salam, etc, and yet they apparently did not absorb the relevant information. The Lie groups which were tried in 1954–60 were always rotation groups, in ever more dimensions. These were the only ones with which physicists were familiar, as stressed by Weyl.

The entire domain of Lie algebras had been charted and classified by the great French mathematician Elie Cartan in 1894. The catalogue includes four infinite series (rotations in any number—odd or even—of dimensions, rotations in any even number of dimensions but with an antisymmetric metric —termed *symplectic*—and unitary or linear transformations in any complex dimensionality), and

five 'exceptional' algebras whose existence is related to that of 'octonions', a number field beyond the complex and quaternions. Unitary transformations preserve a vectorial basis in a complex space, as rotations do in real spaces.

GLOBAL AND LOCAL SYMMETRIES: EMMY NOETHER'S THEOREMS

At this point, it is appropriate to list the type of experimental evidence which can be used to identify the pattern or, once this is done, the type of prediction which can be derived from that pattern. A global symmetry has several implications.

(a) *Classification* in 'families' (these are called '*group representations*'). These groupings sometimes display an empty spot, a species which has not yet been detected, but which has to exist for the classification scheme to be physically correct.

(b) The group describes a symmetry, and one can now *predict ratios between processes involving particles within the same set of families*. These ratios are known as *intensity rules* and generalize the dependence on 'charges' which we apply, for example, in Coulomb's law. In some cases, the intensity rule is just a *no-go* theorem, forbidding some process.

(c) As all this relates to the strong interactions, which dominate the hadron's dynamics, the symmetry implies *energy degeneracy* within a family, i.e. equal masses. In the presence of a weaker interaction, breaking the symmetry, this degeneracy becomes approximate, just as a magnetic field in the z direction breaks the symmetry between the up- and down-spinning electrons. If the symmetry-breaking agent is known, one can chart the direction in which it breaks the symmetry, within the global symmetry's isospace (generalizing the idea of the magnetic field in the z direction, in the case of spin). The same approach relates to any other features induced by the weaker interactions, e.g. electromagnetic mass differences within an isospin grouping (such as the mass difference between π^0 and $\pi^\pm$), or magnetic dipole moments of various hadrons within the same group representation, etc. Should one be able to identify the direction in which the weak interactions break the global isospace, similar predictions would be made about the weak couplings, etc.

(d) In 1918, Emmy Noether, a Göttingen female mathematician interested in general relativity and in D Hilbert's work in physics, proved a powerful theorem *relating every conservation law to a symmetry and vice versa.* (As a woman, the Göttingen University Senate refused to appoint Emmy Noether to a professorship. When Hilbert finally managed to carry through a positive vote, one professor came up with a new objection: with only one toilet in the University Senate Hall, how could a woman be a professor? Noether had to flee Nazi Germany as a Jew in 1933 and died two years later in the USA.) The conservation of linear momentum, for instance, can be derived from the invariance (symmetry) of physical laws under spatial translations, i.e. the requirement that the results of an experiment be the same everywhere, given equal initial conditions. Angular momentum conservation similarly derives from the symmetry of physical laws under spatial rotations, energy conservation from symmetry under time translations, etc. F London showed in 1928 that the conservation of electric charge is related to a symmetry in the mathematical phase angle in the plane of the *complex* quantum amplitude, etc. *Thus every (Lie group) symmetry implies a conservation law and vice versa.*

(e) When the symmetry is *local*, however (i.e. when the system is invariant even under transformations by a different amount at different locations—as in our rubber pipe example), the corresponding conserved 'charge' also becomes the source of a dynamical field (the stresses in the carrying geometry), just as electric charge generates the electromagnetic field. Emmy Noether's second theorem deals with this situation. *The field's coupling strength is given by the amount of conserved charge carried by the relevant source.* This is known as a *universal* coupling, i.e. it is fixed by the measurable value of the conserved charge. Classically, this is the essence of Einstein's equivalence principle in gravity; the *gravitational mass*, which appears as the coupling strength in Newton's law of universal attraction, is equal to the *inertial mass*, which we know to be conserved, as Mc^2, through the law of energy conservation.

An observation of this nature was made in 1955 by S S Gershtein and Ya B Zeldovich in the USSR, with respect to the 'vector' part of the weak interactions. That coupling had been measured in *muon* decay and found to be very close to its value in *neutron* ('β') decay, even though one would have expected the latter's strong interactions (e.g. the action of the pion cloud surrounding it) to affect the result strongly. This is very much like the universality of Newton's gravitational constant, which does not depend on the material's composition and is explained, as we described, by the equivalence principle. It provided a clue to the list of charges conserved by the strong interactions and became known as conserved vector current ('CVC'). A similar (although mathematically somewhat more complicated) observation was made by M Goldberger and S Treiman in 1958, with respect to the other component of the weak interaction currents, the axial vector 'A' component, where in the neutron's case the coupling is indeed larger than that of the muon by some 40%. This was shown to indicate the existence of a larger 'chiral' symmetry, obeyed by the strong interactions, in which they conserve separately left-handed and right-handed currents carrying the same charges (a feature known as *partially conserved axial currents*, PCACs).

Around 1958–60, experiments at Stanford, using powerful photons to probe the structure of nucleons, appeared to contradict the assumed picture of a simple pion cloud. Instead, as first interpreted by Y Nambu, the pions condense in heavier mesons, with $J = 1^-$, i.e. resembling the electromagnetic field, except that they are massive (i.e. short ranged). J J Sakurai then raised the possibility that the strong interactions might also result from electromagnetic-like Yang–Mills vector potentials, induced by locally conserved charges and producing universal couplings. This was termed the vector theory of the strong interactions. Sakurai proposed that the three components of isospin, and the two single-component systems of hypercharge and of baryon number represent local symmetries and induce vector potentials.

UNITARY SYMMETRY IN THE BARYON OCTET VERSION

The present author arrived in London in the last days of 1957, as an army colonel, the Defence Attaché to Israel's embassies in the UK and Scandinavia, intending to do research in theoretical physics. Originally, he had applied for leave from Israel's Defence Forces, intending to study physics at his Haifa *alma mater*, the Israel Institute of Technology ('Technion'), where he had graduated in mechanical and electrical engineering in 1945, and where Einstein's collaborator, N Rosen, had just established a department of physics. It was the Israeli Chief of Staff, General Moshe Dayan, well known for his originality and disregard for conventions, who had suggested instead studies in London, to be combined with the attaché's job. Graduate courses—and political complications—filled 1958–59, with research starting in May 1960. Attracted by the 'order' or global symmetry problem—the negative advice of his thesis adviser A Salam notwithstanding ('you are embarking on a highly speculative project!')— he studied Cartan's classification of the Lie algebras ('but, if you insist on doing it, do not stay with the little group theory I know, which is what I taught you—do it in depth!') in a reformulation by E B Dynkin (Moscow), a translation of which he unearthed as a heliograph in the British Museum library.

Studying Lie transformation groups and their Lie algebras (used to perform an infinitesimal transformation)

made it clear that what one was after was a Lie algebra of *rank 2*, i.e. there should be only two linearly independent additive charges (e.g. I_z and Y), since every reaction allowed by these conservation laws can be seen to be realized. There were five such rank 2 Lie algebra candidates and comparing their predictions with experiment easily excluded three. What was left was a choice between Cartan's exceptional G_2 and the algebra of SU(3). G_2 had several nice features and a six-pointed 'star of David' as its *root diagram*, the diagram describing the quantum numbers of the charges. For SU(3) it is a hexagon, with a double point in its centre, and thus an eight-point diagram, i.e. eight currents, from which the theory got its nickname of *the eightfold way* (figure 4.1). G_2 has a 14-point diagram. For the baryons, G_2 has a seven-member 'multiplet' (family), leaving out the Λ^0, which then appears as a separate entity. In SU(3), the Nagoya group had put p, n, Λ^0 as the 'fundamental' triplet, thus having to assign the three Σ to another multiplet with three unobserved particles. The two Ξ hyperons, in that model, were constrained to have spin $\frac{3}{2}$, with Fermi's four Δ. To the present author, however, one of the nicest features of SU(3) was that it could precisely accommodate the entire set of eight baryons in one *octet* representation (figure 4.2(a)), thus predicting that the Ξ have spin $\frac{1}{2}$, like p, n, etc, and that all eight baryons have the same relative parity, a point that later, in mid-1961, seemed to contradict experiments. The 0^- mesons were also constrained into an octet (figure 4.2(c)), thus predicting a missing eighth meson, the η^0, indeed discovered in 1961, with a mass of 547.5 MeV (fitting the prediction from the 'broken SU(3)' mass formula which we shall further discuss). Assuming SU(3) to act as a *local* symmetry leads to eight 1^- mesons, or nine, if one adjoins the *baryon charge* current, enlarging the *simple* SU(3) into a non-simple (although connected) U(3) with nine currents (figure 4.2(d)). This reproduces the five charges and potentials suggested by Sakurai, together with four additional (K-like) ones, and also predicts the relative strengths, since eight, at least, are now irreducibly related. Indeed, this theoretical prediction fits beautifully with the ratios that Sakurai had evaluated from experiments! All nine 1^- mesons were observed in mass plots in 1961.

M Gell-Mann had investigated the Yang–Mills algebraic structure in 1959. The case of the three charges

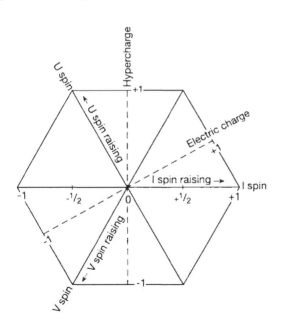

Figure 4.1. The root diagram of the SU(3) Lie algebra. A hexagon with two central points; it gave its name to the 'eightfold way' particle classification scheme.

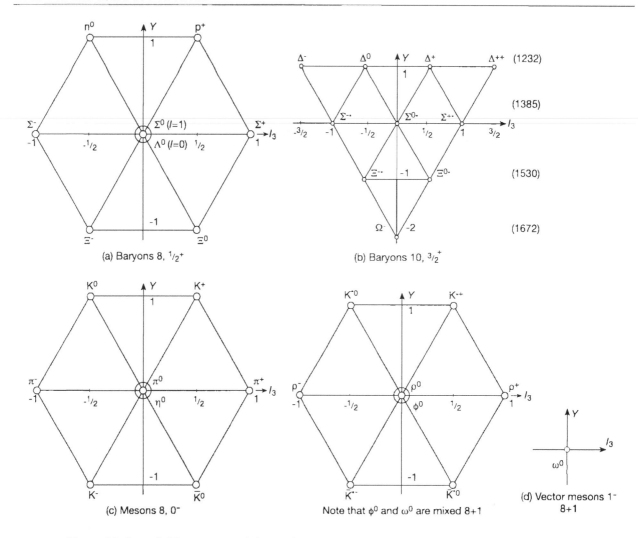

(a) Baryons 8, $1/2^+$

(b) Baryons 10, $3/2^+$

(c) Mesons 8, 0^-

Note that ϕ^0 and ω^0 are mixed 8+1

(d) Vector mesons 1^- 8+1

Figure 4.2. Some SU(3) particle multiplets. I_3 is the third component of isospin, and Y is the hypercharge.

of isospin was that which Yang and Mills had worked out in detail. Trying four charges, Gell-Mann reproduced isospin plus a (hypercharge-like) singlet. Five charges just gave isospin with two singlets; six gave either isospin with three singlets or two 'isospins'. Seven again added a singlet. Gell-Mann concluded that nothing interesting was to be expected from this approach and turned to other issues. Late in 1960 at California Institute of Technology (Caltech) he consulted Feynman, who suggested consulting a mathematician. Gell-Mann climbed one floor in the same building and found the mathematician, who directed him to Cartan's classification. Around the same time as the present author was exploring SU(3), Gell-Mann did too and issued a pre-publication report. He also submitted it for publication, but meanwhile (in June 1961) there were (wrong) experimental results which appeared to show opposite relative parities for the Λ^0 and the three Σ hyperons. Gell-Mann withdrew the eightfold way paper, replacing it in 1962 by a more general article, in which both SU(3) models, Sakata's and the octet, were presented as possible choices.

Throughout 1961–63, new resonances were discovered and filled the SU(3) multiplets. In some cases, the assignment could fit in both the octet and the Sakata models. In the spring of 1962, however, there was at least *one observed process*, the annihilation $p + \bar{p} \rightarrow K_0^1 + K_0^2$, which was *totally forbidden in the Sakata model* and clearly allowed in the octet. Meanwhile, at the XIth International High-Energy Physics Conference at CERN (July 1962), where the present author stressed these results, new Σ and Ξ like $J^P = \frac{3}{2}^+$ resonances were reported, at 1385 MeV and 1530 MeV respectively. On the other hand, the (University of California at Berkeley) Gerson and Sulamith Goldhaber (husband–wife) team reported that there was no resonance in K–N scattering. Fermi's Δ quartet could be assigned to either a **10** or a **27** multiplet, in the SU(3) octet model. The non-existence of an NK resonance invalidated the **27** assignment, leaving the **10** as the only possible one, and thus predicting the Ω^- at 1672 MeV (figure 4.2(*b*)).

This value of the mass was derived from an understanding of the direction of the SU(3) breaking contributions in the SU(3) algebra's space. In this case, it predicts equal spacing, as already displayed by the sequence 1232, 1385, 1530 MeV. Considering its isospin 0, strangeness −3 quantum numbers, the strong interaction could provide for transitions, e.g. $\Omega^- \rightarrow \Sigma^+ + \bar{K}^0 + K^- + \pi^-$, but the Ω^- mass is far too light to allow such a decay. The particle is likewise *stable with respect to any strong interaction decay mode* and can thus only decay via the weak interactions, which is also why it could be photographed in a bubble chamber (its lifetime can be computed to be of the order of 10^{-10} s, allowing for a 1–2 cm flight).

Both Gell-Mann and the present author thus predicted the existence of the Ω^- metastable hyperon. The validation of this prediction in an experiment at Brookhaven early in 1964 put the final stamp of approval on SU(3) in its octet version, as a *global* symmetry. In several ways, mainly through the existence of SU(3) **8** + **1**, spin–parity 1$^-$ vector mesons and by the observed *universality* of their couplings, it even looked as if SU(3) should be considered as a *local* (gauge) symmetry, and yet somehow *broken*, at least through the acquisition of a mass, by gauge bosons which should be massless in a true local gauge symmetry.

We return to the issue of the hadron masses, showing SU(3) to be an approximate (rather than an exact) symmetry. The masses do obey isospin and hypercharge invariance.

In the SU(3) root diagram, the equal-mass levels are horizontal, parallel to the action of the operators raising and lowering the values of I_z, within one isospin (SU(2)) submultiplet. The mass levels are thus measured along a (vertical) axis, that of hypercharge. We can easily identify its direction in the group space as that of an 'eighth component', i.e. the (singlet) middle point, in the root diagram, with quantum numbers like those of the Λ^0: $I = 0$, $Y = 0$. Such a component cannot modify the isospin nor the hypercharge. We identified this direction and Gell-Mann derived a relation for the baryon masses, $m(\Sigma) + 3m(\Lambda) = 2m(N) + 2m(\Xi)$, nicely obeyed by the observed masses. S Okubo derived a general formula, for any SU(3) multiplet. Note that these formulae work perturbatively, i.e. a small correction, in which the first-order contribution is the main one, the higher-order terms adding very little. This is how we compute electromagnetic effects, for instance. In that case, it is the smallness of the coupling constant which guarantees good results from first-order calculations. *What was surprising was the success of perturbative calculations in what seemed to be part of the strong interaction.*

The present author resolved this issue in 1964 by suggesting that the strong interaction proper is fully SU(3) invariant, whereas the symmetry breaking increment is due to another interaction, of a perturbative nature. I coined the term 'fifth interaction' for this effect, but the term is now used in a different context. The 1964 solution of the riddle of the success of the SU(3) mass formulae has been incorporated in the present 'Standard Model' (SM).

Looking at the SU(3) root diagram (figure 4.1), we can also identify the direction of electromagnetism; the electric charge axis is inclined with a 30° slope, and electromagnetic effects are equivalent along lines perpendicular to this direction (e.g. p, Σ^+ have the same charge, or Σ^-, Ξ^-, etc). This direction represents the action of yet another SU(2) subgroup, resembling isospin. H J Lipkin, known for his sense of humour, named the three basic SU(2) subgroups I spin, U spin and V spin, drawing from an ice cream advertisement of the time 'I scream, you scream, we all scream for ice cream'. Returning to the diagram, p, Σ^+ make a U-spin doublet, and so do Σ^-, Ξ^-. We can predict equal magnetic moments for particles in the same U-spin multiplet, etc. From such considerations (plus ensuring that the mass-breaking eighth component is cancelled) it is very easy to derive, for example, the accurate Coleman–Glashow formula for the electromagnetic mass differences,

$$m(\Xi^-) - m(\Xi^0) = m(\Sigma^-) - m(\Sigma^+) + m(p) - m(n).$$

The weak interaction charges and currents also find their place in SU(3) space. N Cabibbo pointed out in 1963 that the weak isospin transition current n → p (e.g. β decay) is somewhat 'deflected' in SU(3) space from the direction of strong interaction isospin. Whereas in the latter the I_z raising current would turn n → p, the same current in weak isospin relates the linear combination $(n \cos\theta + \Lambda^0 \sin\theta)$ → p, where the 'Cabibbo angle' $\theta \simeq 15°$. This explains the difference between the V couplings in μ decay and in n decay (which is about 4% weaker (equal to $\sin^2\theta$) while also predicting a relatively slower decay Λ^0 → p, as compared with n → p (the same 'missing' 4% in the V coupling) and is confirmed by a number of other experiments. Moreover, Gell-Mann had pointed out in 1958 that this is linked to the existence of two neutrinos. With a single neutrino, one would have equally fast transitions for μ, e, Λ^0 and n. What is observed is indeed equal rates for μ and e but very different rates within the Λ^0, n set. With two neutrinos, one has $\mu \to \nu_\mu$, $e \to \nu_e$, $(n \cos\theta + \Lambda^0 \sin\theta) \to p$.

QUARKS

There was one surprising element in SU(3) in its baryon-octet version. Whereas in the Sakata version or in G_2 (which had strong partisans until 1964) the nucleons and some hyperons are assigned to the group's *fundamental* representation (a three-component one in SU(3) and a seven-component one in G_2), *the octet is not a fundamental object* in SU(3). In July 1962, at the XIth International High Energy Conference at CERN, for instance, D Speiser expressed his strong scepticism, for just this reason of 'non-elementarity', with respect to the SU(3) octet model, as compared with the Sakata **3** or to the G_2 **7**. The **8** in SU(3) correspond to compounds, a system of three fundamental triplets for the baryons, or three antitriplets for antibaryons, or a triplet–antitriplet system for the mesons. The present author noted the structure in mid-1961, still in London and concluded that indeed the nucleon, until then believed to be 'simple', might well be a composite object. Returning

to Israel in the summer of 1961 (to take on the scientific directorship of the Israel Atomic Energy Commission's Soreq Nuclear Research Establishment) and working with H Goldberg, we elucidated the mathematical structure necessary to 'make' the baryons from three 'copies' of some fundamental triplets and the mesons from a triplet and an antitriplet. The surprising feature, however, is that these basic bricks have *fractional electric charges!*

In *Quarks for Pedestrians*, H J Lipkin writes. 'Goldberg and Ne'eman then pointed out that the octet model was consistent with a composite model constructed from a basic triplet with the same isospin and strangeness quantum numbers as the sakaton, but with baryon number $\frac{1}{3}$. However, their equations show that particles having third-integral baryon number must also have third-integral electric charge and hypercharge. At that time, the eightfold way was considered to be rather far fetched and probably wrong. Any suggestion that unitary symmetry was based on the existence of particles with third-integral quantum numbers would not have been considered seriously. Thus, the Goldberg–Ne'eman paper presented this triplet as a mathematical device for construction of the representations in which the particles were classified. Several years later, new experimental data forced everyone to take SU(3) more seriously. The second baryon multiplet was found, including the Ω^-, with spin and parity $\frac{3}{2}^+$. Gell-Mann and Zweig then proposed the possible existence of the fundamental triplet as a serious possibility and Gell-Mann gave it the name of quarks.'

The Goldberg–Ne'eman paper appeared with a one year delay; it had been lost and then found again in the journal, on the first page of the first issue of *Nuovo Cimento* in 1963, still a year before the Ω^- results, and at a time when the octet's chances indeed appeared dubious. It had, however, been disseminated a year earlier as an Israel Atomic Energy Report and the mathematical idea of making an **8** through the triple product of triplets $\mathbf{3} \otimes \mathbf{3} \otimes \mathbf{3} = \mathbf{1} \oplus \mathbf{8} \oplus \mathbf{8} \oplus \mathbf{10}$ started to spread among the devotees. In a seminar at Columbia University in March 1963, Gell-Mann's thinking was further stimulated by a question posed by R Serber, namely could the octet be 'made of' Sakata-like triplets? By the fall of 1963, Gell-Mann's idea had matured into a real physical model, beyond the purely algebraic content. Firstly, there was the possibility that

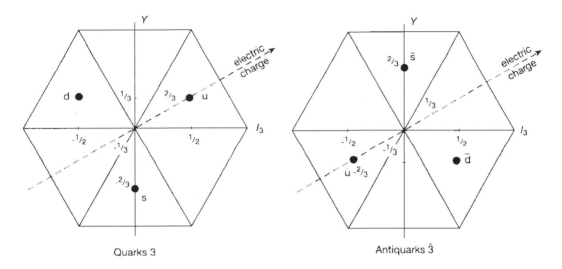

Figure 4.3. The fundamental triplet SU(3) representation did not correspond to any known particles but supplied the quark idea.

the triplets with their fractional charges were present in the soil, the sea or the atmosphere. Secondly, the *quarks* (the term was borrowed from Joyce's *Finnegan's Wake*: 'three quarks for Master Mark') were embodied by *quantum fields* in a Lagrangian RQFT, displaying their role in weak and electromagnetic interactions. Thirdly, the possibility that they might nevertheless be dynamically *confined* was envisaged ('instead of purely mathematical entities, as they would be in the limit of infinite mass'; returning to the Goldberg–Ne'eman picture, although now with a *physical* interpretation). Lastly, it was suggested that the breaking of SU(3) might be due to a difference between the mass of the s quark and that of the u, d doublet, a view which has indeed been incorporated in the present QCD paradigm.

G Zweig, then at CERN, independently conceived his '*aces*', with the same quantum numbers as the two previous versions, but mostly as physical particles, also calling for searches.

Denoting the first two quarks u (up) and d (down), with the *isospins* of the sakaton's p and n respectively, one 'makes' a proton from uud and a neutron from udd. The only way that the appropriate electric charges can be obtained is for u and d to carry charges $\frac{2}{3}$ and $-\frac{1}{3}$ respectively. The Λ should be made of uds (s is the *strange* quark, an isospin singlet and the s should have the same charge as d, namely $-\frac{1}{3}$). u, d and s should all have $B = \frac{1}{3}$,

and by the Gell-Mann–Nishijima rule $Q = I_z + Y/2$, u and d have $Y = \frac{1}{3}$ and s has $Y = -\frac{2}{3}$ (figure 4.3).

The Ω^- experiment (see the chapter by Samios) had a considerable impact on the quark idea. The **10** assignment is realizable as a three-quark combination, as can be seen in figure 4.4, whereas the **27** can only be reached with four quarks and one antiquark. Experiments thus appeared to say that nature indeed favoured three-quark structures for baryons.

At this stage, F Gürsey and L Radicati, G Zweig and B Sakita applied a static algorithm borrowed from an idea of E Wigner in nuclear structure. One assumes that the kinematics conserve spin angular momentum separately from any 'orbital' motion. One then postulates symmetry between the six basic states, u, d and s, each with spin up and spin down, i.e. an SU(6) symmetry. The first interesting result is represented by the assignments for baryons and mesons in SU(6). The sixteen components of the baryon octet (2 spins × 8) and the 40 components of the $J = \frac{3}{2}$ baryon resonances (including the Ω^-) in the **10** (4 spins × 10) fit together in a 56-component structure corresponding to a symmetric arrangement of the quarks (hence, for example, $\Delta^{++} = $ uuu, $\Delta^+ = $ uud and $\Omega^- = $ sss). We return to this point in the following. The mesons fit precisely together in a 35-component structure (3 × (8+1) for 1^-, + 8 for 0^-). The applications of this static

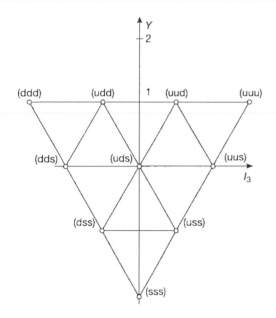

Figure 4.4. More SU(3) representations: (*a*) the ten-fold three-quark combination, and (*b*) the 27-fold pattern given by four quarks combining with an antiquark.

symmetry, with these assignments, produced numerous beautiful results. One example is the resulting ratio between baryon magnetic moments, assuming the quarks have normal Dirac equation values. Both the proton and the neutron display large anomalous magnetic moments. For the proton-to-neutron ratio, the SU(6) prediction is $\mu_p/\mu_n = -\frac{3}{2}$. The measured quantities, known since the 1940s to seven or eight decimals, have a ratio of -1.46. For a static model, this is a surprisingly good fit. Many such results indicated that the *non-relativistic quark model* really 'hit it', although there was as yet no understanding as to how *strongly interacting systems could obey such a static symmetry*.

There was an additional puzzle. Fermions should arrange themselves antisymmetrically, to reproduce Pauli's exclusion principle. In SU(6), however, they display a symmetric arrangement! The answer, suggested in different forms by W O Greenberg and by Y Nambu and M Y Han in 1964, was to indicate an additional degree of freedom, in which the same quarks fall into an antisymmetric arrangement. The simplest such solution would be the

existence of yet *another SU(3) set of charges*. Such a system indeed has a totally antisymmetric combination which is a scalar. This new SU(3) is now known as *colour*. Quarks have colour, but hadrons correspond to 'white' sets, in which the colour indices cancel. A modified version of *colour*-SU(3) was suggested in 1972 by Fritzsch, Gell-Mann and Leutwyler, after other indications pointing to the existence of SU(3)$_{colour}$ became known in the late 1960s.

Yet another spectacular quark model result, typical of its application to high-energy scattering, was obtained in 1965 by E M Levin and L L Frankfurt. Remembering that a nucleon is *made of* three quarks, whereas a pion is a quark–antiquark compound, they predicted that *the ratio of total cross-sections in nucleon/nucleon, versus pion/nucleon scattering should tend to* $\frac{3}{2}$. The experimental figures for the largest available energies are now 41 mb and 26 mb respectively, i.e. a ratio of 1.58. Note that other than the quark composition, this ratio would be allowed to take *any value*! A large number of similar predictions followed, all validated by experiment in the work of H J Lipkin and F Scheck and many other groups. *Quark counting* clearly contained essential dynamical information.

In the applications of SU(3) and its SU(6) 'spin extension', with their *chiral* extensions, one often comes across the 'paradox' of two 'origins' for the symmetry. The hadron classification and the mass spectrum are based on a picture of u and d quarks, related by isospin. These *constituent* quarks can be thought of having $\frac{1}{3}$ of the 'mass' of a nucleon, and $\frac{1}{3}$ of its high-energy cross section. At the same time, working on weak or on electromagnetic interactions, one encounters a Cabibbo-deflected partner $d' = d\cos\theta + s\sin\theta$ to the u quark. Calculations based on these show a 'mass' of 5 MeV for the u and 9 MeV for the d'. It became clear that the latter should be regarded as a conceptual entity, *current quarks*, different from the *constituent quarks*.

In the present Standard Model formulation, the fundamental entities correspond to the *current quarks*, which, like the leptons, enter as matter fields, with their *current* masses (for the heavier quarks, these are almost the same as the constituent masses). QCD, the interquark strong interaction binds them in hadrons, adding some 300 MeV per quark to the total hadron mass, beyond the input *current* masses, and ignoring any difference between the *current quark flavours*, i.e. whether it is u, d', s', c, etc. QCD also produces the mixing, e.g. d, s instead of d', s'.

Between 1962 and 1969, a variety of experiments searched for free quarks. The search included cosmic rays, analyses of minerals, etc, in the hope of detecting an $e/3$ residual charge. Two 'finds' were not confirmed and thirty and more years later, the verdict is that they have not been found. This could have two explanations: either quarks do not exist or they are *confined. The verdict came down in favour of confinement, once experiments in 1967–69 managed to detect the quarks, with their fractional charges, by probing inside the nucleon* (see the chapter by Friedman and Kendall).

The return of RQFT after 1971 brought in its wake (1973) the discovery of *asymptotic freedom*, a spring-like force, which vanishes when the quarks are very close and becomes extremely large when they distance themselves mutually. Qualitatively, this explains why hadrons appear to consist of 'almost free' quarks, acting additively. It also provides a model for confinement, although this remains unproven as of 1997.

FURTHER READING

Adler S L and Dashen R 1968 *Current Algebra* (New York: W A Benjamin)

Gell-Mann M and Ne'eman Y 1964 *The Eightfold Way* (New York: W A Benjamin)

Ne'eman Y 1967 *Algebraic Theory of Particle Physics* (New York: W A Benjamin)

Ne'eman Y and Kirsh Y 1995 *The Particle Hunters* 2nd edn (Cambridge: Cambridge University Press)

ABOUT THE AUTHOR

Professor Yuval Ne'eman has been a Professor of Physics at Tel-Aviv University since 1965, holding the Wolfson Distinguished Chair in Theoretical Physics in 1977–97 and serving as the Director of the TAU Mortimer and Raymond Sackler Institute of Advanced Studies in 1979–97. Ne'eman was TAU's President in 1971–75, after having founded and chaired its Department (now School) of Physics and Astronomy (1964–72) and its School of Engineering (1969–71). He is also the founder of the Center for Particle Theory at the University of Texas, Austin. Ne'eman initiated astronomical research in Israel and established the Wise Observatory in the Negev (1971). He has published some 20 books and 300 scientific articles and has been awarded several national and international prizes, including the Israel Prize (1969), the Einstein Medal (Washington, 1970), and the Wigner Medal (Istanbul, Austin Solvay Conference 1982). He has received several honorary degrees from Israeli, American and European universities, is a Member of the Israel National Academy of Sciences and Humanities, a Foreign Associate of the National Academy of Sciences of the USA, a Foreign Honorary Member of the American Academy of Arts and Sciences, etc. Ne'eman has served in a variety of national functions, mostly relating to science and technology. He was Israel's first Minister of Science (1982–84 and again in 1990–92) and Minister of Energy (1990–92). He was a member of Israel's parliament ('Knesset') in 1981–90. He served as a Member of the Israel Atomic Energy Commission (1952–54, 1965–82 and 1990–92) and as its Acting Chairman (1982–84) and was the Scientific Director of the Israel Atomic Energy Commission's Soreq Research Establishment in 1961–63. He was Defence Chief Scientist in 1975–76 and President of the Bureau of Standards (1974–76). From 1977 to 1983, he served as Chairman of the Steering Committee of the Mediterranean–Dead Sea Project. In 1971 and 1975 he was Senior Strategic Advisor to the Minister of Defence.

Professor Ne'eman was born in Tel-Aviv in 1925. He matriculated at 15, studied at the Israel Institute of Technology (Technion) at Haifa, graduated in Mechanical and Electrical Engineering in 1945 and worked in hydrodynamical design at the family pump factory in 1945–46. He had joined the Haganah (Jewish Underground) in 1940 and participated throughout 1941–48 in various military activities relating to the Second World War and later to Jewish immigration. In the 1948–49 War of Independence he fought on the Egyptian front as an infantry field commander at the company, battalion and brigade levels successively. In 1950–51 he served as a Lieutenant Colonel, at Israel's Defence Forces High Command, as Deputy to General Y Rabin, Chief of Operations. In 1951–52 he studied at the French Staff College in Paris. In 1952–55 Ne'eman served as Director of Defence Planning, making important contributions to the shaping of Israel's strategy as later realized in the Six Day War. In 1955–57 he was Deputy Director of Defence Intelligence.

5 FINDING THE INTERNAL STRUCTURES OF THE PROTON AND NEUTRON

Jerome I Friedman and Henry W Kendall

Editor's Introduction: Of the experiments of modern particle physics, few have had such far-reaching implications as the pioneer 1967 study at the then-new 20 GeV two-mile electron linear accelerator at the Stanford Linear Accelerator Center (SLAC) which discovered that hard scattering centres were hidden deep inside protons.

Just as Rutherford's classic 1911 α-particle experiment had discovered that a small solid nucleus lurked deep inside the atom, the SLAC study, using much higher energy, showed that there was a deeper level of substructure, 'partons', inside the proton itself. While Rutherford himself had been able to explain his result, the interpretation of the 1967 findings took more time and effort. However, partons were ultimately identified as being due to the quarks which had been introduced by Murray Gell-Mann several years before.

For their part in the SLAC experiment, the authors of this article, with Richard Taylor, were awarded the 1990 Nobel Prize for Physics. As well as describing their own historic experiment, the authors take particular care to credit subsequent developments which enabled partons to be identified with quarks and with the gluons which carry the inter-quark force.

INTRODUCTION AND OVERVIEW

In late 1967 the first of a long series of experiments on highly inelastic electron scattering was started at the two-mile accelerator at the Stanford Linear Accelerator Center (SLAC), using liquid hydrogen and, later, liquid deuterium targets. Carried out by a collaboration from the Massachusetts Institute of Technology (MIT) and SLAC, these experiments had the objective of studying the large energy loss scattering of electrons from the nucleon (the generic name for the proton and neutron). In such a process, soon to be dubbed deep inelastic scattering, the target particle is violently jolted. Beam energies up to 21 GeV, the highest electron energies then available, and large electron fluxes, made it possible to study the nucleon to much smaller distances than had previously been possible. Because quantum electrodynamics provides an explicit and well understood description of the interaction of electrons with charges and magnetic moments, electron scattering had, by 1968, already been shown to be a very powerful probe of the structures of complex nuclei and individual nucleons. Indeed, Robert Hofstadter and collaborators had discovered, by the mid-1960s, that the neutron and proton were roughly 10^{-13} cm in size, implying a distributed structure.

The results of the MIT–SLAC experiments were inconsistent with the current expectations of most physicists at the time. The general belief had been that the nucleon was the extended and diffuse object as found in the elastic electron scattering experiments of Hofstadter and in pion and proton scattering. However, the new experimental results suggested instead point-like constituents. This was puzzling because such constituents seemed to contradict well established beliefs. Intense interest in these results developed in the theoretical community and, in a programme of linked experimental and theoretical advances extending over a number of years, the internal constituents were ultimately identified as quarks.

Quarks had previously been devised in 1964 as an underlying quasi-abstract scheme to justify a highly successful classification of the many particles discovered in the 1950s and 1960s (see the chapter by Ne'eman). There were serious doubts about their reality. The new experimental information opened the door to the development of a comprehensive field theory of hadrons (the strongly interacting particles), called quantum chromodynamics (QCD), that replaced entirely the earlier picture of the nucleons. QCD in conjunction with electroweak theory, which describes the interactions of leptons and quarks under the influence of the combined weak and electromagnetic fields, constitutes the Standard Model (SM), all of whose predictions, at this writing, are in satisfactory agreement with experiment. The contributions of the MIT–SLAC inelastic experimental programme were recognized by the award of the 1990 Nobel Prize in Physics. This is how it came about.

THE 1968 HADRON

In 1968, when the MIT–SLAC experiments started, there was no satisfactory model of the internal structures of the hadrons. Indeed, the very notion of 'internal structure' was foreign to much of the then-current theory. Quarks had been introduced, quite successfully, to classify and explain the static properties of the array of hadrons, the bulk of which were discovered in the late 1950s and 1960s. Nevertheless, the available information suggested that hadrons were 'soft' inside. Thus the expectation was that electron scattering would infer diffuse charge and magnetic moment distributions with no underlying point-like constituents. Quark constituent models were being studied by a small handful of theorists, but these had very serious problems, then unsolved, which made them widely unpopular as models for the high-energy interactions of hadrons.

Nuclear democracy

The need to carry out calculations with forces that were known to be very strong introduced intractable difficulties. This stimulated renewed attention to an approach called S-matrix theory, an attempt to deal with these problems by consideration of the properties of the initial and final states of the particles in an interaction. Several elaborations of S-matrix theory were employed to help to deal with the challenges of hadron structure as well as hadron–hadron interactions.

Bootstrap theory, one of the elaborations of S-matrix theory, was an approach to understanding hadronic interactions, and the large array of hadronic particles. It assumed that there were no 'fundamental' particles; each was a composite of the others. Sometimes referred to as 'nuclear democracy', the theory was at the opposite pole from constituent theories. Nuclear democracy led in a natural way to the expectation that electron scattering would reveal an underlying structure that was smooth and featureless.

Numerous models were based on this approach and, aside from their applications to hadron–hadron scattering and the properties of resonances, they had some bearing on nucleon structure as well, and were tested against the early MIT–SLAC results.

1964 quark model

The quark was born in a 1964 paper by Murray Gell-Mann and, independently, by George Zweig (see the chapter by Ne'eman). (The word *quark* was invented by Murray Gell-Mann, who later found quark in the novel *Finnegan's Wake*, by James Joyce, and adopted what has become the accepted spelling. Joyce apparently employed the word as a corruption of the word quart. The authors are grateful to Murray Gell-Mann for a discussion clarifying the matter.) For both, the quark (a term Zweig did not use until later) was a means to generate the symmetries of SU(3), the 'eightfold way,' the highly successful 1961 scheme of Gell-Mann and Ne'emann for classifying the hadrons. This scheme was equivalent to a 'periodic table' for the hadrons, being both descriptive and predictive. This model postulated three types of quark: up (u), down (d) and strange (s), with charges $\frac{2}{3}$, $-\frac{1}{3}$ and $-\frac{1}{3}$ respectively, each a spin-$\frac{1}{2}$ particle. Fractional charges were not necessary but provided the most elegant and economical scheme. That quarks might be real particles, constituents of hadrons, was not a necessary part of the theory, although not prohibited, either.

Three quarks, later referred to as valence quarks, were required for baryons, i.e. protons, neutrons and other such particles, and quark–antiquark pairs for mesons. The initial successes of the theory in classification stimulated

Figure 5.1. View of the Stanford Linear Accelerator. The electron injector is at the top, and the experimental area in the lower centre. The deep inelastic scattering studies were carried out in End Station A, the largest of the buildings in the experimental area.

numerous searches for free quarks. There were attempts to produce them with accelerator beams, studies to see whether they were produced in cosmic rays, and searches for 'primordial' quarks by Millikan oil drop techniques sensitive to fractional charges. None of these has ever been successful.

Questions about the quark model

There were serious problems in having quarks as physical constituents of nucleons and these problems either daunted or repelled the majority of the theoretical community, including some of its most respected members. Murray Gell-Mann, speaking at the International Conference on High Energy Physics at Berkeley in 1967, said '. . . we know that. . . (mesons and baryons) are mostly, if not entirely, made up out of one another. . . . The probability that a meson consists of a real quark pair rather than two

mesons or a baryon must be quite small.' The idea was distasteful to the S-matrix proponents. The problems were as follows. First, the failure to produce quarks had no precedent in physicists' experience. Second, the lack of direct production required the quarks to be very massive, which, for quarks to be constituents of hadrons, meant that the binding had to be very great, a requirement that led to predictions inconsistent with hadron–hadron scattering results. Third, the ways in which they were combined to form the baryons meant that they could not obey the Pauli exclusion principle, as required for spin-$\frac{1}{2}$ particles. Fourth, no fractionally charged objects had ever been unambiguously identified. Such charges were very difficult for many to accept, for the integer character of elementary charges was long established. Enterprising theorists did construct quark theories employing integrally charged quarks, and others contrived ways to circumvent the other objections. Nevertheless, the idea of constituent quarks was not accepted by the bulk of the physics community, while others sought to construct tests that the quark model was expected to fail. At the 1967 International Symposium on Electron and Photon Interactions at High Energy at Stanford in 1967, J D Bjorken said, 'Additional data is necessary and very welcome in order to destroy the picture of elementary constituents', while Kurt Gottfried said 'I think Prof. Bjorken and I constructed the sum rules in the hope of destroying the quark model'.

Some theorists persisted, nonetheless, and used the quark in various trial applications. Owing to the difficulties just discussed, quarks were generally accepted as a useful mathematical representation but most probably not real objects.

Thus one sees that the tide ran against the constituent quark model in the 1960s. One reviewer's summary of the views of the 1960s was that 'quarks came in handy for coding information but should not be taken seriously as physical objects'. While quite helpful in low-energy resonance physics, it was for some 'theoretically disreputable'.

THE EXPERIMENTAL PROGRAMME
Deep inelastic scattering

In view of this theoretical situation, there was no consideration given during the planning and implementation of the inelastic electron scattering studies that a possible point-like substructure of the nucleon might be observable. Deep inelastic processes, in which a great amount of energy is transferred to the target, were, however, assessed in preparing the proposal submitted to SLAC for construction of the facility. Predictions were made using a model based on the picture of the proton as an extended and diffuse object. It was found ultimately that these had underpredicted the actual yields by between one and three orders of magnitude.

The linear accelerator that provided the electron beam employed in the experiments was, and will remain into the next century, a device unique among high-energy particle accelerators (figure 5.1). An outgrowth of a smaller 1 GeV accelerator employed by Hofstadter in his studies of the charge and magnetic moment distributions of the nucleon, it relied on advanced klystron technology devised

by Stanford scientists and engineers to provide the high levels of microwave power necessary for acceleration of electrons. Proposed in 1957, approved by Congress in 1962, its construction was initiated in 1963, and it went into operation in 1967.

The experimental collaboration began in 1964. R E Taylor was head of SLAC Group A, and the present authors shared responsibility for the MIT participation. The construction of the facility to be employed in electron scattering was nearly concurrent with the accelerator's construction. This facility was large for its time. A 200 ft by 125 ft shielded building housed three magnetic

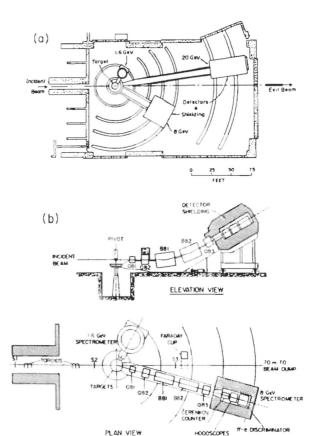

Figure 5.2. (a) Plan view of End Station A and the two principal magnetic spectrometers employed for analysis of scattered electrons. (b) Configuration of the 8 GeV spectrometer, employed at scattering angles greater than 120°.

Figure 5.3. Photograph of the 8 and 20 GeV spectrometers in the experimental hall.

spectrometers with an adjacent 'counting house' containing the fast electronics and a computer, also large for its time, where experimenters controlled the equipment and conducted the measurements (figures 5.2 and 5.3). The largest spectrometer would focus electrons up to 20 GeV and was employed at scattering angles up to 10°. A second spectrometer, useful up to 8 GeV, was used initially out to 34°, and a third, focusing to 1.6 GeV, constructed for other purposes, was employed in one set of large-angle measurements to help to monitor the constancy in density of the liquified target gases. The detectors were designed to detect the scattered electrons and to measure the cross sections (quantities proportional to the probability of scattering) for varied initial and final energies and over a range of scattering angles.

The elastic studies started in early 1967 with the first look at inelastic processes from the proton late the same year. By the spring of 1968 the first inelastic results were at hand. The data were reported at a major scientific meeting in Vienna in August and published in 1969. Thereafter a succession of experiments were carried out, most of them, from 1970 on, using both deuterium and hydrogen targets in matched sets of measurements so as to extract neutron scattering cross sections with a minimum of systematic error. These continued well into the 1970s.

Elastic and inelastic studies

Very-high-energy electrons can interact with both the electric charge as well as the magnetism of the target particle and so it is possible to obtain information on both the distribution of charge and of magnetism within the target. Scattering which leaves the target nuclei intact, so-called elastic scattering, produces information about the average size and distribution of electric charge and magnetism in the target. Deep inelastic scattering, which greatly disrupts the target, can produce information about constituents, if any, that form the target structure. Deep inelastic scattering can be identified experimentally by determining the energies and scattering angles of the scattered electrons.

In actual practice, it was convenient to characterize the scattering by a pair of variables, q^2, the square of the momentum transfer to the target particle, and v, the energy transferred to the target system.

The functions $G_{E_p}(q^2)$ and $G_{M_p}(q^2)$, defined as the elastic electric and magnetic form factors respectively, describe the time-averaged structure of the proton. There are similar functions for the neutron. These are related to the spatial distributions of charge and magnetic moment respectively, and from them the spatial distributions can be found.

Because both the proton and the neutron are extended objects, the elastic scattering cross sections as well as the structure functions $G_E(q^2)$ and $G_M(q^2)$ decrease very rapidly as q^2 increases. This is a general characteristic of the scattering from extended objects. Scattering from point-like objects, on the other hand, exhibits a very small or possibly zero decrease and, as we shall see, this can be used to identify such objects.

For deep inelastic scattering, the structure functions $W_1(q^2, v)$ and $W_2(q^2, v)$ are employed in a similar manner for the proton, deuteron or neutron; they summarize information about the structure of the target particles.

Before the inelastic results had been analysed, J D Bjorken, a young post-doctoral fellow at SLAC, had conjectured that in the limit of q^2 and v approaching infinity, with the ratio $w = 2Mv/q^2$ held fixed, the two quantities vW_2 and W_1 become functions of w only, a property that became known as scaling. At the time that he carried out his study, scaling had never been observed and there were some theoretical reasons to think that it would not be observed experimentally. It is this property that is referred to as 'scaling' in the variable w in the 'Bjorken limit.' The variable $x = 1/w$ came into use soon after the first inelastic measurements; we will use both here.

Inelastic electron–proton scattering: results

Unexpected results became apparent as the first measurements of deep inelastic scattering were analysed. The first was a quite weak q^2 dependence of the scattering at constant W, a quantity related to the energy transferred to the

target internal structure. Examples taken from data of the first experiment are shown in figure 5.4 for increasing q^2. For comparison the q^2 dependence of elastic scattering is shown also, which displays the steep drop-off mentioned above.

The second feature was the phenomenon of scaling. Figure 5.5(a) shows $2M_pW_1$, for a range of values of q^2, plotted for increasing w. Figure 5.5(b) shows vW_2 for a range of values of q^2, plotted against w. It was immediately clear that the Bjorken scaling hypothesis is, to a good approximation, correct.

These two unexpected results had, in fact, fundamental implications, as subsequent events were to demonstrate.

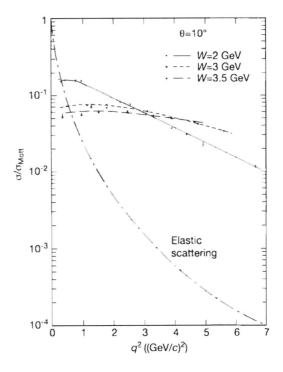

Figure 5.4. Inelastic data for $W = 2$, 3 and 3.5 GeV as a function of q^2. This was one of the early examples of the relatively large cross sections and weak q^2 dependence that were later found to characterize the deep inelastic scattering and which suggested point-like nucleon constituents. The q^2 dependence of elastic scattering is shown also; these cross sections have been divided by σ_{Mott}, the scattering cross section for a hypothetical point proton.

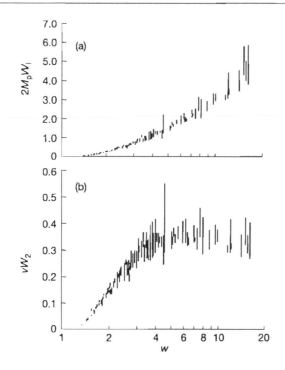

Figure 5.5. The effects revealed by the electron scattering experiments supported the idea of 'scaling': simple dependence on the purely kinematical quantity w.

As noted earlier, the discovery, during the first inelastic proton measurements, of the weak q^2 dependence of the structure function vW_2, coupled with the scaling concept, at once suggested new possibilities concerning nucleon structure. At the 1968 Vienna Meeting, where the results were made public for the first time, the rapporteur, W K H Panofsky, stated in his concluding remarks: 'Therefore theoretical speculations are focused on the possibility that these data might give evidence on the behaviour of point-like charged structures within the nucleon.' This, however, was not the prevailing point of view at this time. The picture challenged the beliefs of most of the community and only a small number of theorists took such a possibility seriously.

Parton model

Theoretical interest at SLAC in the implications of the inelastic scattering increased substantially after an August 1968 visit by Richard Feynman. He had been trying to understand proton–proton interactions at high energy, assuming constituents that he referred to as partons. Deep inelastic electron scattering was an ideal process for the application of his model for in such scattering the electron's interaction and point-like nature were both well understood. However, at that time, in proton–proton scattering, neither the structures nor the interactions were understood. On becoming aware of the inelastic electron scattering data, he immediately saw in partons an explanation of both scaling and the weak q^2 dependence. In his initial formulation, now called the naive parton theory, he assumed that the proton was composed of point-like partons, from which the electrons scattered independently. The partons were assumed not to interact with one another while the interaction with the electron was taking place.

In Feynman's theory, electrons scattered from constituents that were essentially 'free', and therefore the scattering reflected the properties and motions of the constituents. This was based on the assumption of a near-vanishing of the parton–parton interaction during electron scattering, in the Bjorken limit. Feynman came to Stanford again, in October 1968, and gave the first public talk on his parton theory, stimulating much of the theoretical work which ultimately led to the identification of his partons with quarks.

In November 1968, Curt Callan and David Gross at Harvard showed that a quantity R, related to a ratio of the structure functions, depended on the spins of the constituents in a parton model and that its value and kinematic variation constituted an important test of such models. For spin $\frac{1}{2}$, the value assigned in the quark model, R was expected to be small. The small values of R found in the experiment were consistent with the parton spin being $\frac{1}{2}$.

In proposing the parton model, Feynman was not specific as to what the partons were. Were they quarks, were they nucleons or mesons, or were they some wholly unexpected objects? There were, at first, two principal competing proposals for the identity of partons. In one approach, partons were identified with bare nucleons and pions, and in the other with quarks. At that time, one picture of the proton was that of a small inner core (the bare proton) surrounded by a cloud of pions. However, parton models incorporating quarks had a glaring inconsistency. Quarks require a strong final-state interaction that changes their identity as they are knocked out of the proton in order to account for the fact that free quarks had not been observed in the laboratory. Before the theory of QCD was developed, there was a serious problem in making the 'free' behaviour of a constituent during its interaction with the electron compatible with this required strong final-state interaction. This problem was avoided in parton models in which the proton is composed of bare nucleons and pions because the recoil constituents are allowed to decay into real particles when they are emitted from the nucleon.

The first application of the parton model using quarks as constituents (by J D Bjorken and E A Paschos) studied the parton model for a system of three valence quarks in a background of quark–antiquark pairs, often called the sea. A later, more detailed description of a quark–parton model of the proton (by J Kuti and V Weisskopf) contained, in addition to the three valence quarks and a sea of quark–antiquark pairs, neutral gluons, which are quanta of the field responsible for the binding of the quarks.

By 1970 there was an acceptance in some parts of the high-energy community of the view that the proton is composed of spin-$\frac{1}{2}$ point-like constituents. At that time we were reasonably convinced that we were seeing constituent structure in our experimental results, and afterwards our group directed its efforts to trying to identify these constituents and making comparisons with the last-remaining competing non-constituent models. The surprising results that stemmed from the initial deep inelastic measurements stimulated a flurry of theoretical work that gave rise to a number of non-constituent models based on several theoretical approaches to explain the data. After the proton–neutron ratios were determined, by 1972, all non-constituent models were found to be inconsistent with the experimental results.

Electron–deuteron scattering: results

Following the first deep inelastic measurements with hydrogen targets, the series of matched measurements with better statistics and covering an extended kinematic range were carried out with hydrogen and deuterium targets, utilizing all three spectrometers. Neutron cross sections

were extracted from measured deuteron cross sections with a procedure to remove the effects of the internal motion of the neutron within the deuteron. These data sets provided, in addition to more detailed information about the proton structure, a test of scaling for the neutron. In addition, the measured ratio of the neutron to proton scattering cross sections provided a decisive tool in discriminating between the various models proposed to explain the early proton results.

The analysis of these extensive data sets showed that the structures of both the proton and the neutron were consistent with being made up of 'point-like' spin-$\frac{1}{2}$ constituents. The ratio of the neutron to proton inelastic cross sections was found to fall continuously as the scaling variable x approaches unity ($x = 1/w$), as shown in figure 5.6 in which the ratio is plotted as a function of x. From a value of about 1 near $x = 0$, the experimental ratio falls to about 0.3 in the neighbourhood of $x = 0.85$. In the quark model a lower bound of 0.25 is imposed on this ratio. While the experimental values approached and were consistent with this lower bound, models based on the 'bootstrap picture' gave values from 1 to 0.6 in the region of 0.85. The model in which the partons were identified as bare nucleons and mesons predicted a ratio which fell to

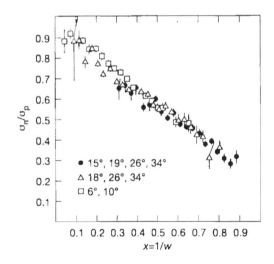

Figure 5.6. Ratio of neutron–electron to proton–electron scattering as a function of x, the scaling variable, for a variety of scattering angles.

about 0.1 at $x = 0.85$. Only the quark model was consistent with this measurement.

Determining the constituent charges

Sum rules played an important role in ascertaining whether the constituents of the nucleon have fractional charges. Sum rules are relationships involving weighted sums of various combinations of scattering cross sections or of quantities derived from them. A theoretical evaluation based on a particular model can be compared with an experimental evaluation to determine the validity of the model.

A particular sum rule that was of great interest was the so-called energy weighted sum rule. As applied to the proton, this was equal to the mean square charge of the charged constituents of the proton times the fraction of the momentum of a rapidly moving proton carried by its charged constituents. In the quark model the proton and neutron are expected to consist of quarks, antiquarks and neutral gluons, all of which carry the proton's momentum. As a consequence, the charged constituents of either the proton or the neutron are not expected to carry its total momentum. The experimental evaluation of this sum rule for both the proton and the neutron was found to be consistent with the quark model if the gluons were assumed to carry half of the total momentum.

CONFIRMATION

Full confirmation of a constituent model of the proton and neutron and the identification of the constituents as quarks took a number of years and was the result of continuing interplay between experiment and theory. While the electron scattering results ruled out competing models, they were not sufficient to establish the validity of the quark model. What was needed was confirmation that the constituents had charges which were consistent with the fractional charge assignments of the quark model. The development of more powerful theories and experiments using different particles soon made important contributions.

Neutrino scattering

The electron scattering data suggested spin-$\frac{1}{2}$ point-like constituents in the nucleon. The experimental evaluation of the sum rule related to the mean square charge of

the constituents was consistent with the fractional charge assignments of the quark model, provided that half of the nucleon's momentum is carried by gluons.

Neutrino deep inelastic scattering produced complementary information that provided stringent tests of the above interpretation. Neutrinos interact with protons and neutrons through the weak interaction, the force responsible for radioactive decay and for energy production in the Sun. Thus charged-current neutrino interactions with quarks (collisions in which neutrinos are transformed into leptons such as electrons and μ mesons) were expected to be independent of quark charges but were thought to depend on the quark momentum distributions in the nucleon in a manner similar to that for electrons. As a consequence, the ratio of electron to neutrino deep inelastic scattering was predicted to depend on the quark charges and the strength of the weak force, with the momentum distributions cancelling out. Because the strength of the weak force was known from other experiments, this ratio could provide information about the quark charges. The ratio of neutrino scattering to that of electron scattering, each averaged over the neutron and proton and properly normalized to take into account the relative strengths of weak and electromagnetic forces, was predicted to be $2/(Q_u^2 + Q_d^2)$, where Q_u and Q_d are the charges of the up and down quarks respectively. Since $Q_u = \frac{2}{3}$ and $Q_d = -\frac{1}{3}$, this ratio was predicted to be 18/5.

The first neutrino and antineutrino total cross sections were presented in 1972 at the XVIth International Conference on High-Energy Physics at Fermilab and the University of Chicago. The measurements were made at the CERN 28 GeV synchrotron using the large heavy-liquid bubble chamber 'Gargamelle'. At this meeting Donald Perkins, who reported these results, stated that 'the preliminary data on the cross sections provide an astonishing verification for the Gell-Mann–Zweig quark model of hadrons'.

The Gargamelle group evaluated the neutrino-to-electron scattering ratio by comparing their results with those from the MIT–SLAC electron scattering measurements and found it to be 3.4 ± 0.7 compared with the value predicted for the quark model, $18/5 = 3.6$. This was a striking success for the quark model. In addition, they evaluated a sum rule which indicated that the quarks and antiquarks in the proton and neutron carry about half of the momentum.

Within the next few years additional neutrino results solidified these conclusions. The results presented at the XVIIth International Conference on High-Energy Physics held in London in 1974 demonstrated that the ratio 18/5 was valid as a function of x. This is equivalent to stating that the electron and neutrino scattering measurements found the same structure. Figure 5.7 shows a comparison of the nucleon (an average of the neutron and proton) structure function, as measured with neutrinos by the Gargamelle group, and 18/5 times the nucleon structure function as measured with electrons by the MIT–SLAC group. These results indicated that both types of particle beam 'see' the same nucleon structure. In addition, the Gargamelle group evaluated the Gross–Llewellyn Smith sum rule, which is unique to inelastic neutrino and antineutrino nucleon scattering as a consequence of parity non-conservation in the weak interaction (see the chapter by Sutton). The value of this sum rule is equal to the number of valence quarks in either the proton or the neutron. The quark model states that this number is 3. The Gargamelle group found this sum to be 3.2 ± 0.6, another significant success for the model.

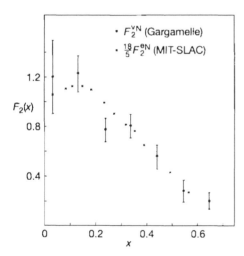

Figure 5.7. Neutrino scattering and electron scattering compared. Early Gargamelle measurements of the structure function determined with neutrinos along with electron scattering results multiplied by a theoretically appropriate factor of 18/5.

Other experiments

There were a number of other important experimental results first reported in 1974 and continuing in the latter half of the decade, that provided further strong confirmations of the quark model and further buttressed QCD. Among these was the discovery of charmonium (see the chapter by Schwitters), a system composed of a charm quark bound to a charm antiquark, and its energy levels could be understood in terms of dynamics based on the quark model and QCD. Its discovery led to the subsequent discovery of the charm quark. Other important developments were the discoveries of quark jets and gluon jets (see the chapter by Wu). These were first seen in electron–positron colliding beam experiments. The electron–positron annihilation produces a quark and antiquark, moving away from one another, and each of them fragments into a jet of collimated particles. When either of them radiates a gluon, this gluon will also fragment into a jet of particles and three jets are observed. The observation of these phenomena provided further validations of the quark model.

Quarks as constituents and quantum chromodynamics

By 1973, the quark–parton model, as it was usually called, satisfactorily explained electron–nucleon and neutrino–nucleon interactions and provided an explanation for the very high-energy 'hard' nucleon–nucleon scattering that had only recently been observed. The experimenters were seeing quark–quark, quark–gluon and gluon–gluon collisions in hadron–hadron collisions.

After the London Conference in 1974, with its strong confirmation of the quark model, the general point of view with regard to the structure of hadrons began to change. The bootstrap approach and the concept of nuclear democracy were in decline, and by the end of the 1970s the quark structure of hadrons became the dominant view for developing theory and planning experiments. An important element in this change was the general acceptance of QCD, a theory which eliminated the last paradox, namely why no free quarks are observed. The infrared slavery mechanism of QCD was hypothesized to lead to confinement and thus provided a reason to accept quarks as physical constituents without demanding the existence of free quarks. The asymptotic freedom property of QCD also readily provided an explanation of scaling, but logarithmic deviations from scaling were inescapable in this theory. While our later results had sufficient precision to establish small deviations from scaling, we did not have a wide enough energy range to verify their logarithmic behaviour. This was later confirmed in higher-energy muon and neutrino scattering experiments at Fermilab and CERN.

The MIT–SLAC deep inelastic electron scattering not only provided the first crucial evidence for the reality of quarks but also provided the initial experimental underpinnings for QCD. The measurements stimulated a wide variety of studies at other laboratories, using complementary techniques, that provided final confirmation of the theoretical advances that accompanied the programme of experiments.

The quark model, with quark interactions described by QCD, became the accepted basis for understanding the structure and interactions of hadrons. QCD successfully describes the strong interactions of the quarks and so can account, in principle at least, for the ground state properties of hadrons as well as hadron–hadron scattering. The weak and electromagnetic interactions of quarks are well described by electroweak theory, itself developed in the late 1960s. Taken together, these comprise the foundations of the Standard Model, a theory which has not been contradicted by any experimental evidence in the intervening years.

FURTHER READING

Riordan M 1987 *The Hunting of the Quark* (New York: Simon and Schuster)

ABOUT THE AUTHORS

Jerome I Friedman and Henry W Kendall were awarded the 1990 Nobel Prize for Physics jointly with Richard Taylor for their 'pioneering investigations concerning deep inelastic scattering of electrons on protons and bound neutrons, which have been of essential importance for the development of the quark model in particle physics.' Their work was carried out at the Stanford Linear Accelerator Center.

Born in Chicago in 1930, Jerome Friedman studied at Chicago, and after research at Chicago and Stanford moved to MIT in 1960, where he became Professor in 1967. At MIT he has served as Director of the Laboratory for Nuclear Science and head of the Physics Department, and in 1991 was appointed Institute Professor. He received the 1989 W K H Panofsky Prize and has also been awarded the Alumni Medal of the University of Chicago. He is a member of the US National Academy of Sciences, the American Academy of Arts and Sciences, and is a Fellow of the American Physical Society and the American Association for the Advancement of Science. He has been a member of numerous scientific advisory committees at major research centres and organizations in the US and overseas. He is currently President-Elect of the American Physical Society.

Currently J A Stratton Professor of Physics at MIT, Henry W Kendall studied at Amherst College and obtained his PhD from MIT. He taught at Stanford from 1956–61 before returning to MIT. As well as continuing to be active in teaching and research, he has served as Chairman of the Board of the Union of Concerned Scientists since 1974 and was a founding member of this organization. He has been active in writing, analysis and public activities in US energy and defence issues and in the global implications of environmental pressures, resource management and population growth. He is co-author of numerous reports, studies and books on nuclear arms, nuclear power and renewable energy. He served for a decade as consultant to the US Defense Department through membership of the Jason Group of the Institute for Defense Analyses and has been an advisor to the World Bank and was Chair of the Bank's Panel on Transgenic Crops, whose report was completed in 1997.

6 GAUGE THEORIES

M Veltman

Editor's Introduction: Gauge theories, and the insights that they have brought to particle physics, are one of the major intellectual achievements of the twentieth century. However, their development has illustrated how science often has to blunder a path through the jungle of ignorance, and eye-witness accounts traditionally reflect this confused and tortuous progress. Only when new physical theories are capable of making reliable predictions do they become valuable, and for future generations this new understanding is ultimately more important than the pedigree of the theory. (Even the name 'gauge' theories reflects misplaced faith. However, as often happens, the unfortunate name has stuck.) Here M Veltman, a major player in the development of these theories, boldly discards the flimsy script of history and reappraises the underlying scientific message. Some candidate pictures, plagued by infinities, are not mathematically tractable and cannot provide quantitative results. Veltman shows how the pre-condition of 'renormalizability'—requiring theories to be mathematically well behaved and capable of yielding reliable predictions—can be used as a tool to shape the theory and ensure its completeness. He shows how many penetrating insights follow simply by demanding that the theory be renormalizable, for Nature, it appears, requires it to be this way.

His treatment relies on 'Feynman diagrams', the graphical method introduced by Richard Feynman in 1949 to depict the intricate mechanisms of subatomic physics. In the same way that these component processes are subtly interwoven into a rich pattern of physics, the diagrams are such an integral part of Veltman's text that independent figure captions would be insufficient to describe their role, and no attempt has been made to supply any. The text and the diagrams are inseparable.

INTRODUCTION

The aim of this article is to give some insight into the notion of renormalizability and the related theories called gauge theories. Gauge theory has very much become the framework of today's high-energy physics. It is a complicated subject, much of it of a technical nature. The main importance of gauge theories lies in their renormalizability, and that by itself is not even well understood on any level other than technical. Why does nature select renormalizable theories? This question is still as much in the forefront today as it was 20 years ago. Yet it is precisely that criterion that has been extremely successful in understanding particles and their interactions. So let us try to explain what this is all about. It seems relevant in understanding Nature.

First of all it must be understood that at present everything, but indeed everything, is understood in terms of particles. The discovery that light consists of photons has changed things fundamentally. Since Maxwell explained that light is nothing but electromagnetic fields, the same electromagnetism that we see in static magnetic and electric fields, we are now forced to understand these same fields and the forces that go with it in terms of particles, photons. That indeed is how we understand all forces today: in terms of particles.

FEYNMAN RULES

Feynman rules are the main tools of the contemporary particle theorist. These rules incorporate the basic concepts of quantum mechanics; most importantly they can be represented in terms of drawings, diagrams, that have a strong intuitive appeal. A few basic concepts must be first understood to appreciate them.

In Feynman diagrams, particles are represented by lines, and interactions between particles by points where these lines join. Such an interaction point is called a vertex. The most obvious example is the interaction of electrons with photons. It is an interaction that we see literally almost permanently: the emission of light by electrons. In Feynman diagram language this interaction is represented in a very simple manner, see figure 6.1.

Figure 6.1.

The electron, represented by a line with an arrow, shakes off a photon and moves on. The arrow is not there to indicate the direction of movement, but rather that of the flow of (negative) electric charge. Later the meaning of the arrow will be changed slightly, but for now this will do. The interaction of electrons with light has been well understood for a long time, and we have a precise quantitative understanding corresponding to this diagram. Typically, the electron may be initially in an excited state in an atom and fall to a state of lower energy, thereby emitting a photon.

The first lesson that we draw from this diagram is that particles can be created in an interaction. Initially the photon was not there, and at some time later it came into existence. The opposite happens when the photon hits the eye; the photon is absorbed by an electron, one that would then move, in an atom, into a higher-energy state. The diagram corresponding to this process is shown in figure 6.2. The difference is that here the photon is incoming, not outgoing.

Figure 6.2.

We just learned another important feature; lines can be ingoing or outgoing. For example, if we bend the line representing the incoming electron to become an outgoing electron, we would get yet another figure, in which a photon changes into an electron pair, see figure 6.3.

Figure 6.3.

One of its members has the charge moving in the opposite way, and we observe it as positive charge. The associated particle is a positron, the antiparticle of the electron. So our rule gets refined: the same line that represents a particle may also represent its antiparticle. Positrons were discovered in the 1930s, and today antiparticles are an almost automatically accepted part of particle physics. Some particles are identical with their antiparticles; the photon is an example. Their lines carry no arrow.

Meanwhile there is another important element to understand. The interactions obey strictly the laws of energy and momentum conservation. An electron in an atom can emit a photon while dropping to a lower-energy state, but a free electron cannot emit a photon. Consider an electron at rest, i.e. with zero momentum; it is then in its lowest-energy state. If it were to emit a photon of finite energy, then an electron with even less energy would be left behind, which is not possible. The same holds then also for a freely moving electron, which one could imagine to be an electron at rest as seen by a moving observer.

Likewise a photon cannot change in mid-flight into an electron–positron pair, even if it is a high-energy photon. This can be understood by realizing that this high-energy photon appears as a photon of lower energy to another observer moving in the same direction as that photon. A photon always moves with the speed of light, and one can never catch up with it, as in the case of a particle with mass; instead, when this observer races along in the direction of the photon, it appears red shifted, i.e. it is perceived as a photon of lower energy. If the observer moves fast enough, the photon energy can become less than is needed to create an electron pair (whose energy is at least twice the rest mass energy of an electron), but in other circumstances where an external agent absorbs some momentum or energy it can happen. In collisions with nuclei a high-energy photon passing through matter will in fact readily convert into an electron–positron pair. An observer moving in the direction of the photon would still see a photon of lower energy, but it would collide with a moving nucleus, and that is where the needed energy would come from. An electron or positron moving through matter may likewise emit a photon, commonly called bremsstrahlung (literally brake radiation).

The next point is one of quantum mechanics. Particles can exist with 'inadmissible' energies, provided that this occurs only for a short time. The more inadmissible the energy, the shorter is its duration. What we mean here by inadmissible is an energy different from the value that one must normally assign to a particle with a given momentum. For example, an electron at rest has an energy corresponding to its rest mass multiplied by the speed of light squared ($E = mc^2$). An electron with zero energy is hence not possible. Yet quantum mechanics allows the existence of zero-energy electrons and even negative-energy electrons, or of electrons with inadmissibly large energies (e.g. a very-high-energy electron at rest), provided that this takes place only for short times. To give another example, a photon in flight can momentarily become an electron–positron pair.

Very quickly the pair must recombine into a photon. This possibility is shown in figure 6.4.

Figure 6.4.

A particle in an inadmissible state of energy and/or momentum is called a *virtual* particle. Because the relation between energy and momentum is not that of a free particle ($E = mc^2$ for a particle of zero momentum) such a particle is said to be 'off mass shell'.

It follows that a photon, for a small fraction of time, can become a virtual electron–positron pair. This can actually be observed by letting another photon cross the path of the first. If the second photon catches the first in a dissociated state, it could be absorbed by one of the two virtual particles, to be emitted again by the other, as shown in figure 6.5.

Figure 6.5.

As we know precisely the quantitative value of the vertices, and we also know the quantum-mechanical behaviour of these particles, the probability of the process can be calculated. It would be observed experimentally as scattering of light by light. The effect is small, so you cannot see it by crossing the beams of two flash-lights. Nonetheless, it has been observed. The classical Maxwell theory of radiation does not allow such a process. It is a purely quantum-mechanical effect.

The effect described here is somewhat similar to the so-called tunnelling effect, well known to students of quantum mechanics. A particle may cross an energy barrier even if it has insufficient energy to go over the top. An electron could cross a mountain even if it had not enough energy to get to the top. It may 'tunnel' through.

One more effect must be discussed, namely interference. Light interferes with itself, and this must therefore be a property of particles as well. The way that this works is that for a given situation there may be more than one path for a particle to get from a given initial state to a given final state. These different possibilities may interfere, either constructively or destructively. That is certainly something in which elementary particles differ from billiard balls or cannon balls! In actual fact, the laws of quantum mechanics apply equally well to macroscopic objects; but in that case the effects become too small to be observable. Imagine two cannons sufficiently precisely synchronized, firing two totally identical cannon balls at the same object, and interfering out precisely at the target! Of course, since the cannon balls have to go somewhere it would be dangerous living next to the target ('next' means at a distance of the order of 10^{-37} m).

In calculations with particles the theorist draws as many diagrams as he can, writes down the corresponding mathematical expressions and sums them. The different possibilities may add up or subtract: they interfere. Only from this sum total can the probability of the happening be calculated (effectively by squaring it). For example, to compute light-by-light scattering one must consider six diagrams, and combine their contributions. Figure 6.6 shows the different possibilities.

Figure 6.6.

All these diagrams correspond to a contribution of possibly different sign, and these contributions interfere. After taking all these contributions together the result must be squared and that is then a real probability.

INFINITIES

Where life becomes difficult is implicit in these diagrams. Not only must they be summed over all the different configurations, but over all the different energy–momentum values of the virtual particles also. Consider for example again the temporary transformation of a photon into an electron–positron pair. The virtual electron–positron pair can have an infinite range of possible energies, when we also include negative energies. The electron may for example be of very high energy, and the positron of very negative energy. The total energy must of course be equal to the energy of the photon, energy conservation being strictly enforced by Nature. One must sum over all the possibilities. That is often a hard calculation. Moreover, sometimes the summation gives an infinite result. It is a question of relative importance, of damping. If the configurations with the electron–positron pair of very high energy (very high negative energy for one of them) keep on contributing as much as barely inadmissible energies, then there is simply no end to the summation. *This, in a nutshell, is the problem of infinities in quantum field theory.*

Several factors affect the occurrence of these infinities. To begin with, the more a particle is away from its correct energy, the shorter is the time that it is allowed to exist in that state. Consequently there is a damping factor associated with the occurrence of any virtual particle; the stronger this factor, the more virtual the particle is. Furthermore, the damping is also a function of the intrinsic properties of the particle (we shall discuss this further below). Another factor is the behaviour of the vertices, i.e. of the coupling, as a function of the energies of the particles involved. By and large these couplings have no strong energy dependence, although there are exceptions.

A difficult point is the behaviour of virtual particles as a function of their intrinsic properties. The main property in this respect is the 'spin' of the particle. One of the very surprising discoveries in the domain of quantum physics was the discovery that particles have an angular momentum, as if they were spinning around an axis. For a body of some size, such as a billiard ball, this is easy to imagine but, for a particle that for all we know has no size, i.e. is pointlike, that is very hard to imagine. Yet it is there, and every elementary particle has a very definite spin as this intrinsic angular momentum is called (it may be zero). When particles are created or absorbed, the interaction is always such that angular momentum is conserved. If a spinning particle enters an interaction, then the angular momentum is preserved throughout, and it appears in the final state either in the form of other spinning particles, or else through non-spinning particles that revolve around each other, or both. All this is quite complicated, but fortunately we need only a few related facts. Spin is measured in terms of a specific basic unit, and spin is always a multiple of $\frac{1}{2}$ in terms of that unit.

As it happens, no elementary particle observed to date is of the spin-0 variety. Most particles have spin $\frac{1}{2}$, and the remainder have spin 1, except for the graviton (the particle responsible for gravitational interactions and analogous to the photon in electromagnetism) that has spin 2. Here now is the important property relevant to our discussion about virtual particles; as their spin becomes higher, virtual particles are less damped for higher energy. Particles of spin 1 are barely damped at high energy in their contributions to a virtual process. Particles of spin 2 are even worse; the quantum theory of gravitation is in a very poor shape.

Quantum field theory for particles of spin 1 (with the exception of the photon) was not part of our understanding of Nature up to 1970. No one knew how to handle the virtual contributions. They invariably led to infinite summations.

What changed the situation was the discovery that the worst effects in individual diagrams can be cured by introducing new interactions and particles (and hence new diagrams) in such a way that in the sum total the bad parts cancel out. For the photon a similar mechanism had been partly understood since 1948, and quantum electrodynamics produced finite numerical results that could be compared with experiment. One of the most important results is the magnitude of the magnetic moment of the electron. The predictions of quantum electrodynamics concerning this magnetic moment have been verified to a truly fantastic degree of accuracy. However, before discussing this we must first fill in some gaps, and explain about perturbation theory.

PERTURBATION THEORY

As we have pointed out, one must sum all possibilities when considering any process. That includes summing over all energy–momentum distributions of the virtual particles. However, also further emissions–absorptions of virtual particles must be taken into account. Figure 6.7 shows an example; the virtual electron emits a photon which is absorbed by the virtual positron.

Figure 6.7.

Here we have another complication of quantum theory: there is no end to this chain. One can exchange one photon, two photons, whatever number, and also any of these photons can momentarily become an electron–positron pair, etc.

As luck has it, there is in many cases no need to consider all these possibilities. The reason is that there is a factor associated with any vertex and that factor, at least for quantum electrodynamics, is quite small: the electric charge. The emission or absorption of a photon by an

electron (or positron) is proportional to the electric charge of the electron. Indeed, if the electron had no charge, it would not interact with the electromagnetic field. For this reason, a diagram as shown above, with an additional photon exchanged between electron and positron gives a contribution that is down by a factor e^2, where $-e$ is the electric charge of the electron. In practice there are some additional factors, and the relevant dimensionless quantity is what physicists call the fine-structure constant $\alpha = e^2/\hbar c$. Numerically $\alpha \approx \frac{1}{137}$, so that a diagram with an extra photon exchange indicates a contribution of the order of 1% compared with that of the diagram without that photon exchange. So, if we restrict ourselves for a given process to diagrams with the least number of vertices, we may expect an answer that is accurate to 1% and, if that is not enough, we can include diagrams with two more vertices and get an accuracy of 0.01% (i.e. 1 part in 10^4).

Here we see a fact of field theory: it is a *perturbation theory*. Never can we compute things exactly but we can approximate them to any desired precision. That is of course true assuming that we can find our way though the maze of summations (over energy–momentum distributions) that arises when considering a diagram with many virtual particles. The calculation of the magnetic moment of the electron is in practice perhaps the most advanced example. The electron, possessing spin, has a magnetic moment, like any rotating charged object. In other words, not only does the electron have a charge, but also it acts as a tiny magnet. That part of the interaction of an electron with the electromagnetic field is also subject to quantum corrections, and figure 6.8 shows the lowest-order diagram and a next-order (in e, the electric charge) diagram.

Figure 6.8.

In suitable units the magnetic moment of the electron, disregarding quantum corrections, is 1. The second- and higher-order contributions alter that magnetic moment by a tiny amount; to give an idea about the accuracy achieved we quote here the theoretical result for this alteration (including fourth- and sixth-order contributions as well):

$$0.5 \left(\frac{\alpha}{\pi}\right) - 0.328\,478\,965 \left(\frac{\alpha}{\pi}\right)^2 + 1.181\,241\,456 \left(\frac{\alpha}{\pi}\right)^3$$
$$- 1.4 \left(\frac{\alpha}{\pi}\right)^4 = 0.001\,159\,652\,201 \qquad (6.1)$$

compared with the experimental value of 0.001 159 652 188. Note that $\alpha/\pi \approx 0.002\,32$. The errors in both the theoretical and the experimental values are of the order of the difference between the values quoted here. In other words, the agreement is excellent.

The sophistication involved in both theory and experiment is impressive. The calculation of the coefficient of α^3 has taken some 20 years, involving some 72 diagrams, while the calculation of the α^4 term (891 diagrams) has been done only by numerical approximation methods, using up years of super-computer time. Any experiment achieving an accuracy of one part in a thousand is already difficult, let alone the experiment relevant here, having an accuracy of order of one part in 10^6. This most spectacular experiment is based on measurements performed on a single electron, caught in an electromagnetic trap.

Here a remark should be made concerning the way that theorists talk about these things. They usually classify the subsequent orders of perturbation theory by means of loops. Without going into details, the lowest-order diagram is called a tree diagram (no loop), the next-order diagrams have one loop, the next order three loops, etc. Figure 6.9 shows examples of a two-loop and a three-loop diagram.

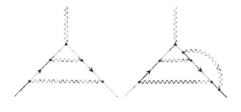

Figure 6.9.

The amazing agreement between theory and experiment, involving these very complicated quantum effects,

must be seen as a strong support for the theoretical insights as well as for the validity of perturbation theory. Many theorists would have liked a formulation of the theory not involving approximations, but so far perturbation theory is all that we have. In certain instance, one has been able to sum the contributions of some classes of diagrams to all orders, but we do not have any general non-perturbative version of the theory. This is a fact of life. Let us be happy with the notion that we know at least the basic ingredients that underlie the quantum-mechanical calculations.

RENORMALIZABILITY

Another problem relates to the non-convergence of the summations over all possible distributions of energy–momentum of the virtual particles. In certain cases these sums do not converge, i.e. the result is infinite. That stopped progress for quite some time, until, in about 1948, the idea of *renormalization* solved the problem at least on a practical level. The subject theory was quantum electrodynamics, and it was noted that the infinities occurred only in some well defined instances. For example, in the calculation of the magnetic moment of the electron discussed above they did not occur, but these very same diagrams, which alter the magnetic properties of the electron, will equally well alter the electric charge of the electron, as they simply affect the way that a photon interacts with the electron. That change to the electric charge turns out to be *infinite*. Here then is the big idea; the electric charge as actually observed is the sum total of the basic electric charge (as occurring in the tree diagram) plus the contributions of all higher orders, but we have no idea how large that charge (the basic charge) is without the radiative corrections. So, let us choose the value of the basic charge such that the total charge comes out equal to the experimentally observed value. In other words, we give the basic charge a value that has an infinite part as well, but opposite to that part of the higher-order corrections, making the sum come out equal to the observed value!

This trick, hard to swallow at first, works very well indeed. The basic observation, somewhat more concisely stated, is that the infinities occur only in conjunction with the free parameters of the theory. These parameters must be obtained from experiment anyway. The electric charge is such a parameter. It is an input to the theory, and not something that we can compute. Another such parameter is the mass of the electron. It is not known from any basic principle, and its value must be obtained by measurement. That gives us an opportunity to hide an infinity.

This scheme for getting rid of infinities is called renormalization. It is by itself far from satisfactory. No one thinks that the basic quantities are actually infinite. Rather we believe that the theory is imperfect, but that this imperfection can be isolated and at least for the moment be hidden. The miracle is that for quantum electrodynamics all infinities can be absorbed into the available free parameters. So, apart from these infinite corrections to free parameters, everything else (such as the magnetic moment of the electron quoted above) is finite and, insofar as it has been checked, agrees well with experiment.

So here we are. We have a theory imperfect on several counts. First, the theory is perturbative. Second, infinities occur, even if they can be isolated and swept under the rug. In spite of these imperfections, all this leaves us nonetheless with a scheme that makes accurate predictions that can be compared with experimental results.

There are also theories such that infinities occur not only in conjunction with free parameters. Such theories cannot make solid predictions. They are called *non-renormalizable*. For a long time, theories involving vector particles (spin 1) were thought to be of that type and, as such, useless. This picture has changed, and we now also have renormalizable theories involving spin-1 particles. These theories are called *gauge theories*. In fact, almost all interactions seen in experiments are of that type, gravitation being the exception. Quantum effects in gravitation have so far not been understood. That casts a shadow on that theory and its consequences, for example black holes.

WEAK INTERACTIONS

Weak interactions constitute a different type of interaction, one that does not produce long-range forces such as those in electrodynamics. A photon has zero mass and can hence have arbitrarily low energy. For this reason a virtual photon of zero energy and small momentum is only slightly 'off mass shell', and little damping is associated with the

exchange of such a virtual photon. It is this type of zero-energy low-momentum photon that is responsible for long-range electromagnetic interactions. For the same reason the graviton will also give rise to a long-range force. These, however, exhaust the list of long-range forces that we experience in daily life. The weak interactions have a very short range; let us discuss them in some detail.

Weak interactions made their entry into physics through the discovery, by Becquerel in 1896, of β radioactivity. Experimentally and theoretically it took really a very long time before these interactions were understood, even on a purely phenomenological level. Since it is not our purpose to present here the history of the subject, we shall straight away describe things as they are understood today.

Consider the most fundamental nuclear β decay: that of a neutron into a proton and an electron (plus an antineutrino). The neutron contains two down quarks and one up quark (denoted by d and u respectively), and the proton one d and two u quarks. As a first step, one of the d quarks decays into a u quark and a negatively charged vector boson, denoted by W^-. Figure 6.10 shows the diagram representing this decay.

Figure 6.10.

It contains one of the basic vertices of weak interactions. The associated coupling constant ('weak charge'), usually denoted by g, is somewhat larger than the corresponding value for electromagnetism. Experiment shows that $\alpha_w = g^2/\hbar c = \frac{1}{32}$. At this point we have a notational problem, because all particles in this reaction are charged (their charges are $-\frac{1}{3}$, $+\frac{2}{3}$ and -1, in terms of a unit such that the charge of the electron is -1, for d, u and W^- respectively), and the arrow no longer represents the flow of negative electric charge. Instead it will be used to distinguish between particles and antiparticles, where an arrow pointing opposite to the flow of energy indicates an antiparticle. Here there is a choice: is the W^- a particle or an antiparticle? Historically, there was the regrettable mistake of having defined the charge of the electron as negative. We shall not do that here for the W, and define the W^- to be the antiparticle. That is why the arrow in the W line points inwards.

The W is very massive (80.3 GeV; compared to the proton mass, 0.938 GeV), and given the low mass of the d quark (about 10 MeV = 0.010 GeV) it must be very virtual (way off mass shell) in the actual process of d decay. In a second step it hence immediately transforms into an electron and an antineutrino, an interaction which is another basic vertex of the theory. Figure 6.11 shows the complete diagram for d decay. (Note that an antineutrino is not the same as a neutrino, despite the fact that the neutrino has zero electrical charge.)

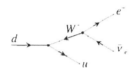

Figure 6.11.

As noted before, the existence of a negatively charged W^- (an antiparticle) implies the existence of a particle with opposite charge, the W^+. It would for example be involved in the decay of an antineutron into an antiproton, a positron (antielectron) and a neutrino; that reaction is simply the same reaction as the one discussed above, with all particles replaced by their antiparticles (reversal of all the arrows, see figure 6.12).

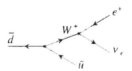

Figure 6.12.

The W has spin 1 and its interactions generally lead, owing to the absence of damping at high energies, to a non-renormalizable theory as we discussed before. We shall discuss this point in a systematic way, showing how the situation can be salvaged, and how new particles and interactions must be introduced in order to achieve a tolerable high-energy behaviour.

COMPTON SCATTERING

Exploring high-energy behaviour can conveniently be done by considering certain simple processes. One has to estimate the energy dependence of a process where particles of very high energy are scattered. The simplest and most important example is Compton scattering, that is the scattering of a photon by an electron. In lowest order, only electromagnetic interactions are of relevance here, and figure 6.13 shows the two diagrams that contribute at the tree level.

Figure 6.13.

Strictly speaking much of what follows below is true only for non-zero mass photons, but we shall ignore this subtle point for the sake of simplicity.

The high-energy behaviour of the diagrams shown can be assessed as follows. An incoming or outgoing photon (vector particle; spin 1) contributes a factor proportional to the energy E of that photon. A virtual photon contributes no energy dependence, i.e. it must be counted as a constant. A virtual electron or, generally, a virtual spin-$\frac{1}{2}$ particle behaves as $1/E$. An incoming or outgoing electron (spin-$\frac{1}{2}$ particle) must be counted as $\sqrt{E}$. A virtual scalar particle (spin 0) must be counted as $1/E^2$; an incoming or outgoing spinless particle contributes a constant. It must be noted that in special cases the energy dependence might be different from the dependence that one would deduce by counting with these rules; whether or not it does depends on details of the actual couplings which sometimes compensate the aforementioned energy dependence related to the magnitude of the spin. A case in point, one that we shall meet in the following, is the coupling of a vector boson to two real (i.e. not virtual) particles. In the case of gauge theories the vertices are always such that spin effects are neutralized in that instance, i.e. for a (possibly virtual) vector boson coupling to two real particles. An example is the decay of the down quark as depicted in the previous section. There the virtual W^- is on both ends coupled to real particles. In that case the energy dependence relating to that virtual W is as that for a scalar particle, i.e. as $1/E^2$. We summarize the various factors given above in the following table.

Spin	In/out	Virtual	Ends*
0	1	$1/E^2$	$1/E^2$
$\frac{1}{2}$	$\sqrt{E}$	$1/E$	$1/E$
1	E	1	$1/E^2$

*Real particles at the ends

In a renormalizable theory the probability that a process occurs must, as a function of energy, either decrease or at worst tend to a constant value. Even on a purely intuitive level a probability increasing indefinitely as the energy of the incident particle increases is hard to accept. Note that the probabilities (cross sections) are obtained, as mentioned earlier, by squaring the contributions of the diagrams. Counting the expected behaviour for the diagrams shown above for Compton scattering, we arrive at a behaviour increasing with increasing energy, in fact as E^2 (so the cross section would go as E^4). This is unacceptable for a renormalizable theory. However, the second diagram shows also a leading dependence proportional to E^2 but with the opposite sign, and the sum of the two actually behaves as a constant. A somewhat simplistic explanation for this behaviour is that the intermediate (virtual) electron in the first diagram has a large positive energy, and in the second a large negative energy. Thus the factors $1/E$ for the intermediate electron have opposite signs for the two diagrams.

This is the idea behind gauge theories, of which quantum electrodynamics is the simplest example. Individual diagrams give unacceptable energy behaviour, but everything is arranged in such a way that in the end the bad behaviour cancels. Let us see whether we can make this work for weak interactions that involve charged vector bosons.

NEUTRAL VECTOR BOSONS

In the following we shall ignore electromagnetic interactions, giving rise to diagrams containing a virtual photon in some of the situations discussed below. They are not essential to the reasoning.

Let us first examine W^+ scattering off an electron. The corresponding lowest-order diagram is the first one in figure 6.14. There is a virtual neutrino mediating this process. The behaviour at high energy assessed by power counting as specified above is bad, namely as E^2 (E for each of the W particles, $\sqrt{E}$ for each of the electrons and $1/E$ for the intermediate neutrino). As the W particles are connected to vertices of which one of the particles (the neutrino) is *virtual* there are no special compensating effects.

Figure 6.14.

Recalling the case of Compton scattering we might think that the situation can be cured by another process, that shown in the second diagram in figure 6.14. Since now the incoming electron emits a positively charged W, the intermediate particle (denoted X^{--} in the diagram) cannot be a neutrino because of charge conservation (a law that holds rigorously) requires the intermediate particle to have charge -2, but no such particle is known. The diagram does not exist. What now?

Figure 6.15.

The solution is to introduce another vector boson, this time one without electric charge (figure 6.15). It is coupled to the charged vector bosons and the electrons in such a way that the high-energy behaviour of the diagram shown cancels the bad behaviour of the first diagram in figure 6.14. The vertex must behave like E and, given that the intermediate vector boson is coupled on both ends to real particles we have indeed the required behaviour. E^2 (E for each of the charged W particles, $\sqrt{E}$ for each of

slightly modify this relation. In this equation, M and M_0 are the masses of the W (80.3 GeV) and the Z^0 (91.2 GeV) respectively. To explain the angle θ_w appearing here would require a detailed discussion about the interplay of weak and electromagnetic interactions, because wherever the Z^0 occurs coupled to charged particles on both ends the photon can take its role. Experimentally one finds that $\sin^2 \theta \approx 0.2315$, and we conclude this discussion with the observation that ρ comes out to the predicted value so that there is no need to have more than one Higgs particle.

It is interesting to note that the higher-order corrections to the equation $\rho = 1$ involve, among others, the mass of the top quark in a most peculiar way; the correction becomes larger as the top quark mass is heavier. Here we have a quantum effect that increases if the intermediate state is energy-wise further away! Many years before the top quark was actually observed, the measured magnitude of the quantum corrections was used to predict the top quark mass. This prediction agrees quite well with the experimental value.

The reason that the radiative correction grows with the top mass is a very typical consequence of a gauge theory structure. The top quark has a function in the scheme for, if it is not there, certain diagrams grow in an intolerable way. So, if you try to eliminate the top quark from the theory (by making it very heavy), you are left with an infinity. Figure 6.21 shows the relevant diagrams, which concern momentary dissociation of the W^+ and Z^0 into a quark–antiquark pair. Such diagrams are called self-energy diagrams. The effect that we discuss involves the first diagram minus $\cos^2 \theta_w$ times the sum of the second and third diagram (since the first diagram gives a correction to M^2, and the other two to M_0^2). The top quark is now essential; without the top quark, only the second diagram would be there, and this diagram all by itself gives an infinity.

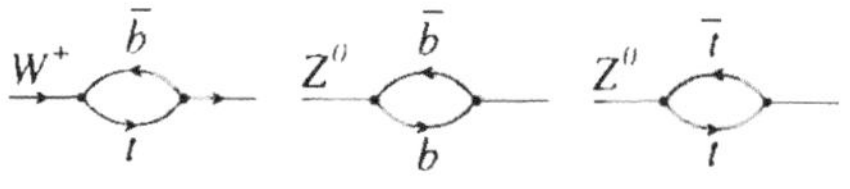

Figure 6.21.

Is this the end of the story? Not quite. First of all, there remain many little problems of the nature sketched above,

but it would carry us too far to enter here into a detailed discussion. Suffice it to say that the Higgs particle must also be coupled to the neutral vector boson (the Z^0) and to the quarks, etc, as well. In short, it must be coupled to *any* particle having a mass. Moreover, the coupling must always be proportional to the mass of the particle to which it is coupled.

To date the Higgs particle has not been observed experimentally. Unfortunately the theory has nothing to say about its mass, except that it should not be too high (less than, say, 1000 GeV), or else its compensating actions set in too late. The present experimental lower limit is roughly 100 GeV. The new collider being built at CERN (the Large Hadron Collider (LHC), colliding protons each with an energy of 7000 GeV) might give us information on this Higgs particle. It will not be easy to produce Higgs particles, because the proton contains only u and d quarks, and these, because of their low masses, couple only weakly to this Higgs particle. Higher order processes, involving virtual (heavy) W and Z^0 particles are needed to produce this particle.

The demonstration that all bad energy behaviour can be made to vanish with only those particles discussed above, and no others is usually referred to as the proof that this theory is renormalizable. On reading the previous discussion, one may easily have the impression that there is no end to new hypothetical particles that must be introduced, but no, this is it! The Higgs particle is the last particle needed.

It is perhaps necessary to state explicitly to what extent the discussion above reflects the historical development. We have sketched a theory involving many particles, with their interactions so orchestrated and tuned as to have a renormalizable theory. The result is a theory possessing a high degree of symmetry. The historical development was quite the opposite of that suggested by our treatment. The symmetry was discovered and investigated some 20 years before its consequence, a renormalizable theory, was finally understood.

SPECULATIONS

Because this Higgs particle seems so intimately connected to the masses of all elementary particles, it is tempting to think that somehow the Higgs particle is responsible for these masses. Up to now we have no clue as to where masses come from: they are just free parameters fixed by experiment. It requires no great imagination to suppose that the Higgs particle might have something to do with gravitation and, indeed, theoretical models suggest a strong involvement of the Higgs particle in the structure of the Universe, otherwise thought to be shaped by gravitation. Many theorists believe today that the Higgs particle does not really exist, but that it somehow mimics a much more complicated reality, involving gravitation in a fundamental way.

These are very exciting and interesting questions and speculations. We are looking forward to LHC experiments, noting that so far theorists have not been able to come up with any credible theory that answers all or some of these questions, including questions concerning the magnitude of the masses, the Cabibbo angle, the existence of all these quarks, the grouping of these particles into families, and so on. There is clearly so much that we do not know! Even so, we have certainly made enormous advances in understanding the structure of the interactions between the elementary particles.

ACKNOWLEGMENT

I would like to express my gratitude to Professor V L Telegdi for advice which led to substantial improvements in the presentation of the material.

FURTHER READING

Cao T Y 1997 *Conceptual Developments of 20th Century Field Theories* (Cambridge: Cambridge University Press)
Veltman M 1997 'The Path to Renormalizability' in *The Rise of the Standard Model* ed L Hoddeson, L Brown, M Riordan and M Dresden (Cambridge: Cambridge University Press)
't Hooft G 1997 *In Search of the Ultimate Building Blocks* (Cambridge: Cambridge University Press)
Weinberg S 1995 *The Quantum Theory of Fields* (2 volumes) (Cambridge: Cambridge University Press)

ABOUT THE AUTHOR

Martinus Veltman was educated in the Netherlands and received his doctorate from the University of Utrecht in 1963. His thesis advisor was Léon Van Hove. He was at CERN from 1961–66, after which he became Professor of Theoretical Physics at the University of Utrecht. In 1981 he became MacArthur Professor of Physics at the University of Michigan, Ann Arbor. To facilitate onerous field theory calculations, he was one of the first to apply the computer to the evaluation of Feynman diagrams. In 1993 he was awarded the European Physical Society's Europhysics High-Energy Prize for his key role in applying field theory to weak interactions, and in 1996 shared the Dirac Medal of the International Centre for Theoretical Physics, Trieste, Italy, for his pioneering investigations on the renormalizability of gauge theories.

7 UNEXPECTED WINDFALL

Roy F Schwitters

Editor's Introduction: One of the greatest and most far-reaching discoveries of modern particle physics, that of the 'J/ψ' in November 1974, was almost totally unexpected. Even more remarkable, the discovery was made simultaneously by two experiments using totally different techniques at two different laboratories: Brookhaven National Laboratory by a team led by Sam Ting and the Stanford Linear Accelerator Center (SLAC) by a team led by Burton Richter. In 1976, Richter and Ting shared the Nobel Prize for their discovery which confirmed that quarks, rather than coming in three types, up, down and strange, as had long been thought, in fact came in four, the new one earning the name 'charm'. In this chapter, Roy Schwitters, a leading member of the SLAC experiment, relates how that experiment stumbled onto a major discovery, how they learned that Ting had found the same particle, and how the new particle could be explained as an atom-like system of a charmed quark and its corresponding antiquark.

In the world of experimental high-energy physics, the subatomic particle called J/ψ is often at the centre of attention, be it as part of formal reports at international conferences or lunchroom conversations regarding the performance of a piece of apparatus. It has been this way since November 1974, when two research groups announced the discovery of the same new particle, called ψ by one group and J by the other.

The J/ψ is an object that weighs slightly more than three times the mass of a proton, is electrically neutral and lives for about 10^{-20} s. It is easily observed in debris of very-high-energy particle collisions, often providing incisive clues for understanding new phenomena. Its distinctive properties and ubiquitous nature are routinely employed to calibrate and test the large detectors used to explore high-energy physics.

More important than finding just another new particle, the surprise discovery of the J/ψ, and related discoveries that soon followed, consolidated thinking about the basic forces and building blocks of matter and changed approaches to performing experiments in high-energy physics. It is understandable that the J/ψ discovery is commonly called the 'November Revolution' of particle physics. Physicists of the time still recall vividly when, where and how they learned of the discovery. This is an account by one who was fortunate enough to be present at the revolution.

FRASCATI AND CAMBRIDGE

I first heard of 'funny events' being detected in experiments at a particle accelerator in Frascati, Italy, from my thesis advisor while I was a graduate student at Massachusetts Institute of Technology (MIT) in the late 1960s. It seemed that there were too many recorded instances, called 'events', where three or more particles were created from the energy released in a head-on collision between an electron and its antimatter counterpart, a positron. The observations were sketchy, limited by available detector technology and few events. Their reception by the community was decidedly mixed; some physicists were interested, while many treated the funny events as mild curiosities or simply ignored them.

Frascati is a delightful town situated in the hills south of Rome, where the pursuit of fundamental science mixes well with the local wines. The physics laboratory there was one of the pioneers (along with groups at Novosibirsk, Russia, Orsay, France, and Stanford University) in mastering the art and science of colliding beams of electrons and positrons for studying particle physics. Colliding beams was a novel approach in the 1960s, in part because of the myriad of technical difficulties encountered in accelerating and controlling high-energy particle beams so that they collide with sufficient frequency to permit meaningful experiments. The more common approach was to collide one accelerated beam, often protons, with a stationary or 'fixed' target, such as a metal foil or liquefied hydrogen.

Frascati's electron–positron collisions were created with a machine called ADONE. Like nearly all such colliders built before and since, ADONE consisted of magnets and acceleration devices arranged in the shape of a race-track to bend high-energy particles onto a closed path and to focus them into beams that could be collided against each other. The beams circulate around the ring at essentially the speed of light, remaining stored for periods of hours. Positrons trace the same path as electrons, but in the opposite direction. Thus, only one ring is needed to contain both counter-rotating beams. At certain locations around the ring, called interaction regions, one beam intersects the other, permitting head-on collisions between individual electrons and positrons. Detectors are placed around the interaction region to collect particles produced in collisions. Information recorded from the debris can be 'played backwards' to deduce the properties and behaviours of the collision products during the infinitesimal time when the electron and positron interact.

High-energy collisions between particles reveal internal structures and details of the forces between them: the higher the energy, the smaller is the structure that can be probed. New particles can be created out of the collision energy, which, according to Einstein's mass–energy relationship $E = mc^2$, require higher energies to create more massive particles. By the late 1960s, dozens of new, unstable and unexplained particles had been observed in fixed-target experiments, and the internal structures of particles such as protons and neutrons and the forces between them were being revealed. Protons and neutrons were found to be extended objects roughly 10^{-15} m in size, whereas electrons and positrons appeared to have no internal structure, behaving like mathematical points. The hosts of newly discovered particles were all 'hadrons', relatives of protons and neutrons that are influenced by the strong nuclear force which binds atomic nuclei. The strong force was a deep mystery in the 1960s; there was no compelling theory, and the prospects of finding one in the foreseeable future seemed remote. Electrons and

positrons were known to be immune to the strong force, interacting via their electric charges according to well understood laws of electricity and magnetism, described by the theory of quantum electrodynamics (QED). From that perspective, electrons and positrons were considered uninteresting. Electron–positron colliders did not appear to be the best way to pursue the most pressing questions.

A few individuals, however, were attracted by the elegance of the electron–positron system. When these particles collide, the most likely outcome is a glancing blow that causes the two to scatter. This process is called 'Bhabha scattering'. Less likely, but much more interesting, is the *annihilation* of the electron and positron into pure energy which materializes into quanta of light (photons) or other particles. Production of a pair of oppositely charged muons is the prototypical reaction involving particles. Muons, discovered in cosmic rays in the 1930s, are heavier forms of electrons, carrying the same electric charge, behaving like point particles and eschewing the strong force. Their production rate in electron–positron collisions could be calculated precisely from QED.

A last possibility for the outcome of an electron–positron annihilation is direct conversion into a single particle. For conversion to occur, the total energy of the electron–positron system must exactly match the mass of the particle and the particle must share certain properties with photons. At the time of the new Frascati results, there were only three such particles known, the ρ, ω and ϕ, collectively called 'vector mesons'. The Frascati experiments searched in vain for more at energies up to 3 GeV. (The conventional measure of collision energy is the 'gigaelectronvolt' (GeV) or 10^9 eV. In a head-on collision, the total energy is given by the sum of the individual beam energies. For example, the mass–energy of a proton is about 1 GeV; to produce two particles of this mass requires 2 GeV. Thus, a collider such as ADONE with a beam energy of 1.2 GeV would yield a collision energy of 2.4 GeV, slightly above the minimum required to produce a proton–antiproton pair.)

The Frascati experiments verified that QED well described Bhabha scattering, annihilation into photons, and production of muon pairs, but they showed something else; there were too many 'multiprong' events (collisions where several particles were produced) for QED to explain.

Furthermore, the particles in these events seemed to be hadrons, and not electrons or muons. Hadrons were expected to be produced only rarely at these high energies. The detectors, however, covered limited fractions of the space surrounding interaction regions and could not prove the particles were hadrons nor determine their energies.

Meanwhile, at Harvard University, physicists and accelerator experts were planning to convert an aging electron synchrotron, the Cambridge Electron Accelerator (CEA), into an electron–positron collider that would extend the Frascati measurements to higher energies. They had hoped to build a new machine, but funds were not available, in part because the federal agency supporting high-energy physics had chosen a competing proposal by a group at Stanford. The Cambridge group applied great determination and imagination in modifying their accelerator to be able to collide beams, making inventions that are central to modern colliders.

Physicists at Harvard and MIT were responsible for CEA's detectors. The plan was to build a 'non-magnetic' detector, similar to those employed at Frascati to make initial surveys of the physics, to be followed by a more elaborate instrument, a 'magnetic' detector, that would permit measurement of the energies of particles created by the collisions. My advisor, Louis Osborne, was the chief proponent of the magnetic detector. Unfortunately, the CEA proved to be a difficult machine to control. Extraordinary efforts were invested into the complex choreography of accelerators and beams needed to achieve collisions, delaying scientific results. In the end, a magnetic detector was not built.

STANFORD LINEAR ACCELERATOR CENTER

I did not work on colliding beams while in graduate school. My thesis research extended a previous CEA experiment to higher energies at the Stanford Linear Accelerator Center (SLAC). We collaborated with Burton Richter, also a former student of Louis Osborne, and his group. After completing my degree, I accepted a post-doctorate position in Richter's group to work on the Stanford Positron–Electron Asymmetric Ring (SPEAR), the Stanford electron–positron collider that had won the competition for funding over CEA. (In fact, SPEAR is an

obsolete acronym that was retained because it made such a good name.)

SLAC is the creation of Wolfgang Panofsky, called 'Pief' by everyone who knows him. It has nurtured three Nobel Prize-winning experiments and numerous other classic results. The centrepiece is its two-mile-long linear electron accelerator, which provides intense beams of high-energy electrons. SLAC's technical facilities were excellent, its scientific standards were high, the best people were there, and young people were given substantial responsibility. Close interactions between theorists, experimenters, and accelerator experts, who are wont to go their separate ways, were embedded into SLAC's culture.

When I arrived for work in June 1971, the excitement over deep-inelastic electron–proton scattering experiments was palpable. The 'parton' and 'scaling' ideas invoked for interpreting them helped to shape our thinking about electron–positron collisions, including the connection to those funny events from Frascati, but practical demands to build SPEAR and a proper detector to go with it would dominate our efforts for the next few years.

Richter's physics group comprised other post-doctoral researchers, senior physicists and technicians who had worked together on various experiments in the past. Richter had recently invited another SLAC group led by Martin Perl and two groups from Berkeley led by Willi Chinowsky, Gerson Goldhaber and George Trilling to collaborate in designing and building a detector for SPEAR, and, eventually, using it for physics research. Our detector group worked closely with the project team constructing the SPEAR collider, which also reported to Richter.

SPEAR construction was well under way in 1971. It was to cover energies between 2.4 and 8 GeV, overlapping ADONE's top energy. SPEAR was 240 m in circumference and had two interaction regions (figure 7.1).

MAGNETIC DETECTOR

There was never doubt that we should build a magnetic detector; the problem was how to do so since none existed at the time. We wanted to measure QED reactions (Bhabha scattering and muon pair production) to test that theory and to learn as much as possible about the intriguing Frascati

Figure 7.1. SPEAR, the 240 m circumference electron–positron collider at SLAC.

events. This suggested a general-purpose device that would be sensitive to the many kinds of particle being produced over large angular and energy ranges. Emphasis was put on observing all or most of the particles in each event, so that sacred principles of energy and momentum conservation could help to constrain inevitable measurement errors. The magnetic field was needed to bend the paths of charged particles, permitting determination of their momenta and energies.

The basic principles of particle detection have not really changed much over this century, but advances in technology have enabled great improvements. When passing through ordinary matter, such as a volume of gas or piece of plastic, high-speed charged particles ionize atoms along their path. This ionization can be detected by several means. Like a high-flying jet airplane that is only visible by its contrail, high-energy particles can be tracked from their trail of ionization.

We organized the detector into cylindrical layers surrounding the interaction region (figure 7.2). Collision particles would traverse, in sequence, the following: (1) the vacuum pipe containing the beams; (2) a tracking chamber to follow the particle's path in the magnetic field; (3) plastic detectors to precisely time the flight from the collision point; (4) a coil producing the magnetic field; (5) 'shower counters' assembled from plastic detectors interspersed with sheets of lead which respond especially to electrons, positrons and photons; (6) thick plates of iron from the magnet yoke, interspersed with detectors, that would signal muons, the only objects likely to penetrate that far (other than neutrinos, which escape detection in all the layers).

The device had a mass of several hundred tons. Its magnet coil was a cylinder, 3 m in diameter and 3 m long, coaxial with the SPEAR beams (figure 7.3). Assembling this detector was like building a ship in a bottle. All the particle detectors produced electronic signals, requiring many wires running in and out of the bottle, to transfer data to computers for analysis and storage.

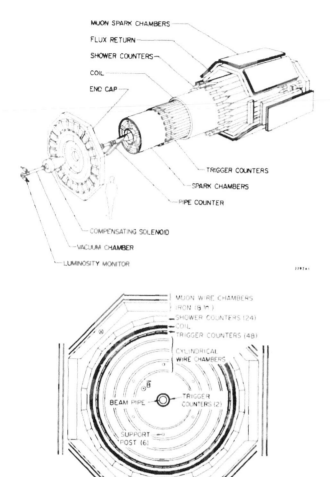

Figure 7.2. labels:
MUON SPARK CHAMBERS
FLUX RETURN
SHOWER COUNTERS
COIL
END CAP
TRIGGER COUNTERS
SPARK CHAMBERS
PIPE COUNTER
COMPENSATING SOLENOID
VACUUM CHAMBER
LUMINOSITY MONITOR

MUON WIRE CHAMBERS
(IRON (8 in.))
SHOWER COUNTERS (24)
COIL
TRIGGER COUNTERS (48)
CYLINDRICAL
WIRE CHAMBERS
BEAM PIPE
TRIGGER COUNTERS (2)
$\vec{B}$
SUPPORT
POST (6)
1 meter

Figure 7.2. Diagram of the concentric layers of the SPEAR magnetic detector which revolutionized the art of building particle detectors.

Figure 7.3. Roy Schwitters, seen at the SPEAR magnetic detector.

Finally, substantial software was developed to reconstruct the particle trajectories and identities from the digitized signals. Each of the detector's layers presented hosts of technical challenges. Coordination of the enterprise was carried out in masterful fashion by Rudy Larsen, a colleague in Richter's physics group, who was spokesman for the collaboration, and by Bill Davies-White, our lead engineer.

The ultimate success of the magnetic detector depended on difficult compromises made between perfor-

mance, cost and space needed for the various component layers. (Modern collider detectors have formal names (and associated 'marketing' trinkets, such as tee-shirts and coffee mugs with catchy logos), usually based on clever acronyms, that are used to identify the device and its parent collaboration. The SPEAR magnetic detector was called just that until after its successor, called 'Mark II' was built a few years later. In retrospect, the SPEAR magnetic detector became 'Mark I'.) We adopted a clear line that emphasized the *area* and *uniformity* of coverage, even if that meant reducing the ultimate performance of some part.

I worked with Harvey Lynch on the tracking system. The technology of choice then was 'spark chambers', planes of fine wires that generate sparks along the ionization trail of particles when a pulse of high voltage is applied across them (figure 7.4). Current from sparks flows down the wires, marking the locations of the sparks. Software,

Figure 7.4. One layer of the cylindrical spark chambers of the SPEAR magnetic detector.

developed by our Berkeley colleagues, Willi Chinowsky, George Trilling and their students, took the spark locations and reconstructed paths. One of our greatest problems was mechanical: fitting square boxes of chamber planes into a cylindrical magnet coil with a beam pipe running down the centre. There would be many blind edges where wire planes ended and the projected coverage was poor. Then, we realized that wires could be strung on cylindrical surfaces, with *no blind edges whatsoever*, as long as their tension was supported by an external frame. This permitted the chambers to be built as a set of cylinders with essentially no dead areas, neatly bypassing a key problem in earlier (and some later) detectors.

Like the 'first light' in a new telescope, the initial detection of colliding beam events is an exhilarating experience. (Also like telescopes, first collisions always seem to occur in the middle of the night!) Our first collisions came in 1973. We shared control room space and computers with the SPEAR operations crew, greatly facilitating communications during the intricate process of loading beams of electrons and positrons and bringing them into collision. When stable beam conditions are achieved, an experimental run begins, usually lasting several hours until the beams need replenishing. During a run, data were logged to magnetic tape for off-line analysis; a small sample of the collision events could be reconstructed by the on-line

computer. Pictures of sampled events were displayed on monitors in the control room. Such technology, so common today, was considered quite advanced at the time.

The detector was triggered to fire its spark chambers and to record data when two or more charged particles emerged from the interaction region. Triggers occurred every few seconds, but only a small fraction were actual collisions. Most were backgrounds caused by cosmic rays or particles lost from SPEAR's beams. The spark chambers generated considerable electrical noise when fired, interfering with the control room's intercom to give an audible 'tick' every time that an event occurred.

I never ceased to be awed watching the 'one-event' display of reconstructed collisions. Matter was being created from energy right before your eyes! Bhabha scattering and muon pair events appeared as two, nearly straight back-to-back tracks. Backgrounds were obvious because there were few tracks and those present failed to point back to the interaction region, but the multiprong events were most beautiful and exciting. Three, four and more tracks arched majestically through the detector from a common origin (figure 7.5). They were unmistakable. It was clear to us from the very beginning that such events were relatively common and that they involved hadrons. The Frascati events were real!

Figure 7.5. Some typical SPEAR electron–positron collisions.

FIRST RESULTS

The first data runs took place in 1973. Our strategy was to collect as many events as possible at a few selected energies, and to 'scan' with brief runs the full energy range available to SPEAR in 0.2 GeV increments to search for possible new vector mesons. The program would take months to carry out. The choice of energy step, crucial to subsequent developments, was based on our use of a single decimal place of precision, 1.2 GeV, 1.3 GeV, etc, to select SPEAR's beam energy. We believed that any new vector meson would be produced over a range of energies greater than 0.2 GeV, a plausible extrapolation from known mesons.

Meanwhile, results at 4 and 5 GeV were announced by the CEA team. Their epic struggle yielded a few hundred multiprong events at the two energies. The results were dramatic; the multiparticle events were five to six times more abundant than muon pairs. Unfortunately, as with the lower-energy data from Frascati, the CEA results were met with scepticism by many physicists because of relatively small numbers of events and concerns stemming from the lack of magnetic tracking. From our preliminary data, we knew that the CEA results were basically correct, but our event reconstruction and data analysis were not yet ready for us to make any claims publicly.

By the early 1970s, the parton picture, introduced to describe the SLAC electron scattering experiments, had focused interest on R, the ratio of the *total* rate for producing hadrons of all kinds to the rate at which muon pairs are created in electron–positron annihilation. This simple picture viewed hadron production as a two-step process: (1) production of parton–antiparton pairs, followed by (2) materialization of hadrons from the partons. Parton production was just like muon production, differing only in rate which should be proportional to the *square* of electric charge. Thus, R should equal the sum of squares of electric charges of all types of parton, *independent of the colliding beam energy*. Partons were introduced as generic point-like objects thought to make up protons, neutrons and other hadrons. It was natural to associate partons with *quarks*, the hypothetical building blocks introduced specifically to describe observed families of hadrons and this was demonstrated conclusively in 1972 (see the chapter by Friedman and Kendall). However, the large values of R

measured at Frascati and CEA seemed to rule out the most plausible quark models.

In December 1973, we took the wraps off our results. Richter was scheduled to talk at a meeting in Irvine, California and gave a 'warm-up' seminar at SLAC. Within our estimated errors, R *doubled* over SPEAR's energy range, from a value near 2.5 at lowest energies to 5 at the highest energy measured, 4.8 GeV. The data agreed with the Frascati and CEA results but had more impact because of the number of energies studied, their statistical power, and the greater capabilities of a magnetic detector. So much for the notion that R was independent of energy! Another curious result was that the fraction of energy carried by charged particles in the hadron events seemed to be diminishing as the beam energy increased. Where was the energy going? This effect came to be called the 'energy crisis', a pun based on the middle-eastern oil crisis taking place at around the same time. QED accurately described our Bhabha scattering and muon pair results.

The Irvine talk created great interest. The theoretical picture was especially confusing because the most compelling new theories could not explain the *increase* in R with increasing energy, not to mention its large values. Within our collaboration, a certain antitheory hubris developed suggesting that not R, but a related quantity called the 'total hadron cross section', had the same value at all energies, implying that R should grow as the *square* of the collision energy. The best way to test this idea would be to run at higher energies to see whether R continued to rise. An upgrade of SPEAR's top energy to permit such measurements was scheduled for summer 1974.

The winter and spring of 1974 were taken up with additional running to accumulate more data. To investigate the energy crisis, a thin steel cylinder was placed inside the tracking chambers. The idea was to measure the energy carried by photons in multihadron events by observing electron–positron pairs created in the cylinder when struck by photons. Unfortunately, relatively few photons converted in the cylinder and our measurement of photon energies was not accurate enough to resolve the 'crisis'. The steel cylinder was soon scrapped.

In June 1974, just before SPEAR was shut down for the energy upgrade, several days were spent running at *low* energies to investigate possible 'bumps' in R.

Superficially, R appeared to be a smooth function of energy, but fluctuations at the level of statistical errors existed in the data near 3 and 4 GeV. The June runs interpolated between energies studied earlier and improved statistical accuracy where measurements already existed.

A key ingredient in these runs was a rapid determination of R, so we could respond if it varied significantly between nearby energies. Marty Briedenbach integrated on-line and off-line computers to accomplish the analysis. The new runs appeared to give the same smooth behaviour for R as seen in earlier data. SPEAR was shut down for its upgrades and most collaboration members went their various ways for the summer.

Burt Richter was the plenary speaker on new electron–positron results at the most important international conference that summer, held in London. He emphasized the idea that R was a smooth increasing function of energy.

OCTOBER 1974

The break in data-taking to upgrade SPEAR gave us a chance to concentrate on analysing the data accumulated over the previous year. In July, we actually prepared a draft paper concluding that R was a smooth increasing function of energy but left several crucial blanks where numbers to be derived from detailed analyses were to be supplied. By October, I could return to the problem of R.

It seemed logical to take all the available data, including the June runs, and to apply a consistent analysis. The analysis was not complicated; count up all the multihadron events and divide by the number of Bhabha events, which was proportional to the muon rate. Then a correction factor had to be applied to account for the small fraction of hadron events that failed to trigger the detector. Adam Boyarski developed a powerful software framework for sifting through events that made this task quite simple.

A shock came when the ratio of the number of hadron events to Bhabha events was plotted versus collision energy (figure 7.6). The ratio jumped all around, most seriously near 3.2 GeV, exactly where we had been looking without success in June. The conclusion was inescapable; either we had serious problems with the reliability of our data or R was *not* a smooth function of energy! Thus began a most interesting quest to resolve this dilemma.

An immediate question was why this had not been observed in June. Checking through the log books revealed the answer: the analysis program running in June assumed that the steel cylinder installed for the energy crisis studies was still in place when, in fact, it had been removed. Events with tracks that appeared to come from the cylinder were placed in a special category and *were left out of the determination of R*. This amounted to a 20% loss of events which just compensated for the larger actual rate, thus completely fooling us!

Burt Richter was away from SLAC that October, delivering a series of Loeb Lectures at Harvard on the subject of electron–positron collisions. I telephoned him to warn that R might not be so smooth after all!

A greater surprise was lurking in the data at 3.1 GeV. These data had an even larger hadron-to-Bhabha event ratio that came entirely from only two of eight data runs taken at that energy. Run 1380, taken on 29 June, had $2\frac{1}{2}$ times as many hadron events for each Bhabha event as the later six runs had, while run 1383 taken less than 5 h later had *five times* as many hadron events as the 'normal' runs! (Runs 1381 and 1382 did not involve colliding beams.) Runs 1384–89 were all quite consistent with a small value of R, the same as that found at 3.0 GeV. This was a worst-case scenario: data taken under seemingly identical conditions appeared grossly inconsistent.

There were checks that could be made. The original data tapes for runs 1380 and 1383 were replayed using the one-event display to scan all events 'by hand' to check the integrity of the data. The data looked fine. Were Bhabha events being lost? This was not likely because various internal monitors of QED processes all agreed with the numbers observed. Just before run 1380, SPEAR had been running at 4.3 GeV, where the normal hadron-to-Bhabha event ratio was almost identical to that found in run 1380. Did we record the wrong energy for this run? The magnetic detector could measure the total collision energy in Bhabha events, giving a clear answer: no! Run 1380 and the others were, indeed, taken at 3.1 GeV. An alert was sent out to the collaboration to use any analysis tools available to determine what was special about runs 1380 and 1383.

By the end of October, the mystery of runs 1380 and 1383 deepened. Everything indicated that the data were correct, but the numbers of hadron events were outside

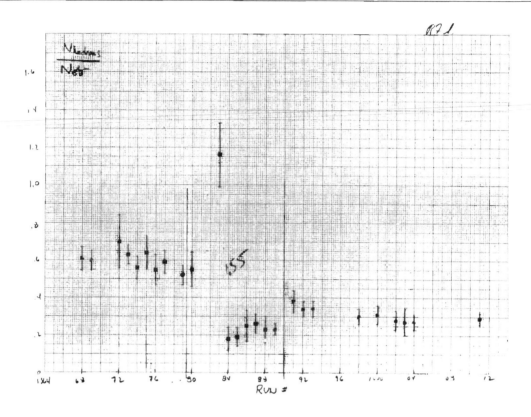

Figure 7.6. Physics history: original SPEAR plot showing the ratio of hadronic to elastic scattering events, and the unexpected burst of activity at a collision energy of 3.1 GeV.

statistical chance. After fighting for weeks to dissolve the bumps in *R*, it became startlingly clear to me that runs 1380 and 1383 were trying to tell us something very important. Even under the best theoretical circumstances, SPEAR could not have produced the number of events seen unless some new physics was responsible. We must have been on the edge of a cliff in *R* with SPEAR's energy ever so slightly higher in runs 1380 and 1383, falling off the peak during the other runs. It was imperative that we go back and look again.

However, the new running period with the upgraded SPEAR was starting and Burt had returned from Cambridge, eager to move on to higher energies. In what must be a classic 'red herring' for physics, our Berkeley colleagues, Gerson Goldhaber and Gerry Abrams, telephoned with the news that there were excess numbers of kaons (common hadrons carrying a property known as

'strangeness') in the crucial runs. The day was Friday, 9 November. In minutes, we would be meeting with Richter and other members of the collaboration to argue for *lowering* SPEAR's energy, then being tuned to 6 GeV, in order to take more data near 3 GeV.

Only the day before, I had read a preprint of a new theoretical paper, soon to be a classic, by Mary K Gaillard, Ben Lee and Jon Rosner, entitled 'Search for charm'. The paper detailed the expected phenomenology for a hypothetical new kind of quark possessing a property not found in ordinary matter, whimsically called 'charm' by its inventors, Sheldon Glashow and 'bj' Bjorken. The quark model of hadron structure had emerged as the most attractive theoretical explanation of much of high-energy physics at the time, including the parton ideas that described the SLAC electron–proton scattering experiments, but the three quarks proposed originally by Murray Gell-Mann gave far too small a value for R. A charmed quark would increase the predicted value. Charm was introduced to restore universality among certain weak decays of hadrons, implying that particles containing the new quark should preferentially decay to particles, such as kaons, that carry strangeness. Furthermore, a new vector meson, composed of a charm quark and an anticharm quark, should exist and it would be directly produced in electron–positron collisions. All this was described in 'Search for charm'.

In our naive groping to understand the runs at 3.1 GeV, the news of kaons coupled with the paper by Gaillard, Lee and Rosner helped to carry the day for those of us who wanted to go back and recheck the old data. We were allotted the weekend; after that, it was on to higher energies. As things turned out, the excess kaons were a statistical fluke!

BUSY WEEKEND

Late that Friday afternoon, while I walked down to the SPEAR control room with Rudy Larsen, a fresh Pacific storm boiled over the hills west of SLAC, creating an ominous looking sky of greens and purples. We joked that the dramatic sunset must be heralding some important Shakespearean-like event. What was to transpire was easily the most thrilling event in our scientific lives. The weekend of 9–10 November 1974 would go into the lore of high-energy physics and into the history of science. What follows are snapshots of some of my most vivid memories; others who participated will have their own. There are carefully documented histories of that weekend and various personal reminiscences in print.

The run plan for the weekend emphasized the need to confirm the 'baseline' value of R at 3.0 GeV before striking out to look for bumps. We had less time than was dedicated in June; care and efficiency were the watchwords if we were to reach any definite conclusions about what was happening near 3.1 GeV.

A few of us showed up early Saturday morning, hoping to sneak a look for hadron events just above 3.1 GeV, before the official run began at 8 o'clock. Ewan Paterson, head of SPEAR operations, agreed to set the energy to 3.12 GeV for our unauthorized run. Scott Whittaker, a Berkeley graduate student involved in much of the data analysis on R, and

I started up the detector, scanning for events right off the computer monitor. After about half an hour, we counted 22 hadron events, about three times higher than the 'official' rate. The excess events were still there; we *knew* that this would be a interesting weekend!

During the next 24 h, the run plan was followed and a peak in the hadron event rate, perhaps three or four times above the baseline level, slowly emerged. Around midmorning on Sunday, we decided to try a new energy, half-way between 3.10 and 3.12 GeV, where the peak was forming. SPEAR had been running at 3.10 GeV and the event rate was quite low, at the baseline value. Then Burt Richter, at SPEAR's control console, nudged the energy slightly higher. The effect was as if a dam had burst; suddenly the SPEAR intercom began madly ticking in response to the rapidly firing spark chambers. Beautiful hadron events were lighting up the computer display as fast as it could handle them. This was not just some peak, but Everest and K2 all wrapped into one!

Over the next couple of hours, an astonishing picture emerged. Within the smallest increment of energy that SPEAR could resolve, the hadron event rate jumped by a factor of 70. Our QED mainstays, Bhabha scattering and muon pairs also increased significantly, a logical response as it turned out, but one that struck me with particular awe, as if we were eavesdropping on Nature's deepest secrets. This was obviously a 'resonance', a particle created from the collision energy with high probability when SPEAR's energy was tuned just right. The particle then rapidly decayed, mainly to hadrons, but also to the excess muons and Bhabha events that we observed. What was astonishing was the narrowness of the range of energies over which the object was produced. This extreme narrowness, by the rules of quantum mechanics, implied the new particle lived for a relatively long time, 100–1000 times longer than anyone expected for such a particle.

As champagne flowed and news of the wonderful discovery went out, the SPEAR control room became mobbed. Gerson Goldhaber and others began drafting a paper to describe the discovery. Plans began to search for more such spikes. Data-taking continued.

Monday morning started early. We wanted to submit a paper to *Physical Review Letters* as soon as possible, and it was traditional to announce important discoveries first to the SLAC staff. A few of us gathered before 8 am to collect information for the draft paper and presentations being scheduled for later in the day. SLAC's Program Advisory Committee (PAC) was meeting that day to consider, among other things, new proposals to run in SPEAR's second interaction region. The first public announcement would be to the PAC. Richter left our gathering to talk to Pief about the PAC meeting and subsequent plans.

In the meantime, a telephone call came in for Burt from one of our colleagues, Gerry Fischer, who was visiting the DESY laboratory in Hamburg, Germany to help with commissioning their new electron–positron collider. I took Gerry's call. He was highly excited about the discovery of a new particle. I was amused by the speed at which rumours of our discovery reached Europe, but then realized we were talking past each other; Gerry was talking about a new discovery made by Sam Ting, a physicist from MIT, working at Brookhaven National Laboratory on a fixed-target proton experiment, while I was talking about yesterday's events. What was going on? In a few minutes, Burt returned from his meeting with Pief with startling news: 'You won't believe this, but Sam Ting is here and has discovered the same particle!'

Burt and I ran down to the PAC meeting, where I showed our new data. There was great excitement, but I shall never forget the stunned figure of Sam Ting slumped in the back of the room. He is a powerful experimenter who had recently joined the PAC. He learned of our discovery the previous night upon arrival in California from the East Coast. After I finished, Sam gave a somewhat halting report on his new discovery, clearly the same object, but produced in a completely different experiment where high-energy protons struck a beryllium target. He and his colleagues had devised a remarkable detector that could accurately measure electrons and positrons coming from the *decay* of hypothetical new particles and had found a peak at 3.1 GeV! At noon, the two of us were to repeat this show, but now to the entire SLAC staff. I was amazed by the audience; SLAC's auditorium was filled to overflowing with scientists, secretaries, technicians and engineers. The full dimension of the weekend's discovery was beginning to set in. Sam gave a brilliant report, in total command of his subject.

Ting soon left the PAC meeting to complete his paper on the discovery. In what must have been a bitter-sweet experience, the ADONE group quickly pushed their energy beyond its design maximum to 3.1 GeV and also recorded the rush of multiprong events. Our three papers were published simultaneously in *Physical Review Letters*.

Naming the new particle was an important and sensitive matter. For complicated reasons, we settled by that Monday on the Greek letter ψ. Later, we learned that Ting chose J. After years of contention, the people who adjudicate such matters suggested the compromise that is accepted now: J/ψ.

ψ-CHOLOGY

Physicists at SLAC and around the world soon mobilized in response to the weekend's news. What was the new particle and what were its implications for physics? Theorists at SLAC began an almost continuous seminar to understand the scientific issues, to suggest new measurements that we should make, and to help to interpret what we found. Haim Harari, a leader of the theoretical effort, distributed a nice compendium of the scientific issues under the title 'ψ-chology'.

An immediate question was whether any more such needles lurked among our haystack of earlier measurements. Within days, Marty Briedenbach was able to tie the SPEAR control computers, the detector and powerful off-line machines together to sweep SPEAR's energy in tiny increments automatically and to trace out the hadron event rate. The very first night that the system was used for real, 21 November, a second narrow spike was found near 3.7 GeV! The starting point for this scan was actually chosen by Marty on suggestions from Harvard theorists who speculated that the J/ψ was an atom-like system of a charmed quark and antiquark, which they dubbed 'charmonium'. Like an atom, charmonium should have higher-mass excited states. Amazingly, the formula that predicted the mass was the same as that used to describe positronium, the atomic system formed by an electron and positron at very low energies. The next day, the SLAC auditorium was again jammed as Marty presented the news of our second new particle, which we called ψ' (figure 7.7).

Several days later Gary Feldman found that a large fraction of ψ' decays yielded a J/ψ and two other

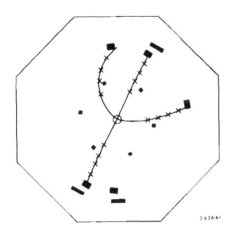

Figure 7.7. An example of a ψ'.

charged particles. Another packed audience in the SLAC auditorium heard about this result. While listening to Gary's presentation, I realized that some of these decays would leave tracks in our detector having a pattern much like the letter ψ, showing that our wonderful new particles could even write their own names! Later, we found such events and immortalized them on some of the first physics tee-shirts to be produced.

The events of early November called out for new measurements in addition to the high-energy studies already planned. Were the J/ψ and ψ' states of a new kind of atom, and, if so, was it composed of charmed quarks? Would charmed quarks combine with ordinary quarks, forming new kinds of particles having the charm property? What else might be happening?

PARTICLE BONANZA

We were in a unique position to follow up on the growing list of important questions raised by the new particles. ADONE could not run above the J/ψ energy, effectively removing Frascati from the hunt. DESY's new electron–positron collider, DORIS, was just starting to operate with detectors that were not optimal for some of the studies needed. DORIS soon became a strong competitor, however. Fixed-target experiments operating at the new proton accelerator at Fermilab were swamped by backgrounds of ordinary particles, making it virtually impossible to investigate subtle

differences that would be crucial to understand the new particles. For the next two years, it seemed like we were publishing some new blockbuster result every few weeks.

The charmonium picture of the J/ψ and ψ' was confirmed by summer 1975 when DORIS observed 'cascade' decays of the ψ' to J/ψ, accompanied by the emission of two photons. We were hot on the trail of the same result but had inferior photon detectors, publishing after the DESY group. This was a nice example of Nature repeating a successful model; when an ordinary atom is excited, by an electrical discharge for example, it returns to the ground state by emitting light of characteristic wavelengths by successive decays to specific intermediate states. Exactly the same thing happens when the ψ' decays to J/ψ, except on a much higher energy scale. These results gave strong support for the emerging theory of strong interactions, quantum chromodynamics (QCD).

Proving the quarks of charmonium were charmed took a little longer. By summer 1976, we had accumulated tens of thousands of hadron events at collision energies around 4 GeV. This energy was chosen because, after the effects of the J/ψ and ψ' were subtracted, R appeared two-tiered, rising from one level to another near 4 GeV. The step was attributed to the production of new, openly charmed particles, no longer constrained by energy to remain bound in charmonium. The proof that this was charm relied on the subtle connection between charm and strangeness predicted by Glashow, John Iliopolos and Luciano Maiani in 1970, which was the basis of our (incorrect) excitement over the kaons seen in runs 1380 and 1383. This time, the 4 GeV data revealed examples of *pairs* of particles, each with masses near 1.9 GeV, decaying in a specific way to kaons and other hadrons. When the parent particle was positively charged, the decay kaon was negatively charged, and vice versa, just as predicted! Quark chemistry was born. The charmed quarks each have a mass equal to about half the J/ψ mass, 1.5 GeV. Add another ordinary quark and you have a charmed particle, of mass 1.9 GeV, which decays according to the recipe.

Data taken at high energies, 5–8 GeV, confirmed the basic quark picture within a year of the J/ψ discovery. Gail Hanson examined these data for 'jets', a clustering of particles along a line that represented the original direction of a pair of quarks produced by the electron–positron annihilation in the same way that muon pairs are created, following a procedure outlined by Bjorken (see the chapter by Wu). Gail found that the correlation increased with increasing beam energy, as predicted. Another stroke of good luck permitted us to measure accurately the angular distribution of the jet axis. SPEAR's beams had become spontaneously polarized at 7.4 GeV, an effect that accelerator experts from Novosibirsk predicted would happen. When the beams were polarized, more jets were produced in the horizontal plane than in the vertical. This implied that the parent quarks were spin-$\frac{1}{2}$ objects, a crucial assumption of the quark picture.

SPEAR had one more great surprise for us in 1975. Martin Perl had long been interested in leptons; electrons, muons and neutrinos, none of which feels the strong force between hadrons. In our original proposal to build the magnetic detector, he urged the search for heavier relatives of the electron and muon. With all the excitement over the events of November, I had not followed these searches until Martin requested a special meeting of SLAC's senior physicists at Panofsky's home. He quietly and deliberately described a painstaking analysis of events containing only two charged tracks, concluding that 24 events existed where one particle was a muon, one was an electron, and nothing else was present. Such events were impossible in the framework of known QED or quark physics.

Martin's hypothesis was that a new heavy relative of electrons and muons, a 'heavy lepton', was being produced in pairs at energies above 3.6 GeV. This was too good to be true—another discovery of historic importance! At first, there was scepticism inside and outside our collaboration, but Martin, Gary Feldman and others presented an iron-clad case, albeit one that pushed the performance of the SPEAR magnetic detector to its limits. Within two years, additional data and other experiments would confirm Martin's results; the τ lepton was discovered!

PERSPECTIVE

Nature blessed the energy range covered by SPEAR with remarkable bounty. Our first glimpse in 1973 of a smooth and rising value of R with its energy crisis gave way to an intricate interplay of quarks, charmonium, charmed hadrons, jets and heavy leptons. Electron–positron collisions proved especially powerful for sorting out these varied phenomena because of the fine control of collision energy that was possible and because new quarks and new leptons could be produced in nearly the same quantities as ordinary ones, as long as the collision energy was sufficient. The layered cylindrical magnetic detector proved its great value in this environment. Essentially all modern collider detectors look like scaled-up versions of the SPEAR magnetic detector, and electron–positron colliders have been the most sought-after research tool in high-energy physics ever since November 1974.

In parallel with our experimental bonanza, theoretical understanding of the building blocks of matter and basic forces of Nature emerged by the mid-1970s from the chaos and confusion of the late-1960s to a satisfying description of all known phenomena in terms of quantum fields and gauge symmetries, now known as the Standard Model (SM). Our experiments at SPEAR contributed, along with others, to the development of the SM, but the drama and excitement of *how* these discoveries came to be have linked them inextricably to the fundamental change in thinking about the physical world that is represented by the SM.

FURTHER READING

Riordan M 1987 *The Hunting of the Quark: A True Story of Modern Physics* (New York: Simon and Schuster)

Cahn R N and Goldhaber G 1989 *The Experimental Foundations of Particle Physics* (Cambridge: Cambridge University Press)

Galison P 1997 *Image and Logic: A Material Culture of Microphysics* (Chicago: University of Chicago Press)

ABOUT THE AUTHOR

Roy F Schwitters is the S W Richardson Foundation Regental Professor of Physics at the University of Texas at Austin, where he teaches and conducts research in experimental high-energy physics. From its founding in 1989 until cancelled by Congress in 1993, he was director of the Superconducting Super Collider (SSC) laboratory in Dallas, Texas. Before moving to Texas, he was Professor of Physics at Harvard University. He joined the Harvard faculty in 1979, after being assistant and then associate professor at the Stanford Linear Accelerator Center. He came to Stanford in 1971 as a research associate after receiving his PhD degree in physics from the Massachusetts

Institute of Technology, where he also earned his bachelor of science degree in physics in 1966.

From 1980 until his appointment as SSC laboratory director, Dr Schwitters was co-spokesman and head of construction for the Collider Detector at Fermilab (CDF) in Batavia, Illinois. From 1978 to 1989, he was Associate Editor of *Annual Review of Nuclear and Particle Science*; from 1980 to 1983, he was Divisional Associate Editor for Particles and Fields of *Physical Review Letters*.

Dr Schwitters is a fellow of the American Academy of Arts and Sciences, the 1980 recipient of the Alan T Waterman Award of the National Science Foundation, the 1996 co-recipient of the Panofsky Prize of the American Physical Society, and a fellow of the American Physical Society and the American Association for the Advancement of Science. He was born and raised in Seattle, Washington.

8 WEAK INTERACTIONS

Christine Sutton

Editor's Introduction: The twentieth century has seen the 'weak nuclear force' play an increasingly important role in our understanding of subnuclear physics. Its subtlety first made it appear as a detail on the subnuclear landscape, but successive insights increasingly showed that this interaction is universal—almost all particles have a weakly interacting facet—and that the weak force is ultimately responsible for life itself.

Curiously, advances in comprehending the weak interaction have called for a series of revolutionary ideas which initially defied everyday experience. The weak interaction, it appears, is very discerning, and these radical proposals eventually rewarded physicists with extraordinary new insights.

The understanding of the weak force has advanced hand in hand with the appreciation of the role played by neutrinos, the most enigmatic of the fundamental particles. Neutrinos are so reluctant to interact that physicists first thought they would be undetectable. It is typical of the dramatic progress of twentieth-century physics that what were first hesitantly introduced as undetectable particles not only have been observed but have gone on to become a major laboratory tool.

The study of neutrinos produced via terrestrial radioactive decay has made incisive contributions to our understanding of fundamental forces. Physicists can also detect extra-terrestrial neutrinos, for example those from the Sun. These have revealed new physics questions (see the chapter by Ellis).

Most introductions to particle physics, whether talks, articles or books, tell you that physicists have identified four fundamental forces or interactions. One of these is the familiar gravitational force, which keeps our feet on the ground, holds the planets around the Sun and pulls matter together to form stars and galaxies. Then there is the electromagnetic force, holding atoms together and binding them as molecules to make matter in bulk. Less familiar are the strong force, which binds quarks together inside the protons and neutrons that make up atomic nuclei, and the weak force, which underlies the radioactive process known as β decay. At this point in a talk someone in the audience will be prompted to comment that, while they can imagine the strong force working like the forces of gravity and electromagnetism, which they have experienced, they find it more difficult to comprehend this thing that we call the weak force. The weak force does not seem to hold anything together, only to break it apart. In physicists'

terms, we do not observe bound states of the weak force. Yet the weak force has a vital role, not only in the radioactive decays that dictate which nuclei exist and which do not, but also in the primal reactions that fuel the Sun and other stars.

So the weak force seems a force apart: one that does not sit quite at ease with the others within a simple intuitive picture. Indeed, since the end of the nineteenth century, the weak force, in one guise or another, has tormented physicists, tempting them to forsake deep assumptions, but rewarding them eventually with a remarkable unification with the processes of electromagnetism.

Interwoven with the surprising story of the weak force has been the story of neutrinos, arguably the most intriguing of the fundamental particles. Parcels almost of nothingness, neutrinos have proved to be surprisingly robust probes of fundamental physics. Unable to interact through the electromagnetic or strong forces, the neutrinos provide a unique and valuable mirror on the weak force.

Table 8.1.

Name	Symbol	Charge	Electron number	Muon number	Tau number	Mass (GeV/c^2)
Electron	e	-1	1	0	0	0.0005
Electron-neutrino	ν_e	0	1	0	0	~ 0
Muon	μ	-1	0	1	0	0.105
Muon-neutrino	ν_μ	0	0	1	0	<0.0003
Tau	τ	-1	0	0	1	1.771
Tau-neutrino	ν_τ	0	0	0	1	<0.031

EARLY CONTROVERSIES AND DISCOVERIES

The weak force caused controversy almost from the start. In the 1920s, Lise Meitner and James Chadwick for a while disputed the shape of the energy spectrum of electrons emitted in β decay—the process whereby an atomic nucleus transmutes to the nucleus of the next element in the periodic table by spitting out an electron. Meitner suspected that the electrons had discrete energies, giving a spectrum of lines, while Chadwick demonstrated, correctly as it turned out, that the spectrum was continuous, i.e. the electron could take a whole range of energies. However, a continuous spectrum was contrary to the single line expected from energy conservation if only two participants, namely the electron and the nucleus, were involved in the process.

No less a person than Niels Bohr advocated abandoning energy conservation in β decay, but in 1930 Wolfgang Pauli daringly proposed that an unseen player, the neutrino, emerges from the process, together with the electron. The neutrino would share the energy released and so allow the electron a continuous spectrum of energies, as observed. Pauli's intuition, to keep the conservation laws and 'let the stars shine in peace', inspired Enrico Fermi in his 'tentative theory of β decay', published in 1934, which was to become the basis for ideas of a universal weak force. The first intimations that β decay is but one manifestation of some deeper fundamental interaction came during the 1940s from experiments which led to the discovery of the muon (see the chapter by Lock).

Both electron and muon are electrically charged, and neither participates in the strong interactions that bind quarks together. We recognize them now as related particles, i.e. members of the family known as leptons (and distinct from the family of quarks that make up protons, neutrons and other strongly interacting particles such as pions). A third charged lepton, the tau, and three neutral neutrinos bring the number of family members to six (table 8.1). In addition, there are six corresponding antiparticles. It appears that in any interaction a lepton can be created (or can disappear) only together with an antilepton. This empirical rule of 'lepton conservation' became apparent in the early 1950s; it implies, for example, that it is an antineutrino that accompanies the electron in β decay (figure 8.1).

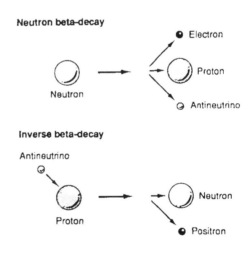

Figure 8.1. In β decay, a neutron turns into a proton, at the same time emitting an electron and an antineutrino. In the related process known as inverse β decay, a proton absorbs an antineutrino and emits a positron (antielectron) as it changes into a neutron.

In the late 1940s, soon after the muon had been separately identified as a new particle, similarities between its interactions and the well known process of β decay began to tantalize a number of physicists. When the decay or capture of a muon was treated in the same way as β decay in Fermi's theory, the coupling constants for the three different processes proved to be remarkably similar. The coupling constant G_F in Fermi's theory is analogous to Newton's gravitational constant G in that it scales the product of properties of the two interacting systems. In Newton's theory, we have two masses, m_1 and m_2, and an interaction that varies (in part) as Gm_1m_2. In Fermi's theory of β decay we have two interacting fields: one for the heavy nuclear particles (neutron and proton) and one for the lightweight particles (electron and neutrino). The probability of β decay then depends on the product of the interacting fields, scaled by the constant G_F. The agreement between the coupling constants for β decay, muon decay and muon capture led to the idea of a 'universal Fermi interaction' and, as this idea took shape, experiments began to reveal more and more new particles with similar weak interactions.

STRANGE BEHAVIOUR AND THE DOWNFALL OF PARITY

During the early 1950s, the number of particles grew steadily, as different types were found in cosmic rays and in laboratory experiments using beams from particle accelerators. Many of the new particles were classed together as 'strange' particles, as they behaved in a puzzling manner. While they were readily produced in collisions between other particles, they did not easily decay. Indeed, their decay times were typical of the weak Fermi-type interactions. Murray Gell-Mann and Kazuhito Nishijima independently explained this behaviour by proposing that the strange particles must possess a new property, namely 'strangeness', which is conserved in strong interactions but not in weak interactions (see the chapter by Samios).

While the strangeness scheme explained many observations, one puzzle remained, concerning two strange particles named θ and τ. The puzzle was that, although the θ decayed to two pions and the τ decayed to three pions, the two particles appeared to have not only the same mass, but also the same lifetime. Could the θ and τ in fact be one and

the same particle? At first sight the answer appeared to be 'no', for the θ and τ decayed to systems of opposite 'parity'. Parity says something about the symmetry of a system, specifically the effect of inverting the system completely in space. This is often described in terms of reflection although as figure 8.2 shows, reflection alone is not sufficient to invert in space (remember that you do not appear upside down in a mirror, only reversed left and right). Figure 8.2 shows that complete inversion is equivalent to a reflection plus a rotation (you need to rotate your image to bring your reflected head to be opposite your real feet if you are to invert yourself completely!). Note that during the inversion a right-handed system changes to a left-handed system. The parity of particles depends on the spatial symmetry of their wavefunction, their quantum-mechanical description. In some cases, spatial inversion changes the sign of the wavefunction, and the parity of the system is said to be odd; in other cases, the sign does not change, and the parity is said to be even. The θ decayed to two pions, which have even parity, while the τ decayed to three pions, a system with odd parity. So it seemed that the θ and τ could not be one and the same thing, for the basic assumption was that

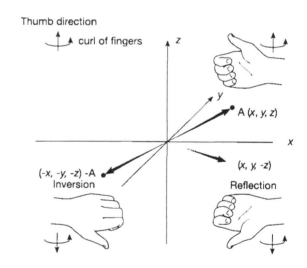

Figure 8.2. The parity operator inverts an object completely (back to front, side to side, and top to bottom) by changing coordinates (x, y, z) to $(-x, -y, -z)$. This is equivalent to a reflection (here in the x–y plane) together with a rotation through $180°$ (here around the z axis).

the parity of a system must remain the same; parity must be conserved.

During the spring of 1956, C N Yang and T D Lee began to consider these arguments more closely. Suppose, they reasoned, that θ and τ were a single particle, created in strong interactions in a state of definite parity (even), which decays through a weak interaction that does not conserve parity. The particle could then decay either to an even state (two pions) or an odd state (three pions). To test this hypothesis, Lee and Yang turned to the best-known weak interaction of all, the familiar process of β decay. However, after searching through the many measurements that had been made, they came to the surprising conclusion that no-one had ever made a measurement that tested parity conservation in β decay.

In June 1956, Lee and Yang submitted a paper to *The Physical Review* in which they suggested ways of testing for parity conservation in weak interactions. One idea was to measure the angles at which electrons in β decay are emitted by nuclei with spins oriented in the same direction. An asymmetry in this angular distribution would be proof of parity violation (non-conservation). Other suggestions included the chain of weak decays from pion to muon to electron. Here, parity violation in the pion's decay would lead to the production of muons with their spin axes polarized in their direction of motion, and this in turn would lead to an asymmetry in the angular distribution of electrons produced when the muons decayed.

Little more than six months later, the long-held notion of parity conservation tumbled. In an experiment at the National Bureau of Standards in Washington, C S Wu had joined forces with Eric Ambler and colleagues to measure the angular asymmetry in electrons produced in the β decay of polarized cobalt-60 nuclei. The nuclear spins were oriented in a magnetic field at temperatures only a few thousandths of a kelvin above absolute zero. After several months of hard work with a difficult experimental set-up, they found a large asymmetry (figure 8.3). By 16 January 1957, after a press conference the preceding day, the whole world knew that β decay does not conserve parity, for the story made headlines on the front page of the *New York Times*. At the same time, results from two experiments studying the pion–muon–electron decay chain were also showing parity violation in these different weak

interactions. With parity violation, the weak interaction—and with it that unwelcome child, the muon—had shaken theoretical precepts. The stage was now set for steps towards a universal weak interaction, and a closer look at the most feebly interacting of all known particles, the neutrino.

NEUTRINOS, SEEN BUT NOT REFLECTED

Neutrinos, with no electric charge and no sense of the strong force, have little probability of interacting with matter. They are famous for being able to travel through the Earth without stopping; indeed, they can travel light-years through water without interacting. So, for many years, the idea of demonstrating that neutrinos really do exist, by creating them at one place and detecting them at another, seemed an impossible dream. However, that dream became much closer to reality with the discovery in 1938 of nuclear fission. Fission provided the possibility of a prolific source of neutrinos, namely a nuclear reactor, where the millions

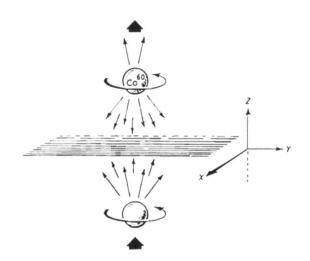

Figure 8.3. In the cobalt-60 experiment, Wu, Ambler and colleagues aligned the nuclei in a magnetic field at very low temperatures so that the spins were nearly all in the same direction. They found that the β decay electrons were emitted asymmetrically, mainly in one direction. When inverted in space this system retains the same direction for the nuclei but reverses the direction of the electrons. Parity is violated as the inverted system is manifestly different from what is observed.

upon millions of fissions per second produce neutrinos as an invisible byproduct.

Around 1951, Clyde Cowan and Fred Reines decided to set themselves the 'supreme challenge' of making these neutrinos, or at least a few of them, visible. (They originally considered detecting neutrinos produced in the explosion of a fission bomb. Later they realized that their technique would allow them to recognize neutrino interactions against many other background signals, so that they could do their experiment at a nuclear reactor.)

Reines and Cowan chose to look for the neutrinos—or, more precisely, antineutrinos—through the process of 'inverse' β decay, where an antineutrino interacts with a proton to produce a positron and a neutron (figure 8.1). They realized that the positron produced in this way would give itself up almost immediately, annihilating with an atomic electron to produce two γ-ray photons, which would shoot off back to back. The neutron would also produce a tell-tale signal of γ-rays, but only after it had wandered around before being captured by an atomic nucleus. The nucleus would be excited by the capture and would lose its excess energy in the form of more γ-rays (figure 8.4). The key factor here was the time that it took, on average, for the neutron to be caught: 5 μs or so in the cadmium-doped water that Cowan and Reines used. (The cadmium served to increase the chance for neutron capture.) The γ-rays themselves were detected in tanks of liquid scintillator placed either side of, and between, two water tanks, to make a three-layer sandwich. The signal for inverse β decay (the 'footprint' of a neutrino in one of the water tanks) was a pair of coincident signals in the scintillator on either side of the water tank, followed 5 μs later by a second coincident pair of signals. The dream became reality in 1956, when Cowan, Reines and their colleagues were finally fully convinced that they had detected more 'delayed coincidences' of this kind when the reactor was on than when it was off. For the first time, an experiment had detected the interactions of neutrinos away from their point of origin.

Coincidentally, this was the same year that Lee and Yang triggered the overthrow of parity conservation and set in train new ideas about the neutrino. The neutrino, like the electron (and many other particles) has an intrinsic angular momentum, or 'spin', of $\frac{1}{2}$ (in units of Planck's constant divided by 2π). So, like the electron, the neutrino can be

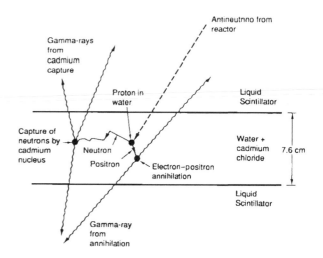

Figure 8.4. Reines and Cowan detected antineutrinos from a nuclear reactor through a three-step process. First, the antineutrino interacted with a proton in the water to produce a neutron and a positron. Second, the positron very quickly annihilated with an electron to emit two back-to-back γ-rays. Third, some 5 μs later, the neutron was captured by a cadmium nucleus added to the water, and the excited nucleus then emitted further γ-rays. The γ-rays from the positron annihilation and the later neutron capture were detected in scintillator tanks on either side of the water.

described by P A M Dirac's theory for a spin-$\frac{1}{2}$ particle. The theory considers all possible states for the particle and its antiparticle and, as both particle and antiparticle can spin in either direction (clockwise or anticlockwise) about the direction of motion, the theory contains four terms altogether. However, if the particle has no mass, the number of terms reduces to two, with the particle in one spin state (clockwise about the motion, say), and the antiparticle in the other.

We can see why this should be in the following way. If a particle has any mass all, it is possible in principle to overtake it and to look back at it and also to see its original spin state (clockwise about its motion, like a right-handed corkscrew) transformed to the opposite state (anticlockwise, like a left-handed corkscrew). With zero mass, on the other hand, the particle will always travel at the speed of light, never to be overtaken, and so it must always

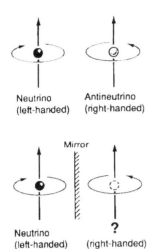

Figure 8.5. The neutrino spins one way about its direction of motion (like a left-hand screw), and the antineutrino the opposite way. On reflection, the spin directions change, and the neutrino becomes right handed—something which is not observed.

appear to be in the same spin state. So a 'two-component' version of Dirac's theory, rather than the standard four-component version, represents massless spin-$\frac{1}{2}$ particles, but it also naturally violates parity, and for this reason had been rejected by Pauli in the 1930s. By 1956, however, Lee and Yang had opened the door to parity violation, and with it the two-component theory of a massless neutrino. Reflect a neutrino in a mirror (or, better still, invert it completely in space) and its spin state changes. This is because the neutrino's direction in space changes on inversion, but its spin direction does not. So the neutrino becomes something that does not exist: a neutrino with opposite 'handedness'. To restore the picture an additional operation is required: the neutrino must be changed to an antineutrino, now with the correct combination of direction and spin (figure 8.5).

V − A AND THE LEFT-HANDED NEUTRINO

While the two-component neutrino meshed neatly with the observation of parity violation in weak interactions, the question of a universal weak interaction remained open. One difficulty concerned the underlying symmetry of the interaction. The theoretical calculations use wavefunctions that describe the initial and final particles in a reaction,

and these wavefunctions must be combined correctly if the theory is to yield the observed probability for the reaction. Fermi, in 1934, had assumed that the wavefunctions combine as they do in electromagnetic theory, behaving like a vector under spatial inversion. Other combinations were possible, however, which transform in different ways when inverted, like a scalar, a tensor or an axial vector (such as spin).

This underlying symmetry determines the possible angles between the final particles: scalar (S) and tensor (T) couplings give one distribution of angles, while vector (V) or axial vector (A) give a distinctly different angular distribution. So, in principle, experiment should reveal the correct way of combining the wavefunctions. By the mid-1950s, measurements from muon decay suggested a (V, A) coupling, but β decay appeared to couple like (S, T), and this difference was a strong argument against an underlying universal weak interaction. However, a theory based on the V and A couplings held considerable appeal, as it naturally gave rise to a two-component neutrino.

So, in 1957, several theorists put forward what Murray Gell-Mann has called the universal Fermi interaction's 'last stand', i.e. that there is a universal interaction, that it is (V, A), but that some of the experimental results on β decay must be wrong.

The end of two momentous years for weak interactions and neutrinos came in December 1957, with the completion of an experiment that tested the coupling in β decay by measuring the spin state of the neutrino. Was it like a right-handed corkscrew, with its spin clockwise relative to its direction in space, or was it like a left-handed corkscrew, spinning anticlockwise about its direction? The cobalt-60 experiment had already demonstrated that electrons emitted in β decay are left handed, but what of the antineutrinos emitted at the same time? If the coupling was (S, T), they would also be left handed, and neutrinos, according to the two-component theory, would be right handed; a (V, A) coupling would instead yield right-handed antineutrinos and imply left-handed neutrinos.

Neutrinos had proved difficult enough to detect at all, but Maurice Goldhaber, Lee Grodzins and Andrew Sunyar succeeded in measuring their handedness in an experiment that Yang later described as 'devilishly ingenious'. The idea was to measure all they could, and then to use conservation laws to lead them to the unknown property that they sought. They chose to study the capture of an atomic electron by a nucleus of europium-152, which yields a nucleus of samarium-152 and a neutrino. The samarium nucleus created in this way has too much energy, which it loses by emitting a γ-ray. The really ingenious part of the experiment was to measure the polarization of the γ-rays emitted in the same direction as the recoil of the samarium nuclei. Conservation of angular momentum implied that these γ-rays would have the same handedness as the neutrinos (figure 8.6).

The experiment demonstrated conclusively that neutrinos are exclusively left handed (they spin anticlockwise about their direction of motion) while antineutrinos, by contrast, are right handed. This confirmed that the underlying interactions must be (V, A) and not (S, T) and moreover suggested that the explicit structure is V − A ('V minus A'). Further experiments confirmed this relative mixture of the V and A components. The results fully vindicated the prescience of the theorists who had pursued this structure against a background of contrary experimental evidence. As they had proposed, the experiments that suggested an (S, T) coupling had been wrong.

THE FIRST HIGH-ENERGY NEUTRINO EXPERIMENT

However, still all was not right in the world of neutrinos and the weak interaction. What Murray Gell-Mann and E P Rosenbaum had referred to in 1957 as the 'unwelcome baby on the doorstep'—the muon—continued to be difficult. Experiments had shown that the muon decays to an electron, a neutrino and an antineutrino, and not into an electron and a photon. But why not? One possible explanation was that the muon was not exactly a heavy electron but carried some property of 'muon-ness', which the electron did not possess. The electron would instead carry 'electron-ness', and both 'muon-ness' and 'electron-ness' would have to be conserved in any interaction. So the muon's decay products had to include a neutrino endowed with the same property. At the same time, to keep the net amount of 'electron-ness' zero, the antineutrino produced in the decay would have to carry the opposite 'electron-ness' to the electron (figure 8.7). (Nowadays we refer to these properties as different lepton 'flavours'.)

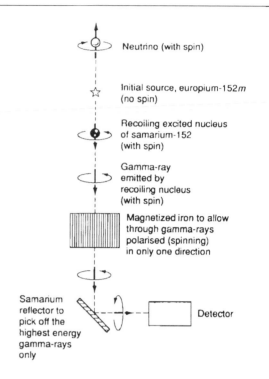

Figure 8.6. The experiment to find the handedness of the neutrino involved measuring the polarization (or handedness) of γ-rays emitted when europium-152m captures an atomic electron and transmutes to samarium-152. Conservation of linear and angular momentum mean that the neutrino and the samarium nucleus head off in opposite directions and with opposite spin, i.e. the same handedness. The samarium shakes off excess energy as γ-rays and transfers to them its unwanted spin. The cunning part of the experiment was to use a reflector that would pick out only the highest-energy γ-rays. These would therefore be the γ-rays emitted in the same direction as the nuclear recoil, and they would therefore have the same handedness as the samarium, and hence the same as the neutrino.

A second difficulty at the time concerned the weak interaction itself. Fermi's theory, in which the particles interact at a single point in space and time, showed that the probability for a neutrino to interact (the cross-section) increases as its energy increases. This was good news as it implies that neutrino experiments become 'easier' at higher energies (although nothing to do with neutrino experiments is ever easy!). However, it was bad news in that the

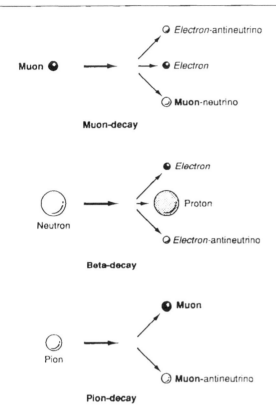

Figure 8.7. The neutrino and the antineutrino emitted in the decay of a muon carry different properties or 'flavours' which must be conserved. The neutrino conserves the muon's 'flavour', while the antineutrino has the 'anti-flavour' of the electron, to cancel the electron's flavour. Similar arguments show that electron-antineutrinos and muon-antineutrinos arise in β decay and pion decay respectively.

probability cannot increase with energy *ad infinitum*, for that would lead to nonsense—a probability greater than 1 (see the chapter by Veltman).

One way to ease this problem was to replace Fermi's interaction at a point with the exchange of a 'field particle' between particles at slightly different places. This idea dated back to Yukawa and his game of quantum 'catch' and owed much to the great success of quantum electrodynamics (QED), the quantum field theory for the electromagnetic interaction. The field particle in QED, exchanged between interacting charged particles, is the massless photon.

However, the field particle for the weak interaction could not be massless. Fermi's 'contact' interaction worked well at the low energies typical of β decay and similar processes, and this implied that any weak exchange had to be short range. This in turn meant that the weak exchange particle, assigned the letter W, had to have some mass, although theory provided little insight as to how much.

In 1959, Bruno Pontecorvo and Melvin Schwartz independently came to the same conclusion about the best way to make progress in understanding both neutrinos and the high-energy behaviour of the weak force. The key was to use a beam of neutrinos, created in the decays of pions produced when a high-energy proton beam struck a target. The probability for neutrinos to interact rises as their energy increases, so, to be useful, the neutrinos would have to be the products of the highest-energy pions available, which in turn would require high-energy protons to make them.

A new proton accelerator being built at the Brookhaven Laboratory in New York offered Schwartz, and colleagues Leon Lederman and Jack Steinberger the opportunity to discover whether they could take the step beyond the pioneering work of Cowan and Reines and turn neutrinos into a research tool. The beauty of using neutrinos to study the weak interaction is that they do not feel the strong force, nor do they respond to the electromagnetic force, as they have no charge. The weak force, with all its peculiarities, dominates their interactions. Indeed, the neutrino is often the calling card of a weak interaction; if it involves a neutrino, it must be weak! The very weakness of the interactions makes neutrino experiments fiendishly difficult, but not impossible, as Reines and Cowan had demonstrated. The experiments require ingenuity, as many neutrinos as possible, as large a detector as possible, and a great deal of patience. Neutrinos, as Maurice Goldhaber has commented, 'are remarkable particles; they induce courage in theoreticians and perseverance in experimenters'.

With the highest-energy particle accelerator in the USA at their disposal, a 10 t spark chamber to detect neutrino interactions, and armour-plate from an old battleship to shield the detector from unwanted particles, Lederman, Schwartz and Steinberger were ready to embark on the world's first high-energy neutrino experiment. One of their first goals was to test the 'two-neutrino' hypothesis. Do the neutrinos created together with muons when pions decay carry a label of 'muon-ness', which distinguishes them from the neutrinos emitted together with electrons in β decay? If there is no difference between the neutrinos involved in the two processes, then the neutrinos born of pions should be able to create electrons when they interact, as often as they create muons; but, if indeed they do carry 'muon-ness', then they should produce only muons. The test of the hypothesis would be to wait patiently for the rare neutrino interactions, and to see how many times electrons were produced, and how many times muons. Fortunately, electrons and muons would produce distinctly different 'footprints' in the spark chamber. Muons would leave long straight trails of sparks; electrons would produce a broader shower of sparks. Over a period of about eight months in 1962, the spark chamber was exposed to some 100 million million neutrinos, and the physicists were rewarded by a total of 29 clean neutrino 'events', where a neutrino had interacted in the detector. In each case, the photographic record showed the long straight track characteristic of a muon; there was no clear sign at all of electrons. 'The most plausible explanation', the team concluded, 'is that there are at least two types of neutrino', which we now call the electron-neutrino and the muon-neutrino. (We now know that there must also be a third neutrino. This is the tau-neutrino, which is associated with the heaviest charged lepton, the tau, which together carry the distinguishing property, or flavour, of 'tau-ness'. However, the distinct nature of the tau-neutrino has yet to be demonstrated in an experiment showing that only neutrinos born of taus can create taus.)

A GIANTESS AND A NEUTRAL CURRENT

The success at Brookhaven demonstrated the potential for neutrino experiments, but further progress, and in particular the search for the W particle, required more intense neutrino beams. One problem with neutrinos is that they have no electric charge, so it is impossible to focus a neutrino beam with magnetic fields, in the way that works routinely with beams of charged particles such as electrons or protons. However, in 1961 at CERN, the European centre for particle physics near Geneva, Simon van der Meer invented a means of focusing the parent particles, i.e. the pions (and

kaons) that decay to produce the desired neutrinos. The pions themselves are produced when a proton beam hits a target, and the problem is that they emerge over a range of angles.

Van der Meer's inspiration was to see how to focus the pions into a parallel beam, rather as a torch produces a beam from a light-bulb that on its own would spray light in all directions. His 'magnetic horn' was based on two metal cones. A current passing through the cones set up a magnetic field shaped appropriately to focus particles produced in a target in the horn's neck. With a magnetic horn in place, by 1963 CERN could boast the most intense neutrino beam in the world.

Along with large numbers of neutrinos (and plenty of patience), neutrino experiments require large detectors. By 1970, CERN's neutrino beam was directed at a giant of a detector, i.e. a bubble chamber, appropriately named Gargamelle, after the mother of the giant Gargantua in the satirical writings of Francois Rabelais. Gargamelle was like a flattened cylinder, 4.8 m long and 1.85 m wide, and held 18 t of liquid Freon. The liquid would stop the occasional neutrino and make visible the tracks of charged particles produced, as if from nowhere, by the neutrino's weak interaction. Eight 'fish-eye' lenses captured these rare events on film, for analysis in the laboratories of the various countries collaborating on Gargamelle.

One such image (on the face of it a rather unprepossessing one) was to prove historic. A single electron straggling through the heavy liquid bore witness to an aspect of the weak force that had slowly come to light in the previous decade. During the 1960s, theoreticians had slowly pieced together a picture that united the recalcitrant weak force with the better behaved electromagnetic force. The unified 'electroweak theory' not only overcame the high-energy crisis in Fermi's theory by including weak force carriers but also exhibited the desirable property of 'gauge invariance' (so the theory remains unchanged under transformations in space).

However, this success came at a price; electroweak theory required the existence of a weak interaction that had never been seen. Until 1973, all weak interactions had involved an exchange of electric charge. For example, in the exchange picture for β decay, a neutron (charge 0) throws out a negatively charged carrier, W$^-$, and turns into a proton

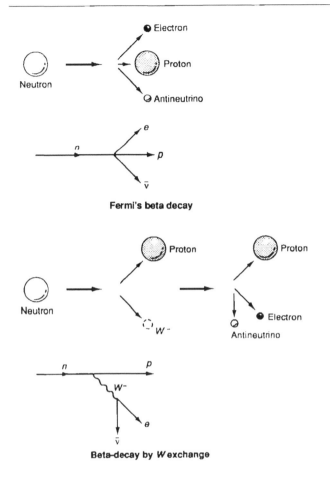

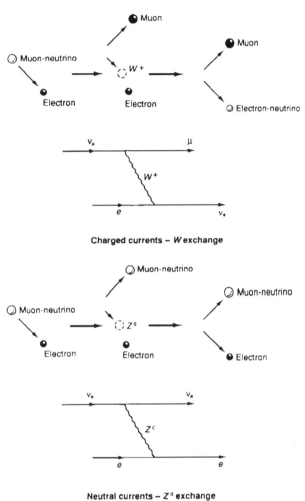

Figure 8.8. In Fermi's theory the weak interaction between particles occurs at a single point. However, the theory is bound to fail at high energies when the predicted probabilities become impossibly high. The problem is relieved if the weak interaction instead involves the exchange of a carrier particle, a W, in this example of β decay.

Figure 8.9. In a weak neutral current interaction no charge changes hands between the participants, as a neutral Z^0 particle is exchanged.

(charge $+1$). The W^- almost immediately changes into an electron (charge -1) and an antineutrino (charge 0) (figure 8.8).

Similarly, electron capture, muon decay, muon capture and other weak interactions can be thought of in terms of the exchange of a charged particle. W^+ or W^-. The additional element that electroweak theory required was a neutral exchange particle, labelled Z^0, which would give rise to a different class of weak interaction, i.e. a 'neutral current'

(figure 8.9), rather than the 'charged current' associated with the exchange of a charged W particle.

Early in 1973, the Gargamelle team at Aachen spotted the historic event in film taken while the chamber had been exposed to a beam of muon-antineutrinos (chosen by selecting negative parent particles for focusing in the magnetic horn). An electron appeared to leave its characteristic track, originating as if from nowhere (see figure 8.10). The best possible explanation was that it had

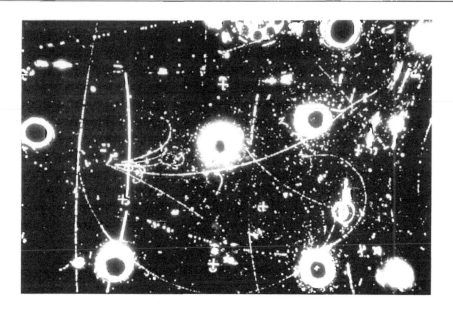

Figure 8.10. A moment of history captured in the Gargamelle bubble chamber. A spray of particles appears as if from nowhere, on the left of the picture, and travels towards the right. This was one of the first examples found of the interaction of a neutrino (unseen) through the weak neutral current.

been knocked from an atom by an incident antineutrino in an 'elastic' collision, i.e. with the particles simply scattering like billiard balls, rather than changing nature as in a more complex interaction. As the antineutrino has no charge, the only way that this could have happened was through the weak force, but with no charge changing hands, the interaction had to be through the exchange of a Z^0.

The discovery helped to convince the Gargamelle team that they were observing 'neutral currents', and that other events involving interactions with nuclei rather than electrons in the liquid were also due to this effect. However, there were some difficult times ahead. Later in 1973, a neutrino experiment at the new high-energy proton accelerator at the Fermi National Accelerator Laboratory (Fermilab) in Illinois initially failed to find similar evidence. Only in 1974 did the team in the USA begin to discern neutral current events. Together, the experiments had demonstrated the value of neutrinos as a probe of new physics, in this case the remarkable unification of weak and electromagnetic interactions, 40 years after Fermi's 'tentative theory' of the weak interactions of β decay.

MODERN TIMES

Since the pioneering days at Brookhaven and CERN, neutrino beams have become a standard tool of particle physics. During the 1970s and 1980s, neutrinos yielded vital information in a variety of studies. They helped to establish the reality of quarks; they revealed some of the first signs of the gluons that transmit the strong force between quarks, and they played a unique part in consolidating electroweak theory. Experiments with high-energy neutrinos, the highest energies being at Fermilab in the USA, have continued into the late 1990s, especially as there are still fundamental questions about neutrinos that remain unanswered.

Can we make the tau-neutrino reveal its flavour, i.e. its 'tau-ness', in interactions where tau-neutrinos create taus, but not muons or electrons? Do neutrinos have some mass after all, albeit very small? Can they (as they might if they have mass) change from one type to another, 'oscillating' as they propagate through space? During the same decades, studies of the weak interaction have moved on from the low energies of β decay, where the story began in the early part of the twentieth century, to the high energies of modern particle accelerators, where the electroweak theory is being tested with great precision. In a sense, one era is over, as the weak force has become fully comprehended within the context of electroweak theory, but as with the neutrino, questions remain; in particular, will we discover how to bring the strong force into the fold, and unite the strong, electromagnetic and weak interactions within one grand unified theory? The answers will come as particle physics continues to progress in the twenty-first century.

FURTHER READING

Brown L M, Dresden M and Hoddeson L (eds) 1989 *Pions to Quarks* (Cambridge: Cambridge University Press) covers the 1950s, from the $\theta-\tau$ puzzle and parity violation, through detection of the neutrino, to the $V-A$ structure of the weak interaction

Brown L M and Hoddeson L (eds) 1989 *The Birth of Particle Physics* (Cambridge: Cambridge University Press) includes reminiscences about the discovery of the muons and its weak interaction

Pais A 1986 *Inward Bound* (Oxford: Oxford University Press) provides excellent coverage of the early days of β decay physics, up to the 1940s

Sutton C 1992 *Spaceship Neutrino* (Cambridge: Cambridge University Press) is a non-specialist introduction to many aspects of neutrino physics

Winter K (ed) 1991 *Neutrino Physics* (Cambridge: Cambridge University Press) provides a comprehensive overview, with expert reviews on all areas of modern neutrino physics

ABOUT THE AUTHOR

Dr Christine Sutton gained her PhD at the University of Sheffield, in 1976, after graduate work on resonance photoproduction at the electron synchrotron, NINA, at the Daresbury Laboratory. She then joined a team working on kaon–nucleon scattering on the NIMROD proton synchrotron at the UK Rutherford Laboratory, and worked briefly on Carlo Rubbia's UA1 experiment at CERN before leaving to become Physics Sciences Editor for *New Scientist*, from 1979–83. After four years as a freelance science writer, she returned to the Rutherford Appleton Laboratory in 1987 to work on the CCD vertex detector for the SLD experiment at the SLAC Linear Collider. She then joined the Crystal Barrel team at Karlsruhe University, and worked at CERN on this experiment for a year from 1988–89. On returning to the UK she joined the Physics Department at the University of Oxford in 1990. Here she has worked on the ZEUS experiment at DESY, but has also become increasingly involved in activities in the public understanding of science, not only at Oxford but on behalf of the UK particle physics community as a whole. At Oxford she is Lecturer in Physics at St Catherine's College, and in addition teaches regularly at the university's summer school for adults, while also giving numerous lectures about particle physics to a variety of audiences. She continues to write frequently about particle physics, in particular for *New Scientist* and *Encyclopaedia Britannica*.

Books published: *The Particle Connection* (Hutchinson, 1984); *Building the Universe* (ed) (Blackwell and *New Scientist*, 1985); *The Particle Explosion* (with Frank Close and Michael Marten) (Oxford University Press, 1987); *The Hutchinson Encyclopaedia of Science in Everyday Life* (ed with Martin Sherwood) (Hutchinson, 1988); *Spaceship Neutrino* (Cambridge University Press, 1992).

9 HUNTING THE CARRIERS OF THE WEAK FORCE

Carlo Rubbia

Editor's Introduction: In the quest to understand the weak force, the realization of its subtle relationship with electromagnetism and the discovery of the particles which carry it provide one of the major advances of twentieth-century physics, related here by Carlo Rubbia of CERN, who shared the 1984 Nobel Physics Prize with Simon van der Meer, then also at CERN, for 'their decisive contributions to the large project which led to the discovery of the field particles W and Z, communicators of the weak interaction'.

In the mid-1930s, Hideki Yukawa in Japan put forward a radical new picture of nuclear forces (see the chapter by Lock) while Enrico Fermi in Italy postulated a framework for the weak interactions (see the chapter by Sutton). Eventually both turned out to be only approximations to the true picture but, at the outset, these bold theories provided physicists at least with a rudimentary sketch map into the unknown. One thing which was known for sure was that the other natural force acting on the microscopic scale, electromagnetism, was carried by intermediate particles: photons.

As initially proposed, the theories of Yukawa and Fermi were not analogous. Yukawa's strong nuclear forces were carried by intermediate particles: mesons. However Fermi's picture of the weak interactions assumed a 'point interaction' with the incoming particles immediately transforming into the outgoing particles. There were no carrier particles.

In 1938, in an obscure paper, Oscar Klein suggested that the weak interactions too might be carried by intermediate particles, in the same way that electromagnetism was carried by photons. Few took any notice at the time, but his words were curiously prophetic: 'The idea of these particles, and their properties, being similar to those of photons, we may perhaps call them "electrophotons".'

In the late 1940s, this view was reinforced by the discovery that the weak interactions of electrons and muons had the same strength. Lee, Rosenbluth and Yang, and Tiomno and Wheeler proposed that the weak interaction could be mediated by electrically charged heavy intermediate particles.

The similarities of electromagnetic and weak interactions were taken up fervently by Schwinger. Both forces were carried by vector (spin-1) fields and both had some universal coupling force, or charge. In the late 1950s, Schwinger developed a unified formalism in which the weak interaction was carried by heavy short-range, electrically charged W particles, of which the massless photon was somehow the electromagnetic counterpart. This ambitious work came before the parity-violating nature of weak interactions had been fully assimilated. At this time, one of Schwinger's postgraduate students at Harvard was Sheldon Glashow, and Schwinger assigned him the thorny problem of working out the details of how the weak interaction and electromagnetism could interrelate. To his credit, Glashow took this mighty problem and showed how it could be solved. However, in trying to build a unified theory of electromagnetism and weak interactions, there were still many obstacles to be overcome.

In the early 1960s, the availability of high-energy neutrino beams from particle accelerators gave physicists a new tool for studying the mechanisms of weak interactions. For the first time they were able to look for signs of the charged W particles. However, the rudimentary theory available at the time gave them no indication of where they should look, and the kinematical region probed by the first neutrino beams was not very extensive. No W particles were seen, and all that could be said was that the elusive particles were heavier than about 2 GeV.

At the same time, powerful unitarity arguments, basically saying that the sum of everything that can possibly happen cannot exceed 100%, dictated that the Fermi interaction broke down above 300 GeV. The W hunters now had an upper limit and a lower limit, albeit widely separated. Perhaps it would be discovered at the new super proton synchrotrons being constructed at the Fermi National Accelerator Laboratory (Fermilab) and at CERN, but there was a cloud on the horizon. A weak interaction built up only in terms of charged W particles is not renormalizable (see the chapter by Veltman), throwing up infinities which make its predictions unreliable. However, introducing a heavy, electrically neutral particle which somehow takes part in weak interactions could take care of the renormalization problem.

The classic varieties of weak interaction, mediated by the charged W particles, involved a permutation of the electric charges; they were 'charged currents'. In the 1930s, Gamow and Teller, and Kemmer had looked at the idea of an weak interaction which did not switch electric charges. Such a phenomenon, a so-called 'neutral current', carried by a Z particle, an electrically neutral partner for the W, had never been seen, although they had been searched for. Physicists tended to discount neutral currents and Glashow, in his attempts to unify electromagnetism and the weak force, had not even considered them.

Also, at the beginning of the 1960s, theorists, in their continuing attempts to unify the different forces of physics were homing in on ideas of spontaneous symmetry breaking, in which a local field theory is gauge invariant but not necessarily all its representations. The problem was that these theories invariably seemed to say that the resultant carrier particles had to have zero mass. The Nambu–Goldstone theorem dictated that, if a manifestly Lorentz invariant field theory exhibits spontaneous symmetry breaking, it will contain zero-mass bosons. This was no good for the short-range interactions which had to be mediated by heavy particles.

New hope came in the mid-1960s. Anderson had pointed out that in a superconductor, if a local gauge symmetry is broken, symmetry-breaking effects give rise to massive particles. For particle physics, Higgs, Brout and Englert, and Guralnik, Hagen and Kibble showed how spontaneous breaking of a local symmetry could sidestep the Goldstone argument. Using what is now universally known as the 'Higgs mechanism', they showed how masses could be generated. With an initial theory containing massless particles, additional scalar (spin-0) fields are coupled to the massless particles. These new background fields can interact with themselves so that the lowest-energy state (the vacuum) has a physical content. Although the interactions are gauge invariant, these new fields are not, thus breaking the symmetry.

In 1967, Steven Weinberg and Abdus Salam pounced on this new Higgs mechanism to write down a prescription which could at last achieve the quest to unify electromagnetism and the weak force. Their prescription has two electrically neutral states, which under symmetry breaking mix to give firstly the massless photon coupled to the conserved charge of electromagnetism and secondly a massive Z. This mixing is governed by a parameter, the 'Weinberg angle', but the theory itself could not predict what it should be.

To illustrate this spontaneous symmetry breaking, Kibble drew the analogy of a pencil balanced on its tip, an unstable but very symmetrical condition. When the pencil falls, the symmetry is broken. So it is with the Higgs phenomenon; it is as though there are 'pencils' throughout space, even in the vacuum. These pencils are all coupled together; so they all fall together, giving the vacuum a preferred direction, that of the photon. The W and Z particles, moving in other directions, have to move the pencils and are more sluggish, acquiring mass.

However, the Weinberg–Salam picture (people did not then call it a theory) necessarily said that the weak force would be accompanied by a neutral current, carried by the Z, which did not alter the electric charges of the participating particles. The problem was that nobody had ever seen such an effect. In addition, the new Weinberg–Salam picture, despite its inclusion of the Z, still did not seem to be renormalizable. Initially it fell on infertile ground and few took any notice.

Then two major things happened. In 1971, Gerard 't Hooft showed that the Weinberg–Salam picture was, after all, renormalizable. Physicists sat up and took notice.

After doing some calculations on neutral currents, Weinberg realized that they could easily have been overlooked in all experiments thus far and urged a new search. In 1973 the weak neutral current, the effect mediated by the Z, was dramatically discovered in neutrino beam experiments using the new Gargamelle bubble chamber at CERN (see the chapter by Sutton). After 40 years of struggle, it looked as though the unification of electromagnetism and the weak force had finally arrived. However, to clinch the theory, the W and Z carrier particles would have to be uncovered.

New experiments at Fermilab and at CERN using higher-energy neutrino beams for the first time were soon able to measure the Weinberg angle. Injecting this value into the equations showed that the mass of the W particle was about 80 GeV, and that of the Z about 90 GeV. Although the physicists now knew where to look, the W and Z were out of reach of the generation of neutrino experiments then getting under way. Something more ingenious was needed.

COLLIDING BEAMS

When particles from a synchrotron hit a stationary target, most of the kinetic energy is transferred to the recoil of the target particles. In the mid-1950s, the Midwestern Universities Research Association (MURA) led by Donald Kerst, a group led by Gerald O'Neill at Princeton, and another by Gersh Budker at Novosibirsk began to look at machines to collide contra-rotating beams (figure 9.1). In these head-on collisions all the kinetic energy could be tapped.

The MURA group was looking at bunched proton beams, while the other two were more concerned with electrons. In their seminal 1956 paper, the MURA group pointed out that the number of particle bunches which can be successively accelerated without leading to excessive beam spread is governed by Liouville's theorem in statistical mechanics (which says that all elements of phase space have equal *a priori* probabilities). For this valuable insight, the MURA group acknowledged the help of Wigner.

The first colliding beam machines were developed at Stanford (by a Princeton–Stanford group) and at Novosibirsk to handle electrons. They used a figure-of-eight configuration, with adjacent circles touching at a common tangent, where the collisions occurred. While

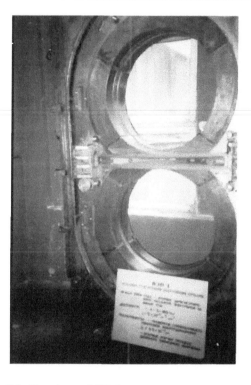

Figure 9.1. The pioneer VEP-2 electron–electron collider, built by Gersh Budker at Novosibirsk.

these machines were being built, Bruno Touschek, working at Frascati, realized that electrons and their antiparticles, positrons, could be collided together in the same ring. As well as having a single storage ring instead of two, this also meant that the colliding particles, with their mutually opposite quantum numbers, could annihilate, forming new states made of quarks and antiquarks bound together. The first electron–positron colliders soon followed (see the chapter by Wilson), opening up a new realm of physics and the dramatic discovery of new particles (see the chapter by Schwitters).

As well as quark–antiquark states, W and Z particles also can be manufactured in such electron–positron annihilations, but the W and Z masses meant that in the 1970s the weak interaction carriers were far out of reach of any contemporary electron–positron collider. Nevertheless the search for the W and Z particles was the physics objective which had priority. In a 1976 paper, the present author and Peter McIntyre (Harvard) and David

Cline (Madison) suggested harnessing proton–antiproton colliders to look for the long-awaited W and Z particles. These colliders would exploit one or other of the new proton synchrotrons then beginning operation at Fermilab and at CERN.

The colliding protons and antiprotons contain quarks and antiquarks respectively, and the W and Z particles could be produced by quark–antiquark annihilations. With the quark momentum distributions of quarks inside the proton (structure functions) known from other experiments, to produce the W at its expected mass of about 80 GeV would require a total proton–antiproton collision energy of about 450 GeV.

Antiprotons can be readily produced as secondary particles, and experiments had been done using such beams. However, with their wide momentum spreads, getting enough antiprotons and fashioning them into a workable beam would be very difficult. To overcome the Liouville theorem restrictions, the proponents of the ambitious scheme pointed out that new beam cooling schemes which had been investigated at CERN and at Novosibirsk might be exploited to reduce the phase space spread of the antiproton beams. Achieving proton–antiproton collisions first called for some serious accelerator physics research and development work.

BEAM COOLING

Liouville's theorem, although far reaching in its implications, is only applicable to forces derived from a standard Hamiltonian. However, it is possible to introduce external factors to affect the phase space distributions of particle beams. This is what is meant by 'non-Liouvillian' methods. The first approach, electron cooling, had been suggested by Budker and Skrinsky in 1966. Their idea was to take a circulating 'hot' beam of particles with a wide momentum spread and bring it into 'thermal contact' with a 'cold' beam of electrons with a much smaller momentum spread but with the same average velocity. The unwanted 'heat' is thus absorbed by the electrons. It was successfully demonstrated by Budker's team at Novosibirsk in 1974.

Another non-Liouvillian technique had been suggested by Simon van der Meer at CERN in 1968 to damp

the betatron oscillations of proton beams in the Intersecting Storage Rings (ISR). Van der Meer proposed detecting fluctuation noise in a particle beam by a pick-up. While the particles continue on their circular orbit, this pulse is fed diametrically across the ring, traversing on the way a high-gain wide-band amplifier. Contact is re-established with the original particle at a kicker which delivers a signal designed to correct for the deviation of the particle from an equilibrium orbit. At the same time, other particles produce other pulses. These are not infinitely narrow, and particles will feel many kicker impulses, but the pulses are short compared with the revolution time, and each particle is influenced by a small but continuously changing feedback signal.

In coherent betatron oscillations, the bunches of particles in the beam behave as single particles and these oscillations can be damped by pick-up–kicker feedback. The small fraction of statistical fluctuations that happens to be coherent can be damped and, after some time, the feedback eventually reduces statistical fluctuations and increases the beam density. After development of the appropriate fast electronics, the technique was demonstrated at the ISR in 1974.

As well as the transverse cooling to reduce betatron oscillations, it is also important to apply cooling to control the longitudinal energy spread. For this, the 'notch filter' method developed by Lars Thorndahl at CERN proved invaluable. Particle momentum information is obtained via its relationship with the revolution frequency. A filter in the feedback chain conditions signals to accelerate or decelerate towards a specific rotation frequency (momentum). A shorted transmission line whose length corresponds to half the rotation period has zero input impedance for all harmonics of the rotation frequency, giving a series of 'notches' in a voltage divider.

Cline, McIntyre and I suggested exploiting these new cooling techniques, still in their infancy, to control antiprotons and enable proton synchrotrons to act as antiproton storage rings. Both CERN and Fermilab followed up the suggestion. At CERN, the Initial Cooling Experiment (ICE) ring was built to test the schemes (figure 9.2), where it was found that stochastic cooling is relatively independent of the beam energy but better suited to lower intensities. On the other hand, electron cooling

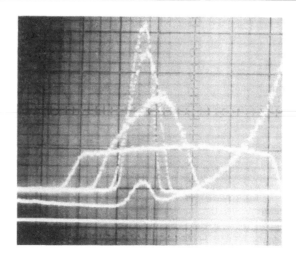

Figure 9.2. The technique of stochastic cooling, invented by Simon van der Meer at CERN, showed how particle beams could be controlled. A proton beam stored in CERN's ICE storage ring initially has a wide spread of momenta, and as cooling is applied, the momentum spread decreases, giving successively sharper peaks.

Figure 9.3. The specially built Antiproton Accumulator, the heart of CERN's antiproton project.

is more suitable for low-energy beams. The conversion of CERN's Super Proton Synchrotron (SPS) into a proton–antiproton collider was put on the fast track. Although Fermilab continued to work on an antiproton scheme, the development and construction of a superconducting magnet ring to double the output energy from the laboratory's Main Ring were considered the priority.

Stochastic cooling at CERN was initially investigated using the ISR and was achieving cooling rates of 89% per hour by 1977. With the new development work using the ICE ring, cooling times were soon down to 15 s and momentum density increases were a factor of 20. All these tests were carried out with protons. However, before an antiproton project could be formally approved, it remained to show that these antiparticles were sufficiently stable to be stored in a ring; at the time, the longest that anyone had seen an antiproton was 140 μs. In the summer of 1978 the ICE ring comfortably held antiprotons for several days, the observed losses being consistent with scattering on residual gas molecules in the ring vacuum. With antiprotons thus shown to be sufficiently stable (in principle, they should live for as long as protons, at least 10^{33} years), the project

to convert CERN's SPS into a proton–antiproton collider was formally approved.

ANTIPROTON FACTORIES

To achieve proton–antiproton collisions first meant building an intense source of antiprotons and a means of accumulating the precious particles until they were ready to collide with protons. CERN's antiprotons came from a 150 mA proton beam accelerated to 26 GeV in CERN's proton synchrotron. From the secondary particles released at a production target, a focusing system (a magnetic horn also devised by van der Meer) selected antiprotons of around 3.5 GeV for injection into a new machine, the specially built Antiproton Accumulator (AA) (figure 9.3), the heart of CERN's antiproton project.

For each pulse of 10^{13} protons hitting the production target, 10^7 antiprotons were obtained. It therefore took several days for the AA to build up stacks of 6×10^{11} antiprotons, using a complicated procedure with precooling applied to the incoming particles, with their wide momentum spread. After this pre-cooling, the antiprotons could be transferred to the stored stack, separated from the incoming raw beams by a mechanical shutter. The AA produced its first antiproton stacks in 1980, and the first collisions in the SPS, at 270 GeV per beam, were achieved one year later. (Later a separate Antiproton Collector ring was built to complement the AA and to take over the job of pre-cooling the antiprotons.)

A similar antiproton source was built at Fermilab, where the superconducting Tevatron came into action in 1983, and its first proton–antiproton collisions, at 800 GeV per beam, were achieved in 1985.

THE DETECTORS

Having suggested how to convert large proton synchrotrons into proton–antiproton colliders, it remained to conceive a special detector to observe the results of these carefully engineered collisions. The new physics conditions required a detector to match. I led the first detector collaboration.

Previous physics discoveries had been missed by 'keyhole' detectors which sampled only a small slice of produced particles. The first objective was therefore to surround the collision region completely to catch as many of the produced particles as possible; the detector had to be 'hermetically sealed'. Within its sealed volume, the detector would have to be built up from different detector elements, each picking up and measuring different kinds of particle. The resultant detector would thus resemble a series of concentric boxes, each box doing what the previous one could not do.

A W particle would be produced when an up quark in a proton hit a down antiquark in an antiproton (or vice versa) and would decay into an electron or a muon, accompanied by the appropriate neutrino. Because the W is so massive, its decay products would carry high momentum. At the same time the accompanying quark–antiquark interactions would produce many hadrons, but the characteristic signature of the W, the simultaneous emission of a high-energy lepton and an accompanying neutrino, would have to be unscrambled from the mass of accompanying hadronic debris. Apart from the challenge of producing proton–antiproton collisions at the required intensity, many physicists were sceptical that the subtle W signal could be detected at all amidst so much hadronic clutter.

In parallel, CERN was preparing plans for the Large Electron–Positron Collider (LEP), a 27 km ring to collide electrons and positrons, to mass-produce Z particles and to make precision measurements, while the Stanford Linear Accelerator Center (SLAC) pushed a scheme to modify its 2 mile electron linear accelerator into an electron–positron collider. However, designing and building such large machines would take about a decade. In the meantime, CERN's ambitious proton–antiproton collider hoped to make the discovery.

To catch W particles meant that the detector for CERN's proton–antiproton collider had to be able to 'see' the otherwise invisible neutrinos. This can be achieved by the 'missing mass' method. All the outgoing energy due to other particles is carefully measured. In principle this outgoing energy has to balance around the detector; any apparent mismatch of outgoing energy in a particular direction transverse to the beam means that an otherwise invisible particle (a neutrino) has escaped via that route. As well as good angular coverage of the whole collision region so that as many of the emerging particles as possible were intercepted, this required fine-grained calorimetry (energy measurement) to pinpoint where particles emerged. The detector was therefore well suited to the study of 'jets', the narrow clusters of particles produced at wide angles by quark interactions deep inside the proton–antiproton collisions.

The electrically neutral Z would be produced by quark–antiquark annihilation, producing an electron–positron or a muon pair. In proton–antiproton collisions, Z production was expected to be about a tenth of W production. Again it would a challenge to pick out from the accompanying hadronic debris, but at least this Z signal did not include neutrinos.

In designing the detector, it was vital that it should be carefully optimized with respect to the expected W and Z production rates, which in turn depended on the proton–antiproton collision rates. Codenamed UA1, construction of our large new detector was approved in 1978 (figure 9.4). The UA stood for 'underground area', referring to the large new pit which had to be excavated around the SPS ring to house the new detector. The general-purpose magnetic detector marked a new dimension in instrumentation for particle physics, with an 800 t conventional magnet

Figure 9.4. The UA1 general-purpose detector marked a new era in instrumentation for particle physics.

supplying a dipole field of 0.7 T in a region of 85 m^3, covering the solid angle round the collision point down to just 0.2° around the beam pipe.

At the centre of the detector, the innermost box, was the tracking chamber, a drift chamber 6 m long with 6000 sense wires with image readout, resembling an 'electronic bubble chamber', able to reconstitute the complex tracks flying out from the high energy proton–antiproton collisions. Equipped with computer-aided technology, the tracker would be able to analyse the three-dimensional collision patterns and present them as seen from all angles.

Catching the photons which eluded the central tracker and better analysis of the electrons were the tasks of the second UA1 box, the electromagnetic calorimetry using 48 semicylindrical lead–plastic scintillator 'gondolas' each divided into four segments to monitor the longitudinal development of showers. Neutral pions and other hadronic material were absorbed in the next layer of calorimetry, using scintillators mounted between the slabs of the magnet yoke and each divided into two compartments. The pattern of energy deposition in these successive layers of electromagnetic and hadronic calorimetry allowed high-energy electrons to be readily distinguished from pions and other hadronic particles. The forward zones of the detector were closed by additional plugs ('bouchons') of calorimetry. The whole detector was enclosed by large slabs

of drift chambers for muon detection, covering a total area of 500 m^2. Of total mass 2000 t, this detector kept busy a team of 130 physicists from 13 research centres, the largest particle physics collaboration at the time. It also became the role model for many subsequent total solid-angle collider detectors, most of which, however, have chosen to use solenoidal rather than dipole field configurations. Apart from the geometry and the construction of the detector, its electronics and the operating software required a major effort. Sophisticated triggers were developed to filter off the more interesting physics, while fast off-line processing selected the cream of the events for subsequent analysis.

Initially, the detector had been designed as an 'electronic bubble chamber' complemented by calorimetry and enclosed by muon detectors. However, under the high-energy conditions of the experiment, the calorimeter signals were highly localized and could be used to help to analyse what was recorded in the central tracker, rather than the other way round.

Complementing UA1 was another large and highly segmented detector, UA2, in a second underground collision area, concentrating on electron signals and using no central magnetic field. Beyond the central vertex detector of proportional and drift chambers was the central calorimeter, covering the range of polar angles from 40° to 140° to the beam with 240 towers divided into an electromagnetic and two hadronic compartments. This was supplemented by forward–backward detectors closer to the beam pipe direction fitted with magnetic tracking and additional electromagnetic calorimetry.

In the initial 1981 running, the UA2 underground area was used by the UA5 experiment, using two large 7.5 m streamer chambers to record the tracks produced by the proton–antiproton collisions.

THE DISCOVERY AND AFTER

To detect the W meant that the CERN proton–antiproton collider had to attain luminosities of the order of several 10^{28} cm^{-2} s^{-1} and to maintain coasting colliding beams for days at a time. Naturally with such a revolutionary new technique it took some time to gain experience and to perfect the complicated operations procedures. However, soon the highly segmented UA2 detector was

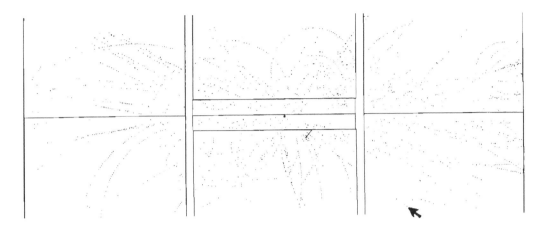

Figure 9.5. One of the proton–antiproton collisions recorded by the UA1 detector in late 1982. The high transverse energy emerging electron (arrowed) was found to be produced back-to-back with 'missing mass', indicative of the emission of an invisible neutrino. The electron and the neutrino are the decay products of the W particle. (See colour plate 1.)

able to provide important new insights on high-transverse-momentum hadron jet production, reported at the 1982 Paris International Conference.

By late 1982, the collision luminosities had edged towards the critical region needed to see W particles in the detectors, and the integrated luminosity was piling up, with one antiproton shot lasting for 42 h. From this data, UA1 found six candidate W events, identified by both the high-transverse-momentum electrons (muons were yet to come) and the corresponding missing-mass neutrino signals (figure 9.5). The paper was submitted for publication on 23 January 1983. In a paper submitted on 15 February 1983, UA2 reported four W events producing electrons.

In principle the Z should have been easier to see, with no missing mass to have to worry about. However, in proton–antiproton collisions Z production is a rarer phenomenon. In the 1983 run, the collider luminosity climbed above 10^{29}, and soon the first Z candidates were announced. UA1 had five, made up of four electron–positron pairs and one muon pair. UA2 followed soon afterwards. At the end of the 1983 run, UA1 and UA2 together had about a dozen Zs, and several hundred W particles.

In 1989, LEP at CERN and the Stanford Linear Collider (SLC) at SLAC began to mass-produce Z particles.

What had been a discovery just six years before had become production line physics.

FURTHER READING

Cahn R and Goldhaber G 1989 *The Experimental Foundations of Particle Physics* (Cambridge: Cambridge University Press)

Cline D 1997 *Weak Neutral Currents* (Reading, MA: Addison-Wesley)

Fraser G 1997 *The Quark Machines* (Bristol: Institute of Physics Publishing)

Sutton C 1984 *The Particle Connection* (London: Hutchinson)

Watkins P 1986 *The Story of the W and Z* (Cambridge: Cambridge University Press)

ABOUT THE AUTHOR

Carlo Rubbia is currently a Senior Physicist at CERN, Geneva, where he was Director General from 1989–93.

After his doctorate at the Scuola Normale Superiore, Pisa, he was research fellow at Columbia University, New York, from 1958–59, where his experiments at the Nevis cyclotron marked the beginning of a long series of investigations of weak interactions. He was lecturer at Rome from 1960–61. He subsequently carried out research at CERN, Brookhaven and Fermilab and in 1971 was appointed Higgins Professor of Physics at Harvard, a position he held until 1988.

At CERN he carried out research at all the laboratory's major machines—the Synchro-Cyclotron, the Proton Synchrotron, the Intersecting Storage Rings and the Super Proton Synchrotron (SPS). He proposed converting the SPS into a proton–antiproton collider to search for W and Z particles and led one of the experimental teams which made the historic discovery in 1983. For this he was awarded the Nobel Prize jointly with Simon van der Meer in 1984.

Subsequently he has played a leading role in the development of the large ICARUS underground experiment searching for proton decay and for neutrino oscillations, and most recently has turned his attention to the concept of an Energy Amplifier, a novel means to produce nuclear energy using accelerator technology.

He is a member of the Papal Academy of Sciences, the American Academy of Arts and Sciences, Accademia dei XL, Accademia dei Lincei and numerous other scientific academies and advisory bodies. He holds honorary doctorates from more than 20 universities throughout the world. He has been awarded numerous prizes, including the Accademia dei Lincei's President of the Republic Award and the Gold Medal of the Italian Physical Society in 1983, the Lorenzo il Magnifico Prize for Sciences 1983, the Achille de Gasperi Prize for Sciences 1984, the Leslie Prize (Harvard Faculty) for exceptional achievements 1985, the Castiglioni di Sicilia Prize 1985, the Carlo Capodieci Gold Medal 1985, the Jesolo d'Oro 1986 and the Sanremo Primavera Prize 1988. He also holds the Italian Knight Grand Cross, is an Officer of the French Legion of Honour and has been awarded the Polish Order of Merit.

10 QUARK MIXING AND CP VIOLATION

Ahmed Ali and Boris Kayser

Editor's Introduction: Quarks, which bind together under the action of the strong 'colour' force, decay under the action of the weak force. Studying these processes showed that different quark decay processes have different strengths. Quark decays (the simplest of which is the well known nuclear β decay) also revealed subtle asymmetries. The first was parity violation (see the chapter by Sutton), in which the weak interaction appears to change when reflected in an ordinary mirror, and charge–parity (CP) violation. Searching for a good 'mirror' of weak interactions, physicists proposed the CP operation, which as well as switching left to right also switches particle to antiparticle, and vice versa. However, CP too was found to be violated in weak interactions.

With six quarks, a description of the many different quark decay processes can be accommodated into a constraining 3×3 matrix. However, the weak decays of many of these quark decays are only just beginning to be explored. In this chapter, Ahmed Ali and Boris Kayser analyse the weak decays of quarks and anticipate new experiments which will probe the weak decay sector of the Standard Model in detail.

The results are quite detailed but the emerging message is clear. This is one of the major growth areas in current particle physics. Only when all this information is in place can the underlying quark pattern be completely understood. What underlies the regularities, and the differences, between the six quarks?

Elementary particles carry many additive attributes (quantum numbers) which are conserved in the strong and electromagnetic interactions. These quantum numbers are called flavours and are used to characterize hadrons (strongly interacting particles). If electromagnetism and strong nuclear forces were the only interactions in nature, there would have been no flavour changing reactions seen in laboratory experiments. However, it has been known since the early days of weak interactions that the neutron is unstable and it decays into a proton by emitting an electron and its associated antineutrino $n \rightarrow pe^- \bar{\nu}_e$ with a mean life of about 15 min. On the other hand, to date not a single proton decay has been observed. Laboratory experiments have put the proton lifetime in excess of 10^{32} years, which is some 22 orders of magnitude larger than the age of our Universe!

In quark language, the two lightest quarks, called up (or u), having the fractional electric charge $+\frac{2}{3}$, and down (or d), having the fractional electric charge $-\frac{1}{3}$, are at the base of the neutron β (electron-emitting) decay and the stability of the proton. One can imagine that the u and d quarks form a doublet and the charged weak interaction causes a transition from the heavier d to the lighter u component. Then, a neutron, which consists of two d and one u quarks (n = ddu) turns into a proton (p = uud), which consists of two u and one d quarks with the charged weak interaction causing the transition $d \rightarrow ue^- \bar{\nu}_e$. The lightest of the quarks, the u quark, is then stable, as ordinary charged weak interactions do not allow the transition of a quark into a lepton. The consequence of this is that a proton, being the lightest known baryon (a hadron with spin $\frac{1}{2}$), remains stable. This example shows that charged weak interactions allow transitions between hadrons (or quarks) with different flavour quantum numbers. Here the quark flavours are up and down.

Back in 1933, Enrico Fermi wrote an effective (i.e. low-energy) theory of charged weak interactions by introducing an effective coupling constant G_F, the Fermi coupling constant. In the example given above, the mean lifetime of the neutron determines the strength of G_F and present-day experiments have measured it very precisely: $G_F = 1.166\,392(2) \times 10^{-5}\,\mathrm{GeV}^{-2}$ in units used by particle physicists in which the reduced Planck constant $h/2\pi$ and the velocity of light are both set to unity. There are other known reactions in which charged weak interactions are at work. An example is the decay of a muon into an electron, a neutrino and an antineutrino: $\mu^- \rightarrow e^- \bar{\nu}_e \nu_\mu$. The decay rate (and hence the lifetime of the muon) is also determined by the Fermi coupling constant. For a long time, until experimental precision improved, it was generally accepted that the Fermi coupling constants in neutron β decay, G_n, and in muon decay, G_μ, were one and the same. However, as the experimental precision improved and theoretical calculations became more sophisticated, by including quantum corrections as well as nucleus-dependent effects in nuclear β decays, from where most of the information on neutron β decay comes, it was established that indeed $G_n \neq G_\mu$, although the difference is small. Today, thanks to very precise experiments, this difference is known very precisely: $G_n/G_\mu = 0.9740 \pm 0.001$. So, it seemed that experiments on nuclear β decay and muon decay required not *one* but *two* different Fermi coupling constants.

As the particle 'zoo' enlarged, in particular with the discoveries of hadrons which carry a new quantum number called strangeness, it became clear that the decay rates of these newly discovered weakly decaying particles were different. A successful description of the decay widths (a measure of transition rate) of kaons and hyperons (nucleon-like particles with a non-zero strangeness quantum number) required introduction of effective coupling constants which were very different from either G_n or G_μ. A good example is the decay of a charged kaon, K^-, which was found to decay into a neutral pion π^0, an electron and an electron antineutrino, $K^- \rightarrow \pi^0 e^- \bar{\nu}_e$. In this case, the effective Fermi coupling constant was found to have an empirical value of $G_K/G_\mu \simeq 0.22$. In quark language, this transition is induced by the mutation $s \rightarrow u e^- \bar{\nu}_e$, as the charged kaon has the quark content $K^- = s\bar{u}$ and a neutral pion is built up from the linear combination of the up and down quarks and their antiquarks $\pi^0 = 1/\sqrt{2}(u\bar{u} - d\bar{d})$, reflecting its isospin properties. So, experiments seemed to have implied the existence of at least *three* different Fermi coupling constants: G_n, G_μ and G_K. The question in the theory

of weak interactions being asked in the early 1960s was: should one give up the concept of a universal charged weak interaction, as opposed to electromagnetism and the strong nuclear force?

The answer came in the hypothesis of flavour mixing, implying that the quantum eigenstates of the charged weak interactions are rotated in quark flavour space with respect to the mass eigenstates. In other words, the states which have simple charged weak interactions are not the states of definite mass, but linear combinations of them. The concept of 'rotated' charged weak currents (involving the W bosons) in the flavour space was introduced by Nicola Cabibbo in 1963, following an earlier suggestion by Murray Gell-Mann and Maurice Levy. It solved the two outstanding problems in weak interactions, explaining the strongly suppressed weak decays of the kaons and hyperons compared with the weak decays of the non-strange light hadrons (containing the u and d quarks), and the difference in the strength of the nuclear β decays compared with μ decay. Calling the Fermi coupling constant in μ decay G_F, the coupling constants in neutron β decay and the strange hadron decays in the Cabibbo theory are given by $G_F \cos\theta_C$ and $G_F \sin\theta_C$ respectively. Here, the Cabibbo angle θ_C is the angle between the weak eigenstate and the mass eigenstate of the quarks. A value $\theta_C \simeq 13°$ describes all data involving weak decays of light hadrons with the same Fermi coupling constant, preserving the universality of weak interactions.

The rates for numerous weak transitions involving light hadrons or leptons are adequately accounted for in the Cabibbo theory in terms of two quantities: G_F and θ_C. This was a great triumph. However, Cabibbo rotation with three light quarks turned out to cause havoc for so-called flavour-changing neutral-current (FCNC) processes, which in this theory were not in line with their effective measured strengths.

What are these FCNC processes? One example from the Cabibbo epoch characterizes the processes in question. Consider particle–antiparticle mixing involving the neutral kaon (K^0–$\overline{K^0}$) complex, in which, through a virtual transition, a K^0 ($= \bar{s}d$) meson turns into its charge conjugate antiparticle $\overline{K^0}$ ($= s\bar{d}$). Now, the K^0 meson has the strangeness quantum number S equal to $+1$. The strangeness quantum number of its conjugate antiparticle $\overline{K^0}$ is then $S = -1$. So, in the virtual K^0–$\overline{K^0}$ transition, the electric charge Q does not change, i.e. in this process $\Delta Q = 0$ (hence neutral current), but S has changed by two units, i.e. $\Delta S = 2$. Such transitions, and we shall see several counterparts in heavy-meson systems, are FCNC processes.

Since the quantum number S is conserved in strong and electromagnetic interactions, the K^0–$\overline{K^0}$ states cannot be mixed by these forces. The charged weak force is the only known force which changes flavours; so it must be at work in inducing the K^0–$\overline{K^0}$ mixing. Now, it is known that mixing of two degenerate levels must result in level splitting, introducing mass differences between the mass eigenstates, named K_S and K_L, being the short lived and longer lived of the two mesons. The mass difference $\Delta M_K \equiv M(K_L) - M(K_S)$ in the Cabibbo theory turned out to be several orders of magnitude larger than the observed mass difference, whose present-day value is $\Delta M_K \simeq 3.49 \times 10^{-6}$ eV. (As a fraction of M_K, the average of the K_L and K_S masses, this mass difference is only 7.0×10^{-15}!)

This great disparity in the effective strength of the K^0–$\overline{K^0}$ transition in the Cabibbo theory and experiment remained for a long time a stumbling block in developing a consistent theory of neutral weak currents involving hadrons. For example, it was not at all obvious whether the same weak force which causes the decays of the muon, the neutron and the charged kaon discussed above was at work in K^0–$\overline{K^0}$ transition, or whether a new effective force had to be invented to explain ΔM_K. During this epoch came the seminal papers by Steven Weinberg and Abdus Salam in 1967–68, proposing renormalizable weak-interaction models for leptons unifying weak and electromagnetic interactions (see the chapter by Rubbia), in which the outstanding problem of FCNC hadronic weak interactions was pushed to one side. It took several years after the advent of this electroweak theory before the FCNC problem was solved elegantly through the 'GIM' mechanism, invented by Sheldon Glashow, John Iliopoulos and Luciano Maiani in 1970, using the hypothesis of the fourth (charm or simply c) quark. According to the GIM proposal the charge-changing (W-emitting) weak current involving quarks has the form

$$J = \overline{u}d' + \overline{c}s' \qquad (10.1)$$

where d' and s' are rotated (orthogonal) combinations of the d and s quarks which can be described in terms of the

Cabibbo angle θ_C as:

$$d' = d \cos\theta_C + s \sin\theta_C$$
$$s' = -d \sin\theta_C + s \cos\theta_C. \qquad (10.2)$$

The GIM construction of the charge-changing weak current, involving four quark flavours (u, d, s and c), removed the leading contribution to the K_L–K_S mass difference. Quantum (loop) effects, such as those shown in the box diagram in figure 10.1, with the contribution of the u and c quarks in the intermediate states, give non-zero contributions to the K^0–$\overline{K^0}$ mass difference. The result of the box diagrams can be written as (here m_μ is the mass of the muon)

$$\Delta M_K \simeq \frac{4(m_c^2 - m_u^2)\cos^2\theta_C}{3\pi m_\mu^2} \Gamma(K^+ \to \mu^+\nu_\mu). \quad (10.3)$$

With the other quantities known, ΔM_K could be predicted in terms of the mass difference of the charm and up quark. This led Benjamin Lee and Mary Gaillard in 1972 to estimate a mass of 1–2 GeV for the charm quark in the Cabibbo–GIM four-quark theory. The GIM proposal remained a curiosity until the charm quark was discovered in 1974 by the experimental groups led by Samuel Ting at the Brookhaven National Laboratory, and Burt Richter at the Stanford Linear Accelerator Center (SLAC), through the $c\overline{c}$ bound state J/ψ (see the chapter by Schwitters), with the charm quark mass compatible with theoretical estimates. Subsequent discoveries of the charmed hadrons D^0 (= $c\overline{u}$), D^+ (= $c\overline{d}$), D_s (= $c\overline{s}$) at SLAC, DESY and elsewhere and their weak decays have confirmed the Cabibbo–GIM

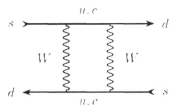

Figure 10.1. The box diagram contributing to the mass difference ΔM_K in the GIM theory. In the six-quark theory, also the top quark contributes whose contribution is small and hence not shown.

current, with again all the decays governed essentially by the parameters G_F and θ_C, thus restoring the universality of charged weak interactions.

We now know that there are not four but six quarks. The charged weak currents then would involve linear combinations of these quarks. We leave the discussion of flavour mixing in the six-quark theory to a subsequent section and discuss first another development in kaon decays which had a profound effect on the theoretical developments. namely charge–parity (CP) violation.

CP VIOLATION IN THE K^0–$\overline{K^0}$ COMPLEX

For every elementary particle, there is a corresponding antiparticle. However, a particle and its antiparticle do not always behave in the same way. For example, in the process $\pi^+ \rightarrow \mu^+ + \nu$, in which a positively charged pion decays into a positively charged muon and a neutrino, the muon emerges with its spin vector *antiparallel* to its momentum. By contrast, in the process $\pi^- \rightarrow \mu^- + \overline{\nu}$, in which every particle in the previous decay has been replaced by its antiparticle, the muon emerges with its spin *parallel* to its momentum. This difference between the two processes shows that the world is not invariant under charge conjugation C, which replaces every particle by its antiparticle. However, one may wonder whether the world is nevertheless unaltered by matter–antimatter interchange in the sense that it is invariant under charge conjugation *combined with a parity (space) reflection P*. Under P, particle spins do not change, but their momenta are reversed. Thus, the CP-mirror image of the process $\pi^+ \rightarrow \mu^+$ (spin antiparallel to momentum) $+ \nu$ is $\pi^- \rightarrow \mu^-$ (spin parallel to momentum) $+ \overline{\nu}$. Experimentally, the rates for these two processes are equal. Thus, CP invariance holds for $\pi \rightarrow \mu\nu$.

However, it has been discovered that CP does not hold everywhere. There are, as already noted, two neutral K mesons, the short-lived K_S (decaying mainly into two pions) and the longer-lived K_L (decaying mainly into $\pi e\nu$, $\pi\mu\nu$ or three pions). If CP invariance held, K_S and K_L would each be its own CP-mirror image. Thus, the CP-mirror image of the decay $K_L \rightarrow \pi^- e^+ \nu$ would be the decay $K_L \rightarrow \pi^+ e^- \overline{\nu}$, with all the momenta in the first decay reversed. If we ask about the rates for these two decays integrated over all possible directions of the outgoing particles, the momentum reversals become irrelevant, and CP invariance would require that the two rates be equal. However, these rates differ by 0.3%. Thus, the world is non-invariant, not only under C, but under CP as well. Non-invariance under the symmetry operation CP is accompanied by non-conservation of the associated CP quantum number, and the first observation that either of these phenomena occurred was the discovery in 1964 that the CP quantum number is not conserved in the decays of neutral K mesons to pairs of pions. The CP quantum number of a system, referred to as its CP parity, can be either $+1$ or -1. If CP were conserved, K_S and K_L would be CP eigenstates with opposite CP parity. The CP parity of the pion pair $\pi^+\pi^-$ (the dominant decay of the K_S) is even. However, in 1964 it was discovered by James Christenson, James Cronin, Val Fitch and René Turlay that the process $K_L \rightarrow \pi^+\pi^-$ also occurs. That is, both K_S and K_L, one of which would have CP $= -1$ if CP were conserved, decay to $\pi^+\pi^-$, which has CP $= +1$. Thus CP is not conserved in neutral K-meson decays, although the observed non-conservation is small: the ratio of the CP-violating amplitude to the CP-conserving amplitude, $|\text{amp}(K_L \rightarrow \pi^+\pi^-)/\text{amp}(K_S \rightarrow \pi^+\pi^-)|$, is only 2.3×10^{-3}. However, much larger effects may be revealed in the future, as we shall see.

Now. CP violation has so far been seen only in the decays of neutral K mesons. Thus, this violation could perhaps be a feature of K^0–$\overline{K^0}$ mixing, rather than of particle decay amplitudes. Then there would be no CP violation in the decays of *charged* K mesons. (The charged K mesons, K^+ and K^-, certainly do not mix, because the conversion of one of them into the other would violate charge conservation.) Several very challenging experimental efforts are being made to see whether decay amplitudes do violate CP. So far, the results are inconclusive. One experiment finds that the quantity 'ϵ'_K/ϵ_K', whose non-vanishing would signal that the decay amplitudes violate CP, is $(23.0 \pm 6.2) \times 10^{-4}$, but another finds that it is $(7.4 \pm 5.9) \times 10^{-4}$, consistent with zero. The experiments continue.

Regardless of the value of ϵ'_K/ϵ_K, the fact that nature violates CP invariance and CP conservation has been established. What is the origin of this CP violation? In addressing this question, we note that, as remarked earlier,

CP-violating effects have thus far been observed only in the decays of neutral K mesons. These decays are known to be due to the weak interaction. Thus, it is natural to ask whether CP violation is also due to the weak interaction, rather than to some (so-far unknown) mysterious force.

Among the discrete symmetries C, CP, T (time reversal) and CPT, the CPT symmetry is considered to be exact as it follows from fundamental principles underlying all field theories. Lately, the invariance of natural laws under CPT has been put in question in the context of the superstring theories of particle physics, in which the particles are described by extended objects in space–time, such as a string, lifting the assumption of locality (point-like nature) ascribed to the particles in field theories (see the chapter by Ross). However, even if CPT invariance should prove to be broken in superstring theory, the effects on the flavour physics being discussed in this chapter would very probably be negligible. Therefore, in what follows, we shall assume that CPT holds exactly. The CPT-invariance principle has a number of implications, such as the equality of the masses and of the lifetimes of a particle and its antiparticle. The best limit on CPT violation stems from the upper limit of the ratio of mass difference to the mass: $m(K^0) - m(\overline{K^0})/m(K^0) \leq 10^{-18}$.

If CP violation is indeed produced by the known weak interaction described by the Standard Model, then it is caused by complex phase factors in the quark mixing matrix. How these complex phases produce physical CP-violating effects will be explained shortly.

THE CABIBBO–KOBAYASHI–MASKAWA MATRIX

We now know that there are six quarks in nature. The fifth and the sixth quarks are called the beauty (or bottom or simply b) and top (or simply t) quarks. The b quark was discovered in the form of its bound state $\Upsilon = (b\bar{b})$ and excited states $\Upsilon', \ldots$ by Leon Lederman's group working at Fermilab in 1977. The discovery of the top quark had to wait until 1994 when two large experimental groups (D0 and CDF) working again at Fermilab finally discovered the top quark in the process $p\bar{p} \rightarrow t\bar{t}X$ and the subsequent decays of the top quarks $t \rightarrow bW^+$ (see the chapter by Shochet). However, indirect evidence of a top quark with

a rather large mass, $m_t = O(100)$ GeV, was found earlier from B^0–$\overline{B^0}$ mixing by the UA1 experiment at CERN's proton–antiproton collider, the ARGUS experiment at DESY's DORIS electron–positron collider, and the CLEO experiment at the Cornell electron–positron collider CESR. The expectation that m_t is large was further strengthened by electroweak precision measurements at CERN's LEP. No top meson has so far been constructed from its decay product, but there exists an impressive amount of data on the properties of the beauty hadrons from experiments at DORIS (DESY), CESR (Cornell), LEP (CERN), SLAC (Stanford) and Fermilab.

Given that there are six quarks, arranged in terms of three 'weak isospin' doublets (u, d; s, c; t, b), the obvious questions are: how are the weak interaction eigenstates involving six quarks related to the quark mass eigenstates, and what does this rotation imply for the decays of the light (containing only u, d, s quarks) and heavy hadrons (containing c, b, t quarks)?

The answers follow if the two-dimensional rotation of equation (10.2) is now replaced by a three-dimensional rotation, where the W-emitting weak current takes the form

$$J = \bar{u}d' + \bar{c}s' + \bar{t}b' = (\bar{u}, \bar{c}, \bar{t}) \, V \begin{pmatrix} d \\ s \\ b \end{pmatrix}. \quad (10.4)$$

Here V is a 3 × 3 matrix in the quark flavour space and can be symbolically written as

$$V = \begin{pmatrix} V_{ud} & V_{us} & V_{ub} \\ V_{cd} & V_{cs} & V_{cb} \\ V_{td} & V_{ts} & V_{tb} \end{pmatrix}. \quad (10.5)$$

Thus, in this case the weak (interaction) eigenstate d' is $d' \equiv V_{ud}d + V_{us}s + V_{ub}b$. (Likewise, the other rotated states s' and b' follow from equation (10.5).) Every weak process involving the W boson is proportional to some product of the elements of V. Now comes a crucial observation: if some of the elements of the 3 × 3 matrix V are not real but complex (so that V is not, strictly speaking, a rotation matrix, but a 'unitary' matrix), then the hadronic weak interactions can violate CP. However, if there were only four quarks, so that we did not have a 3 × 3 quark mixing matrix but only the two-dimensional rotation of equation (10.2) of the

Cabibbo–GIM theory, then making the coefficients in that rotation complex would not lead to any physical effects, as in this case these complex phases in the (2×2) rotation matrix can be eliminated by a redefinition of the quark fields. Not so, if there are six or more quarks.

These facts were first pointed out by M Kobayashi and T Maskawa in 1972, long before the c, b, and t quarks were actually discovered. In fact, the GIM mechanism, put forward to describe FCNC transitions in the $K^0-\overline{K^0}$ complex, was immediately followed by the KM hypothesis to accommodate CP violation in the same $K^0-\overline{K^0}$ system and which predicted the existence of all the three heavy quarks.

Now that all these quarks have been discovered, the six-quark theory of Kobayashi and Maskawa (KM) stands on firm experimental ground, as far as its quark content is concerned. The crucial question now is whether the complex phases in the matrix V are the only source of CP violation in flavour physics. These phases predict CP-violating phenomena in the decay amplitudes (also called *direct CP violation*) of many hadrons. They also predict *indirect CP violation*, which resides in the mass matrix of a neutral meson complex $M^0-\overline{M^0}$, and in particular $K^0-\overline{K^0}$, and can manifest itself only when such mixings are involved. It is widely appreciated that B physics has the potential of providing crucial tests of the KM paradigm.

The 3×3 flavour mixing matrix, which is now aptly called the Cabibbo–Kobayashi–Maskawa (CKM) matrix, plays a central role in quantifying CP-violating asymmetries.

Present status of the CKM matrix

The magnitudes of all nine elements of the CKM matrix have now been measured in weak decays of the relevant quarks, and in some cases in deep inelastic neutrino nucleon scattering. The precision on these matrix elements varies for each entry, reflecting both the present experimental limitations but often also the theoretical uncertainties associated with the imprecise knowledge of the hadronic quantities required for the analysis of data. In most cases, the decaying particle is a hadron and not a quark, and one has to develop a prescription for transcribing the simple quark language to that involving hadrons. Here, the theory

of strong interactions, quantum chromodynamics (QCD), comes to the rescue. Powerful calculational techniques of QCD, in particular renormalization group methods, lattice-QCD, QCD sum rules and effective heavy-quark theory, have been used to estimate, and in some cases even determine, the hadronic quantities of interest, thereby reducing theoretical errors in the CKM matrix elements.

This theoretical development is very impressive although the QCD technology has not quite reached its goal of achieving an accuracy of a few per cent in the determination of the hadronic matrix elements. Nevertheless, it has been crucial in quantifying the CKM matrix elements. Fascinating as these calculational aspects are, their discussion here would take us far from our mainstream interest and we refer to the suggested literature for further reading.

Present knowledge of V comes from a variety of different sources and the present status can be summarized as follows:

$$|V| = \begin{bmatrix} 0.9730\text{–}0.9750 & 0.2173\text{–}0.2219 & 0.0023\text{–}0.0040 \\ 0.208\text{–}0.24 & 1.20\text{–}0.88 & 0.038\text{–}0.041 \\ 0.0065\text{–}0.0102 & 0.026\text{–}0.040 & 1.14\text{–}0.84 \end{bmatrix}.$$

(10.6)

The following comments about the entries are in order.

(i) $|V_{ud}|$. This is based on comparing nuclear β decays $(A, Z) \rightarrow (A, Z+1) + e^- + \bar{\nu}_e$ that proceed through a conserved vector current to muon decay $\mu^- \rightarrow \nu_\mu e^- \bar{\nu}_e$. In the three-quark Cabibbo theory, this matrix element was identified with $\cos\theta_C$.

(ii) $|V_{us}|$. This is based on the analyses of the decays $K^+ \rightarrow \pi^0 \ell^+ \nu_\ell$ and $K^0 \rightarrow \pi^- \ell^+ \nu_\ell$ and β decays of the hyperons. In the Cabibbo theory, this matrix element was identified with $\sin\theta_C$.

(iii) $|V_{cd}|$. This is derived from the neutrino and antineutrino production of charm quarks from d quarks in a nucleon in deep inelastic neutrino nucleon scattering experiments, $\nu_\ell + d \rightarrow \ell^- + c$. In the Cabibbo–GIM current, this matrix element is $\sin\theta_C$.

(iv) $|V_{cs}|$. This comes from the semileptonic decays of the charmed hadrons $D^\pm$ and D^0, involving for example the decay $D^\pm \rightarrow K^0 \ell^\pm \nu_\ell$. Again, in the Cabibbo–GIM current, this matrix element is identified with $\cos\theta_C$.

(v) $|V_{cb}|$. This arises from the semileptonic decays of B hadrons, such as $\overline{B^0} \to D^{*+} + \ell^- + \nu_\ell$, or the inclusive decay of the b quark, $b \to c + \ell^- + \nu_\ell$.

(vi) $|V_{ub}|$. This is obtained from the semileptonic decays of B hadrons into non-charmed hadrons, such as $\overline{B^0} \to \pi^+ + \ell^- + \nu_\ell$, or the inclusive semileptonic decays of a b quark into a non-charm quark, $b \to u + \ell^- + \nu_\ell$.

(vii) $|V_{td}|$. This originates from the measured mass difference between the mass eigenstates in the B^0–$\overline{B^0}$ meson complex. Being an example of a FCNC process, this transition is a quantum effect and in the Standard Model takes place through a box diagram very similar to that shown in figure 10.1 for the K^0–$\overline{K^0}$ system, except that in this case the transition amplitude is dominated by the top quark owing to its very large mass ($m_t \simeq 175$ GeV).

(viii) $|V_{ts}|$. This comes from the measured branching ratio of the electromagnetic process $b \to s + \gamma$, measured by the CLEO experiment at CESR (Cornell) and recently also by the ALEPH collaboration at CERN. Again, an example of a FCNC process, this is also a quantum effect and again in the Standard Model the transition rate is dominated by the top quark.

(ix) $|V_{tb}|$. This is due to the production and decay of the top quark in the process $p\bar{p} \to t\bar{t} + X$ followed by the decay $t \to b + W^+$.

One sees that present knowledge of the matrix elements in the third row of the CKM matrix involving the top quark in equation (10.6) but also of the matrix elements V_{cs}, V_{cd} and V_{ub} is still rather imprecise. A check of the unitarity of the CKM matrix from the entries in equation (10.6) makes this quantitatively clear. Unitarity requires, among other things, that the absolute squares of the elements in any row of the CKM matrix add up to unity. We have at present

$$|V_{ud}|^2 + |V_{us}|^2 + |V_{ub}|^2 = 0.997 \pm 0.002$$
$$|V_{cd}|^2 + |V_{cs}|^2 + |V_{cb}|^2 = 1.18 \pm 0.33 \qquad (10.7)$$
$$|V_{td}|^2 + |V_{ts}|^2 + |V_{tb}|^2 = 0.98 \pm 0.30.$$

This shows that, except for the first row, the information on the unitarity of the CKM matrix is very imprecise. However, all data, within errors, are consistent with the CKM matrix being unitary.

Unitarity triangles

The unitarity of the CKM matrix also requires that any pair of rows or any pair of columns of this matrix be orthogonal. This leads to six orthogonality conditions. These can be depicted as triangles in the complex plane of the CKM parameter space. The constraint stemming from the orthogonality condition on the first and third column of V,

$$V_{ud} V_{ub}^* + V_{cd} V_{cb}^* + V_{td} V_{tb}^* = 0 \qquad (10.8)$$

is at the centre of contemporary theoretical and experimental attention. Since present measurements are consistent with $V_{ud} \simeq 1$, $V_{tb} \simeq 1$ and $V_{cd} \simeq -\lambda$, where $\lambda = \sin\theta_C$, the unitarity relation (10.8) simplifies to

$$V_{ub}^* + V_{td} \simeq -V_{cd} V_{cb}^* \simeq +\lambda V_{cb}^* \qquad (10.9)$$

which can be conveniently depicted as a triangle relation in the complex plane, as shown in figure 10.2. In drawing this triangle, we have used a representation of the CKM matrix due to Wolfenstein, characterized by four constants A, λ, ρ and η. We have also rescaled the sides of the triangle by λV_{cb}, which makes the base of the triangle real and of unit length and the apex of the triangle given by the point (ρ, η) in the complex plane. This is usually called the unitarity triangle. Knowing the sides of the unitarity triangle, the three angles α, β and γ of this triangle are determined, but these angles can also, in principle, be measured directly through observation of CP violation in various B decays. By measuring both the sides and the angles, the unitarity triangle will be overconstrained, which

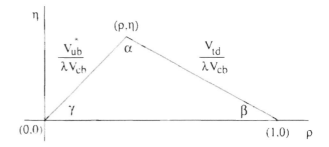

Figure 10.2. The unitarity triangle. The angles α, β and γ can be measured via CP violation in the B system, and the sides from the rates for various CC- and FCNC-induced B decays.

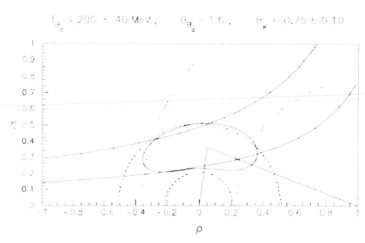

Figure 10.3. Allowed region in ρ–η space obtained by overlaying the individual constraints following from $|V_{ub}/V_{cb}|$ (dashed curves), ϵ_K (solid curves) and ΔM_d (dotted curves), by letting the hadronic quantities vary in the range shown above. The 95% confidence level contour resulting from a simultaneous fit of the data is also shown ('Haggis'-type curve). The triangle shows the best fit.

is one of the principal goals of the current and forthcoming experiments in flavour physics.

In the Wolfenstein representation,

$$\tan\alpha = \frac{\eta}{\eta^2 - \rho(1-\rho)}$$
$$\tan\beta = \frac{\eta}{1-\rho} \qquad (10.10)$$
$$\tan\gamma = \frac{\eta}{\rho}.$$

A profile of the unitarity triangle based on our present knowledge of the CKM matrix is now given from which the CP-violating asymmetries which will be measured in forthcoming experiments in B physics can be estimated. For this, the present experimental input can be summarized as follows:

$$\sqrt{\rho^2 + \eta^2} = 0.363 \pm 0.073$$
$$\left(f_{B_d}\sqrt{\hat{B}_{B_d}}/1\ \text{GeV}\right)\sqrt{(1-\rho)^2 + \eta^2} = 0.202 \pm 0.017$$
$$\hat{B}_K \eta[0.93 + (2.08 \pm 0.34)(1-\rho)] = 0.79 \pm 0.11$$
$$(10.11)$$

which come from the measurements of the CKM matrix element ratio $|V_{ub}/V_{cb}| = 0.08 \pm 0.02$, the mass difference

induced by the B^0–$\overline{B^0}$ mixing, which is measured very accurately, $\Delta M_d = (3.12 \pm 0.20) \times 10^{-4}$ eV, and the CP-violating parameter in the K^0–$\overline{K^0}$ system, $\epsilon_K = (2.28 \pm 0.013) \times 10^{-3}$, which is likewise known very precisely. The quantities f_{B_d}, $\hat{B}_{B_d}$ and $\hat{B}_K$ are various hadronic quantities whose knowledge is needed to analyse data. Present estimates, based mostly on lattice QCD calculations, put them in the range $f_{B_d}\sqrt{\hat{B}_{B_d}} = 200 \pm 40$ MeV and $\hat{B}_K = 0.75 \pm 0.10$. The resulting allowed regions in the (ρ, η) parameter space from each of these constraints individually and the resulting overlap region from all the constraints put together are shown in figure 10.3. The triangle drawn is to guide the eye and represents the currently preferred solution. Two messages are clear. First, current theoretical uncertainties in hadronic quantities translate into rather large uncertainties in the profile of the unitarity triangle. Second, and despite this, a good part of the allowed parameter space is now ruled out by data and the CKM matrix provides a consistent solution only over a limited parameter space.

CP violation in B decays

The paramount interest in B physics lies in that it will test the CKM paradigm of CP violation in flavour-changing weak interactions. The B mesons can decay in many different ways, and a large number of their decay modes are potentially interesting from the point of view of observable CP-violating effects. In some of the decay modes, these effects can yield *clean* information, free of theoretical uncertainties, on the angles in the unitarity triangle of figure 10.2. Since these angles are just the relative phases of various combinations of CKM elements, the clean information on the angles will stringently test the hypothesis that CKM phases cause CP violation. The decay modes which can provide clean information on the angles include the decays of neutral B mesons to final states which are CP eigenstates or at least can come from both a pure B^0 and a pure $\overline{B^0}$, and certain decays of charged B mesons. The decays of neutral B mesons to CP eigenstates provide a particularly good example of how CP violation comes about, and how the phases of CKM matrix elements can be determined. In any decay, for CP violation to be non-zero, there must be interfering amplitudes with clashing phases. Now, in neutral B decay to a CP eigenstate f_{CP}, there are two routes to the final state. If the parent B was born as a B^0, firstly it may decay directly to f_{CP}, or else secondly it may turn via weak mixing into a $\overline{B^0}$, and then this $\overline{B^0}$ decays to f_{CP}. The amplitudes for these two routes must be added coherently and will interfere. If the parent B was born as a $\overline{B^0}$, decay to f_{CP} can again proceed through two routes: $\overline{B^0} \to f_{CP}$ and $\overline{B^0} \to B^0 \to f_{CP}$. As before, the amplitudes for these two routes will interfere. If the CKM matrix elements have complex phases, then these (weak) amplitudes will have different phases from when the B was born as a B^0. As a result, the interferences encountered in (B born as B^0) $\to f_{CP}$ and (B born as $\overline{B^0}$) $\to f_{CP}$ will differ, and consequently the rates for these two decays will differ as well. Since the two decays are CP-mirror-image processes, the difference between their rates is a violation of CP.

Since the rates $\Gamma[$B born as $\overset{(-)}{B^0} \to f_{CP}$ after time $t]$ $\equiv \overset{(-)}{\Gamma}(t)$ depend non-trivially on the time t that the B lives before decaying, experiments will study the time-dependent CP-violating asymmetry

$$a_{f_{CP}}(t) \equiv \frac{\Gamma(t) - \overline{\Gamma}(t)}{\Gamma(t) + \overline{\Gamma}(t)}. \qquad (10.12)$$

When the unmixed B decays, $B^0 \to f_{CP}$ and $\overline{B^0} \to f_{CP}$, are each dominated by one diagram, $a_{f_{CP}}(t)$ is given by the simple expression

$$a_{f_{CP}}(t) = \eta_{f_{CP}} \sin(\phi_{f_{CP}}) \sin(\Delta M t). \qquad (10.13)$$

Here, $\eta_{f_{CP}} = \pm 1$ is the CP parity of the final state, ΔM is the mass difference between the two mass eigenstates of the B^0-$\overline{B^0}$ system, and $\phi_{f_{CP}}$ is the phase of a certain product of CKM elements. Namely, $\phi_{f_{CP}}$ is the relative phase of the product of CKM elements to which the amplitude for $B^0 \to f_{CP}$ is proportional, and the product to which the amplitude for the alternate decay route, $B^0 \to \overline{B^0} \to f_{CP}$, is proportional. Of course, the identity of $\phi_{f_{CP}}$ depends on the choice of f_{CP}.

There are two neutral B systems: $B_d^0(\overline{b}d)$ and its antiparticle, and $B_s^0(\overline{b}s)$ and its antiparticle. For the B_d^0-$\overline{B_d^0}$ system, the mass splitting ΔM_d between the mass eigenstates is already known, as previously mentioned. For the B_s^0-$\overline{B_s^0}$ system, the analogous splitting ΔM_s will no doubt eventually be determined as well. Thus, the ΔM in equation (10.13) for the CP asymmetry may be assumed known. For any chosen final state, the CP parity $\eta_{f_{CP}}$ is also known. Thus, we see from equation (10.13) that, once the asymmetry $a_{f_{CP}}(t)$ is measured, $\sin \phi_{f_{CP}}$ is *cleanly* determined, with no theoretical uncertainties. This makes it possible to test cleanly whether complex phases of CKM matrix elements are indeed the origin of CP violation.

As an example, suppose that f_{CP} is $J/\psi K_S$. In the decays (B born as $\overset{(-)}{B_d^0}$) $\to J/\psi K_S$, each of the unmixed B decays, $B_d^0 \to J/\psi K_S$ and $\overline{B_d^0} \to J/\psi K_S$, is expected to be dominated by one diagram. Thus, equation (10.13) for $a_{f_{CP}}(t)$ should hold. The dominating diagrams are such that, for this final state, $\phi_{f_{CP}}$ is simply 2β, where β is one of the angles in the unitarity triangle of figure 10.2. Thus the decays (B born as $\overset{(-)}{B_d^0}$) $\to J/\psi K_S$ can give us clean information on β. It appears that obtaining information on the other angles in the unitarity triangle will be more difficult but should still be possible. A major experimental effort will be made to determine all the angles of the unitarity triangle.

How large are the CP-violating asymmetries in B decays? They depend in part on the mass-mixing related quantities $x_d \equiv \Delta M_d \cdot \tau(B_d)$ for the B_d^0-$\overline{B_d^0}$ system, which is well measured with $x_d \simeq 0.74$, and on $x_s \equiv \Delta M_s \cdot \tau(B_s)$ for the B_s^0–$\overline{B_s^0}$ system, for which experiments at LEP have provided only lower limits $x_s \geq 16$. Here, $\tau(B_d)$ ($\tau(B_s)$) is the lifetime of the B_d^0 (B_s^0) meson, but the CP asymmetries depend crucially on the angles of the unitarity triangle, which can be estimated from the unitarity fits. With the help of the relations given in equations (10.10), the CP-violating asymmetries in B decays can be expressed straightforwardly in terms of the CKM parameters ρ and η. The constraints on ρ and η discussed above can then be used to predict the correlated ranges of the angles α, β and γ in the Standard Model. Representative of the current theoretical expectations are the following ranges for the CP-violating rate asymmetries parametrized by $\sin(2\alpha)$, $\sin(2\beta)$ and $\sin^2 \gamma$, which at the end of 1997 were estimated by Ali and London in the context of the Standard Model:

$$-1.0 \leq \sin(2\alpha) \leq 1.0$$
$$0.30 \leq \sin(2\beta) \leq 0.88 \qquad (10.14)$$
$$0.27 \leq \sin^2 \gamma \leq 1.0$$

with all ranges corresponding to the 95% confidence level (i.e. $\pm 2\sigma$). The currently preferred solutions of the unitarity fits yield $\rho \simeq 0.12$ and $\eta \simeq 0.34$, which then translate into $\alpha \simeq 88°$, $\beta \simeq 21°$ and $\gamma \simeq 72°$. The central values of the parameters which determine the asymmetries are then $\sin(2\alpha) \simeq 0.07$, $\sin(2\beta) \simeq 0.67$ and $\sin^2 \gamma \simeq 0.89$. These parameters will be measured in decays such as (B born as $\overset{(-)}{B_d^0}$) $\rightarrow$ $J/\psi K_S$, where the CP-violating asymmetry is proportional to $\sin(2\beta)$, (B born as $\overset{(-)}{B_d^0}$) $\rightarrow$ $\pi^+ \pi^-$, which can determine $\sin(2\alpha)$, and (B born as $\overset{(-)}{B_s^0}$) $\rightarrow$ $D_s^\pm K^\mp$ or $B^\pm \rightarrow DK^\pm$, which can yield $\sin^2 \gamma$. The actual asymmetries in the partial rates are expected to be quite large in some of these decays, which will make them easier to measure in the next round of B physics experiments.

Additional decay modes which appear to be promising places to study CP violation include $B^\pm \rightarrow \pi^\pm K$, (B born as $\overset{(-)}{B_d^0}$) $\rightarrow$ $\pi^\pm K^\mp$, $B^\pm \rightarrow \pi^\pm \eta'$, (B born as $\overset{(-)}{B_d^0}$) $\rightarrow$ $K_S \eta'$, (B born as $\overset{(-)}{B_d^0}$) $\rightarrow$ $D^{(*)\pm} \pi^\mp$, (B born as $\overset{(-)}{B_d^0}$) $\rightarrow$ $K^0 \overline{K^0}$, and many others. Moreover, one expects measurable CP violation in the inclusive radiative decays such as $B \rightarrow X_d + \gamma$, where X_d is a system of light non-strange hadrons,

and in exclusive radiative decays such as $B \rightarrow \rho + \gamma$, which are governed by the FCNC process $b \rightarrow d + \gamma$. These processes are similar to the observed decays $B \rightarrow K^* + \gamma$ and $B \rightarrow X_s + \gamma$ but are suppressed by a factor of about 20. Measurements of CP asymmetries in these processes do not directly determine the angles of the unitarity triangle. However, they all depend on the parameters ρ and η and hence their measurement will contribute to determine the unitarity triangle more precisely, and to the understanding of CP violation. However, most of these measurements will require sufficiently many B hadrons that they will probably have to await the second round of experiments in B factories at SLAC (Stanford), KEK (Japan) and CESR (Cornell).

Apart from the CP violation measurements discussed above, some of the anticipated landmark measurements in B physics include: (i) the determination of the mass splitting ΔM_s in the B_s^0–$\overline{B_s^0}$ complex, and (ii) rare B decays, such as $b \rightarrow d + \gamma$, $B \rightarrow \rho^0(\omega) + \gamma$, $b \rightarrow s\ell^+\ell^-$, $b \rightarrow d\ell^+\ell^-$, all examples of FCNC processes, which have been the driving force behind theoretical developments in flavour physics.

Likewise, several planned and ongoing experiments in K physics will measure rare decays such as $K_L \rightarrow \pi^0 \nu\bar{\nu}$ and $K^\pm \rightarrow \pi^\pm \nu\bar{\nu}$, and the CP-violating ratio ϵ_K'/ϵ_K. These important K-system measurements will complement the B-system experiments and will help us to determine the properties of the unitarity triangle and to explore the origin of CP violation.

CONCLUDING REMARKS

The elegant synthesis of seemingly diverse, and in their effective strengths widely differing, empirical observations involving weak interactions in terms of a universal constant G_F and a 3×3 unitary matrix is one of the great simplifications in elementary particle physics. All data on weak interactions can at present be analysed and understood in terms of a few universal constants, and the consistency of the picture is indeed remarkable. With improved theoretical and experimental precision, this consistency will provide in the future one of the most promising search strategies for finding physics beyond the six-quark Standard Model of particle physics. A good candidate in that context is supersymmetry which may contribute to many of the FCNC processes discussed here but whose anticipated effects are quite subtle and their detection would require high-precision data (see the chapter by Ross).

Despite this success, there are many discomforting features which deserve attention. It must be stressed that the parameter ϵ_K, which describes CP violation in K^0–$\overline{K^0}$ mixing, and whose first measurement dates back some 35 years, still remains the only source of information on CP violation in laboratory experiments. This state of affairs is deeply disturbing, in particular as CP violation has a direct bearing on a fundamental phenomenon in nature, namely the observed large-scale preponderance of matter over antimatter in the Universe. The next round of experiments in B (and K) physics will certainly help to fill in some of the numerous blanks. At a deeper level, however, the connection between complex phases in the CKM matrix and the observed matter–antimatter asymmetry in the Universe remains very much a matter of speculation. It is conceivable that fundamental progress here may come from completely different quarters, such as observation of CP violation in the lepton sector and the understanding of baryogenesis at the grand unification scale—all aspects not directly related to the flavour physics of quarks discussed here.

We started this article with the discussion of the Fermi theory postulated some 65 years ago. The physics behind the effective Fermi coupling constant G_F has come to be understood in terms of a fundamental gauge interaction. The question is: are the elements of the CKM unitary matrix also some kind of effective parameter, which some day one would be able to derive in terms of more fundamental quantities? Some ideas along these lines are being pursued enthusiastically in grand theoretical schemes where the CKM matrix elements are derived in terms of quark masses. As theoretical and experimental precision on the CKM matrix improves, many of these relations will come under sharp experimental scrutiny. The emerging pattern will help us to discard misleading theories, and perhaps to single out a definitive and unique theoretical perspective. The flavour problem, namely understanding the physics behind the parameters of the CKM matrix which seem to describe all flavour interactions at present energies consistently, remains one of the most challenging problems of particle physics.

FURTHER READING

Jarlskog C (ed) 1989 *CP Violation (Advanced Series on Directions in High Energy Physics)* vol 3 (Singapore: World Scientific)

Okun L B 1987 *Leptons and Quarks* (Amsterdam: North-Holland)

Stone S (ed) 1994 *B Decays* revised 2nd edn (Singapore: World Scientific)

ABOUT THE AUTHORS

Ahmed Ali is a staff member of the theoretical physics group at the German high-energy physics laboratory DESY in Hamburg. His physics interest is heavy flavour physics and quantum chromodynamics. His early work from the mid-1970s covered charmed hadron and tau leptons, then recently discovered at SLAC and DESY. His quark and gluon jet analysis helped to determine the strong interaction coupling constant at the electron–positron colliders PETRA (DESY) and PEP (SLAC).

Since the discovery of the beauty quark in 1977, he has been involved in theoretical studies of heavy flavour, especially B physics, including studies of particle–antiparticle mixing in the neutral B-meson systems. His more recent work includes applications of QCD in flavour-changing neutral-current-mediated decays of the B hadrons, non-leptonic B decays and CP violation. With David London, he developed a quantitative profile of the quark flavour mixing matrix which has helped the planning of forthcoming precision flavour physics at B factories and hadron machines.

Boris Kayser, Program Director for Theoretical Physics at the US National Science Foundation, works on elementary particle theory close to experiment. In recent years, he has been especially interested in CP violation, and in the physics of massive neutrinos. In the area of CP violation, he and his colleagues have helped to develop the coming experimental test of the Standard Model of CP violation, analysing the question of what there is to measure, and the issue of how it can be measured. In neutrino physics, he and colleagues have explored the nature of neutrinos which are their own antiparticles, the potential neutrino-mass implications of extremely rare nuclear processes, and the quantum mechanics of neutrino oscillation.

11 QUARK GLUE

Sau Lan Wu

Editor's Introduction: For a long time, physicists were convinced that the pion, and perhaps its close relations, were the carriers of the force proposed by Yukawa in the 1930s to account for the extraordinary stability of nuclei. The modern view of the nuclear force is very different. Protons, neutrons and other subnuclear particles are composed of quarks, and the inter-quark force is due to carrier particles: 'gluons'. With the gluon picture concentrating on what happens inside the proton and other subnuclear particles, the forces between protons and neutrons, once thought to be the ultimate substructure and as such the focus of Yukawa's theory, are the gluonic equivalent of molecular forces in chemistry.

This chapter traces the emergence of this gluon picture, the dramatic discovery of the gluon in 1979 and goes on to describe a few examples of modern gluon physics. It is also an introduction to the study of 'jets', the tightly confined sprays of particles which are the key to the inner quark and gluon mechanisms.

QUANTUM ELECTRODYNAMICS AND QUANTUM CHROMODYNAMICS

One of the greatest achievements of twentieth-century physics is the successful marriage of the classical electromagnetic theory of James Clerk Maxwell with quantum mechanics, leading to quantum electrodynamics (QED). The main ingredients of quantum electrodynamics are (i) the photon of Albert Einstein, (ii) the relativistic electron of Paul Dirac, and (iii) the gauge interaction between photons and electrons.

One of the earliest formulations of quantum electrodynamics was that of Enrico Fermi. In QED, there are four parameters: Planck's constant, the velocity of light, the charge and the mass of the electron. From these four parameters, one dimensionless constant can be formed. This dimensionless 'fine-structure constant' denoted as α, specifies the strength of interaction between the photon and the electron; its numerical value is quite small, about 0.007 297 353. Because of the smallness of α, it is natural to apply perturbation methods to QED, expressing various physical quantities as power series in α.

When such perturbation calculations in QED are carried out to low orders, the results are in good agreement with measurements. However, when they are carried out to higher orders, the results become nonsensical, with integrals that fail to converge. It took several decades to solve this problem of divergences through a deeper understanding of the meaning of the mass and the charge of the electron. The resulting theory is referred to as renormalization, which for QED is the cumulative work of many theoretical physicists, including Sin-Itiro Tomonaga, Julian Schwinger, Richard Feynman and Freeman Dyson.

Renormalized QED is the first successful quantum field theory (see the chapter by Veltman) and thus has become the model for all subsequent quantum field theories. Since QED deals exclusively with electromagnetic interactions, it is natural to raise the question of whether there is also a quantum field theory for nuclear, or strong, interactions. There is and it is called quantum chromodynamics (QCD).

Two major events in theoretical physics, ten years apart, mark the beginning of QCD. Both can only be described as strokes of genius.

In 1954, Chen Ning Yang and Robert Mills generalized the notion of gauge theories beyond the electromagnetic case. In QED, the sources of photons, or electromagnetic field, are electrons or other charged particles, but the photon itself does not carry any electric charge. In other words, while electrons can radiate photons, photons themselves cannot radiate any photon. The underlying symmetry group for such a gauge theory is an Abelian group (a group whose operations are commutative: $AB = BA$). Yang and Mills generalized the concept of a gauge theory to the case where a gauge particle (the photon in the electromagnetic case) can itself radiate gauge particles. Let the charge e in QED be generalized to some charge-like quantity, say f; then not only can any particle that carries this f radiate the corresponding Yang–Mills gauge field or particle, but also the Yang–Mills field or gauge particle itself carries this f. An example of such a quantity f is isotopic spin (see the chapter by Ne'eman). The underlying group for such a gauge theory is non-Abelian. Thus the Yang–Mills theory is also referred to as non-Abelian gauge theory.

Shortly thereafter, many attempts were made to apply the Yang–Mills gauge theory to strong interactions. The basic question is: in QCD, what plays the role of the electron? This question was finally answered, ten years later, by Murray Gell-Mann (see the chapter by Ne'eman). He proposed the concept of quarks, whose charges are fractions of the charge e of the electron. For example, a proton consists of two quarks of charge $\frac{2}{3}e$ and one of charge $-\frac{1}{3}e$, while a positively charged pion consists of a quark of charge $\frac{2}{3}e$ and an antiquark of charge $\frac{1}{3}e$. The role of quarks in QCD is similar to that of electrons in QED. Because of complications due to quantum statistics as first discussed by Oscar W Greenberg, the novel feature of the quarks is their 'colours'. There are several pieces of evidence that the number of colours for quarks is 3, the most direct and convincing one being from the work of Stephen Adler on the decay rate of the neutral pion. This decay rate is proportional to the square of the number of colours, and a comparison of Adler's theoretical calculation with the experimental data showed clearly that there are three colours. Indeed, the word 'chromodynamics' in QCD refers to this major feature, the 'colour' of the quarks. In particular, the three quarks in a proton have respectively these three colours, and the quark and the antiquark in a pion must have the same colour.

When these two concepts of the Yang–Mills gauge theory and the Gell-Mann quarks are combined, the result is QCD, as discussed by Harald Fritzsch, Murray Gell-Mann and Heinrich Leutwyler in 1973. More technically, QCD is the Yang–Mills gauge theory for the group SU(3) of 3×3 unitary matrices with determinant equal to 1. In QCD there are eight gauge particles which are the carriers of the strong force between the quarks just as the photon is the carrier of the electromagnetic force between electrons. These Yang–Mills particles are called gluons.

HADRON JETS

Since quarks are charged, they are produced in the annihilation of electrons with positrons. More precisely, the process

electron + positron → quark + antiquark

has a known cross section that depends on the charge of the quark, provided that enough energy is available for this process to proceed.

Yet these produced quark pairs of fractional charge have never been observed directly. This dilemma led to the working hypothesis, due to Richard Feynman and to James Bjorken, that somehow the quarks turn into hadrons through strong interactions. Another way of saying this is that quarks are 'confined' so that the quarks must combine with other quarks or antiquarks to form hadrons. Independent of the mechanism of turning quarks into hadrons, the hadrons are expected to retain some memory of the momentum of the quark. In other words, if a high energy quark is produced in the x direction, the resulting hadrons are expected to have, on the average, larger momenta in the x direction than the y or z directions. From this point of view, the behaviour of quarks is expected to be observed through such 'jets' of hadrons.

Since these hadron jets play a central role in the observation of quarks and gluons, it is worthwhile to describe their formation. In the process under consideration, a quark and an antiquark are produced in the annihilation of the electron and the positron. Since 'colour' is conserved, the quark and antiquark must be of the same colour. Although the 'colour' in QCD has nothing whatsoever to do with our daily experience of colour, it is nevertheless convenient to designate the three colours by ordinary colours, say the primary colours red, green and blue. Let us say that the quark (full circle) and the antiquark (open circle) are both red when produced, as shown in figure 11.1(a). The idea of quark 'confinement' says that the force between the quark and the antiquark does not diminish with increasing distance.

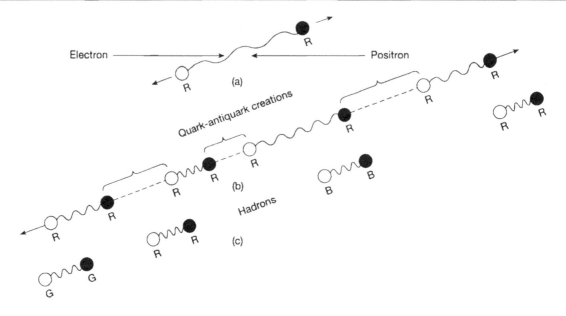

Figure 11.1. Quarks (full circles) and antiquarks (open circles) are permanently confined and come in three colours (red (R), green (G) and blue (B)). (a) After a quark and an antiquark are created in the annihilation of an electron and a positron (anti-electron), (b) more quarks and antiquark pairs are created when they fly apart. (c) The resulting quarks and antiquarks are reassembled to form hadrons, leading to two jets of hadrons in opposite directions.

As the produced quark and antiquark fly apart, the potential energy of the system, similar to a rubber band, increases rapidly and a new quark and a new antiquark are created, as shown in figure 11.1(b). These new quarks and antiquarks combine with those already present to form mesons, as shown in figure 11.1(c). The mesons are colourless, meaning that the colours of the quark and the antiquark forming the meson are the same at any time, but these three colours are equally represented and may change, as shown in figure 11.1(c), where one red pair has turned into a green pair, and another red pair into a blue pair. Note that the mesons so formed may or may not be stable; for example it may be a rho meson, which decays into two pions.

This process of forming mesons, and sometimes protons, may continue, depending on the total energy of the electron–positron annihilation. Thus the process of producing a quark and an antiquark in electron–positron annihilation is observed as two jets of hadrons emerging in opposite directions.

Motivated by this beautiful picture of Feynman and

Bjorken, the jet structure in electron–positron annihilation was looked for and found by the Mark I collaboration (see the chapter by Schwitters) using the colliding accelerator SPEAR at the Stanford Linear Accelerator Center (SLAC). These jets reflect the underlying quark mechanisms.

INDIRECT EVIDENCE FOR THE GLUON

As already mentioned, the strong interactions between quarks are mediated by the gluon, in much the same way as the electromagnetic interactions between the electrons are mediated by the photon. Indirect indications of gluons were first given by deep inelastic electron scattering and neutrino scattering experiments (see the chapter by Friedman and Kendall). The issue here is whether the proton, for example, consists of just the three quarks, or perhaps in addition other particles such as gluons or something else.

Neutrino inelastic scattering using the Gargamelle bubble chamber at CERN showed that only 50% of the momentum of the nucleon is carried by the quarks. Further indirect evidence for gluons was provided by

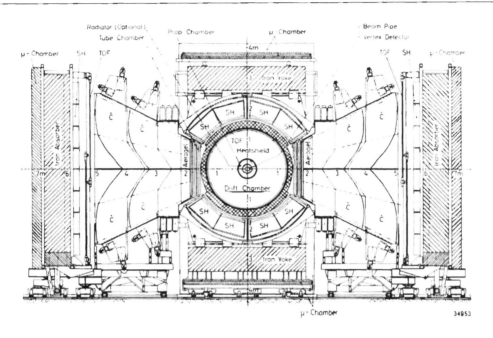

Figure 11.2. The end view of the TASSO detector. The main elements of this detector can be clearly seen: the innermost vertex detector surrounding the beam pipe; the central proportional chamber for tracking, the large drift chamber, and the time-of-flight (TOF) counters; the solenoid magnet with a field of 5 kG; the liquid argon and other shower counters (SH); the two spectrometer arms; the outermost muon chambers.

the observation of detailed behaviour in deep inelastic scattering. The very extensive neutrino scattering data from the BEBC and CDHS collaborations at CERN made it feasible to determine the distributions of quarks and gluons in nucleons compared with what was expected from QCD, and it was found that the gluon distribution function is sizable.

This information about the gluon is interesting but indirect, similar to that for the intermediate boson Z of weak interactions through the measurement of the muon asymmetry in electron–positron annihilation using the Positron–Electron Tandem Ring Accelerator (PETRA) at DESY. Further indirect indication of gluons was given by the PLUTO collaboration using the electron–positron DORIS colliding beam accelerator at DESY. PLUTO established that the decay of upsilon particles cannot be described by a two-jet model with only quarks while remaining consistent with a QCD three-gluon decay but could not establish this three-gluon upsilon decay. This summarizes what we knew about the gluon at the time that

DESY's higher-energy PETRA colliding beam accelerator became operational near the end of 1978.

PETRA AND TASSO

PETRA is located in the German Deutsches Elektronen-Synchrotron (DESY) in a western suburb of Hamburg, Germany. The name DESY refers to the laboratory's first accelerator, a 7 GeV alternating gradient electron synchrotron. The laboratory was established in 1959 under the direction of Willibald Jentschke and has played a crucial role in the re-emergence of Germany as one of the leading countries in physics.

The proposal to construct PETRA was submitted to the German government in November 1974. Under Herwig Schopper, then Director of DESY, approval was granted one year later. Under the leadership of Gus Voss, PETRA construction proceeded very rapidly. The electron beam was first stored on 15 July 1978, more than nine months earlier than originally scheduled. In September 1978, collisions were first observed; a month later, three detectors, PLUTO, MARK J and TASSO (Two-Arm Spectrometer Solenoid) were moved into place. On 18 November 1978, the first hadronic event was observed by PLUTO at a total energy of 13 GeV. The JADE detector was moved in the beam in February 1979. Shortly thereafter, the energy of PETRA increased steadily, eventually reaching 43 GeV. This record electron–positron collision energy was broken only when the SLC accelerator at SLAC became operational in 1989. PETRA was the highest-energy electron–positron machine for more than a decade.

PETRA, housed in a 2.3 km tunnel, is composed of eight straight sections and eight identical curved sections consisting of bending and focusing magnets. Of the eight straight sections, four are long for radio-frequency accelerating cavities, and four are short for experiments: CELLO and PLUTO took turns in Hall NE, JADE in Hall NW, MARK J in Hall SW, and TASSO in Hall SE.

Figure 11.2 shows the end view of the TASSO detector. The main elements of this detector can be clearly seen: the vertex detector; the central tracking detector that consists of the proportional chamber, the large drift chamber, and the time-of-flight (TOF) counters; the solenoid magnet that gives a field of 5 kG; the liquid argon and other shower counters (SH); muon chambers; and the two spectrometer arms.

The excellent performance of the large drift chamber was crucial for the discovery of the gluon. It has a sensitive length of 3.23 m with inner and outer diameters of 0.73 m and 2.44 m respectively. There are in total 15 layers of sense wires, nine parallel to the axis of the chamber and six with the wires oriented at an angle of approximately $\pm 4°$. These six layers make it possible to measure not only the transverse momenta of the produced charged particles but also their longitudinal momenta. This drift chamber was designed and constructed under the direction of Björn Wiik (who became DESY Director in 1993). Figure 11.3 shows him next to the signal cables from his TASSO drift chamber.

GLUON BREMSSTRAHLUNG

In 1977, I moved from MIT to the University of Wisconsin to become a faculty member, and at the same time joined the TASSO collaboration. Besides taking an active part

Figure 11.3. Björn Wiik and the present author relaxing after tedious work on drift-chamber cabling. The cables in the photograph carried the timing information from the preamplifier boxes of the large central TASSO drift chamber and were to be connected to the read-out electronics. Ulrich Kötz of DESY–TASSO took this photograph in 1978.

in the construction of the TASSO Čerenkov counters, I thought a lot about what physics to do. Among the particles that were theoretically expected but not yet observed experimentally, the most interesting included the gluon and the corresponding W and Z for weak interactions. They were especially intriguing because they were new gauge particles. Up to that time, the only known gauge particle was the photon. Furthermore, all three are fundamentally different from the photon in that they have the characteristic self-interactions of non-Abelian gauge particles.

Since the energy available at PETRA was nowhere near high enough to produce the W or the Z, the only possibility at that time was the gluon. One possible effect of the gluon would be the broadening of a quark jet in the two-jet events as the collision energy increases. I felt that the discovery of the gluon required direct observation.

A photon is produced when an electron is shaken, i.e. accelerated. In particular, when an electron and a positron are scattered from each other, sometimes a photon is produced in addition. Such a process is called bremsstrahlung. Similarly, when an electron and a positron are annihilated into a quark and an antiquark as discussed above, sometimes a gluon is produced in addition:

electron + positron → quark + antiquark + gluon.

Such a process is called gluon bremsstrahlung. Gluons, just like quarks, are expected to turn into a jet of hadrons, as pointed out in 1976 by theorists John Ellis, Mary K Gaillard and Graham Ross. Thus gluon bremsstrahlung leads to three-jet events. How could I find such three-jet events?

One of the first worries was whether the energy at PETRA was sufficiently high to produce events where the three jets are clearly separated. At the time of PETRA commissioning in 1978–79, it was widely believed that the initial energy would not be high enough to see three-jet events clearly. Therefore one looked for jet broadening and high transverse momentum phenomena. Figure 11.4 shows a schematic comparison of jet broadening and three-jet events.

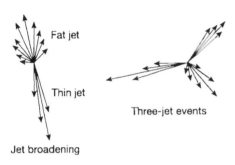

Figure 11.4. Schematic comparison of an event showing jet broadening (left) and a three-jet event (right).

I wanted to know what PETRA energy would be needed to produce clear three-jet events and was encouraged by my rough estimate, carried out early in 1978, that three times the SPEAR energy of 7.4 GeV might be sufficient. The argument was as follows.

Consider the most favourable situation where the three jets make angles of 120 with each other, as shown in figure 11.5. If each pair of jets is taken to have the invariant mass of 7.4 GeV/c^2, as is the case at SPEAR, then the energy of each jet is about 4.3 GeV. The total energy of the three jets is thus 13 GeV, which must be further increased because each jet has to be narrower than the SPEAR jets for them to be clearly separated. This additional factor was estimated to be 180°/120° = 1.5, leading to about 20 GeV. Since this estimate is for the most favourable

configuration where the three jets make angles of 120° with each other, a further increase of perhaps 10% or 20% is called for, resulting in a factor of 3 over the SPEAR energy of 7.4 GeV. Since PETRA was expected to exceed this 3 × 7.4 GeV = 22.2 GeV very soon, I decided to proceed on the assumption that the energy was high enough.

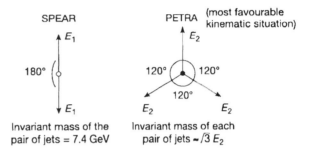

Figure 11.5. Two-jet and three-jet configurations at SPEAR and PETRA respectively.

DISCOVERY OF THE GLUON

Even with my belief that the PETRA energy in 1979 was high enough to produce three-jet events with clearly separated jets, I made a number of false starts until I realized the implications of the following simple observation. Since the total momenta of the quark, the antiquark, and the gluon must add up to zero, the three momenta necessarily lie in one 'event plane'. Therefore the search for the three jets can be carried out in the two-dimensional event plane.

As pointed out above, the large drift chamber of the TASSO detector was capable of measuring all three momentum components of charged particles. Therefore, for each event the vector momenta of the produced charged particles are known. There is a qualitative difference between vectors in three-dimensional spaces and two-dimensional spaces such as the event plane: with a number of vectors in a three-dimensional space, there is no natural way to put them in order; however, vectors in a plane can be naturally ordered in a cyclic way according to their polar angle.

My three-jet analysis therefore began by first defining an approximation to the event plane as the one with the least average perpendicular momenta, and then projecting the measured momenta to this plane. These N projected momenta can be arranged so that

$$0 \leq \theta_1 \leq \theta_2 \leq \theta_3 \leq \cdots \leq \theta_N < 2\pi.$$

With this cyclic ordering, these projected momenta can be split into three groups of contiguous vectors, and these three groups of particles are to be identified as the three jets. There are of course a number of ways of carrying out this splitting, and, with suitable restrictions, the one with the smallest average transverse momentum with respect to the three jet axes is chosen as the best way of identifying the three jets.

This procedure has a number of desirable features. First, all three jets are identified, and the three jet axes are in the same plane. This makes it easy to display any three-jet event, simply by projecting the observed momenta into this event plane. Secondly, particle identification is not needed, since there is no Lorentz transform and hence the particle masses do not enter. Thirdly, the computer time is moderate even for the rather slow computers of that era. Finally, it is not necessary to have the momenta of all the particles produced in the electron–positron annihilation; it is only necessary to have at least one particle from each of the three jets. Thus, for example, the procedure works well even when no neutral particles are included.

This last advantage is important and is the reason why this procedure is well matched to the TASSO detector at the time of the PETRA turn-on in September 1978, when the large drift chamber described above was already working well.

My procedure of identifying the three jets, programmed with the help of my postdoctoral researcher Georg Zobernig, was ready before the PETRA turn-on and published shortly thereafter.

When data were obtained for collision energies of 13 GeV and 17 GeV, Zobernig and I looked for three-jet events. It was not until just before the Neutrino 1979 Conference in Bergen in the late spring of 1979 that data started to come in at the higher collision energy of 27.4 GeV. We found one clear three-jet event from a total of 40 hadronic events at this energy. This event, viewed in the event plane, is shown in figure 11.6. When it was found, Wiik had already left Hamburg to go to Bergen. Therefore, during the

weekend before the Neutrino Conference, I took the display produced by my procedure for the event to Norway to meet Wiik at his house near Bergen. During this weekend, I also phoned Günter Wolf, the TASSO spokesman, at his home in Hamburg and told him of this finding. Wiik showed this event in his plenary talk 'First results from PETRA'. Writing my name next to the event display, he referred questions to me. Donald Perkins took up this offer and wanted to see all 40 TASSO events. I showed him all the events, and, after we had spent some time studying them, he became convinced.

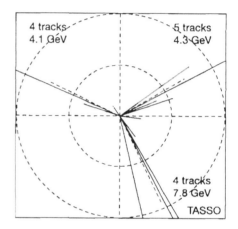

Figure 11.6. This first three-jet event from PETRA was shown during the Bergen Conference, June 1979.

In his talk Wiik said:

'If hard gluon bremsstrahlung is causing the large transverse momentum values in the plane then a small fraction of the events should display a three-jet structure. The events were analyzed for a three jet structure using a method proposed by Wu and Zobernig ... A candidate for a three jet-event, observed by the TASSO group at 27.4 GeV, is shown (viewed in the event plane). Note that the event has (three clear, well separated jets) and is not just a widening of a jet.'

As soon as I returned from Bergen, I wrote a TASSO note with Zobernig, dated 26 June 1979, on this three-jet event. Both in Wiik's talk and in the note, this event was already considered as being due to the hard gluon bremsstrahlung process. This first three-jet event (figure 11.6) with its three clear, well separated jets was considered to be more convincing than a large amount of statistical analysis. Indeed, before the question of statistical fluctuation could be seriously raised, events from the 27.4 GeV run rolled in, and we found more three-jet events. Less than two weeks after the Bergen Conference, four of the TASSO three-jet events were shown by Paul Söding at the European Physical Society (EPS) Conference in Geneva.

CONFIRMATION OF THE GLUON DISCOVERY AT PETRA

With these events, the question was: what are the three jets? Since quarks are fermions, and two fermions (electron and positron) cannot become three fermions, it follows immediately that these three jets cannot all be quarks and antiquarks. Therefore, the presence of three-jet events in electron–positron annihilation implies the discovery of a new particle: a new boson. Similar to quarks and antiquarks, this new particle hadronizes into a jet. Previously known particles, such as the baryons (protons, neutrons, etc) and mesons (pions, kaons, etc) do not metamorphose into jets. For these reasons, most high-energy physicists accepted that the three-jet events are due to hard gluon bremsstrahlung.

With more data, by the end of August the other three experiments at PETRA began to have their own three-jet analyses. At the Lepton–Photon Conference at the Fermi National Accelerator Laboratory (Fermilab) in late August 1979, all four experiments TASSO, MARK J, PLUTO and JADE at PETRA gave more extensive data, confirming the earlier TASSO observation. Fermilab Director Leon Lederman called a press conference on the discovery of the gluon.

SPIN OF THE GLUON

Since QCD is a gauge theory, its gluon must have spin 1. It is nevertheless nice to have this spin of the gluon determined experimentally. The amount of data collected at PETRA up to the time of the 1979 Lepton–Photon Conference at Fermilab was not quite enough for the determination of the spin of the gluon, but the TASSO collaboration carried out this spin determination as soon as there were enough data. As described above, my procedure gives the axes of all

three jets, and they are in the same plane. This ability to isolate individual jets in high-energy three-jet events makes it possible to measure the correlation between the directions of the three jets. Since the spin of the quark is known to be $\frac{1}{2}$ and this correlation depends on the gluon spin, it can be used to determine the spin of the gluon.

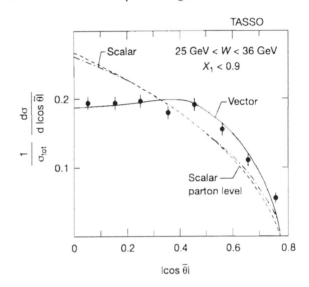

Figure 11.7. Observed distribution of the TASSO data as function of the Ellis–Karliner angle. All curves are normalized to the number of observed events. The data show that the gluon is a vector particle (spin 1).

Without going into details, the TASSO result is shown in figure 11.7. The $\bar{\theta}$ in the horizontal axis is the angle proposed by Ellis and Karliner to discriminate between vector (spin-1) and scalar (spin-0) gluons. The figure clearly shows that the spin of the gluon is indeed 1. With this determination of the spin of the gluon, QCD was put on a firm experimental footing.

QUANTUM CHROMODYNAMICS EFFECTS AT HIGHER ENERGIES

During the two decades since the experimental observation of the gluon, a number of higher-energy accelerators have become available. Among electron–positron colliders, the energy has been increased nearly seven times, from 27.4 GeV at PETRA for the discovery of the gluon to

189 GeV at the CERN Large Electron–Positron Collider (LEP). At present, the highest-energy accelerator is the Tevatron proton–antiproton collider at the Fermi National Accelerator, where the maximum energy is 1800 GeV.

With this increase in the available energy, QCD effects have been observed in numerous processes. Already at PETRA, the dimensionless coupling constant α_s of QCD, corresponding to the fine-structure constant α of QED, was measured and found to be about 0.1. While this is 15 times larger than α, it is still fairly small so that perturbation methods can be applied in QCD. At higher energies, α_s has been found to decrease, giving experimental confirmation to the idea of asymptotic freedom, proposed in 1973 by theorists D Gross, G 't Hooft, H Politzer and F Wilczek.

The development of perturbative QCD in recent years has been very impressive. Two examples will be given here. Since the gluon was discovered at PETRA, an electron–positron colliding accelerator, these two examples will be taken from proton–antiproton and electron–proton colliding accelerators. These high-energy machines have increased greatly the resolution power of unravelling the nature of the proton and hence look much deeper into its structure.

The first example is from the Fermilab Tevatron collider at 1800 GeV. Figure 11.8 shows a direct comparison

between the QCD prediction of S Ellis, Z Kunszt and D Soper with the data obtained by the Central Detector Facility (CDF) collaboration. As explained in the figure, the points are the CDF data while the curve is the theoretical prediction. Unlike the situation with electron–positron colliders such as PETRA, in proton–antiproton colliders most of the hadron jets are produced near the directions of the incident beams. The number of hadron jets decreases rapidly as a function of the 'jet transverse energy' (the horizontal axis of figure 11.8), which is the component of the energy of the hadron jet along the direction of the incident beams.

Up to a jet transverse energy of 200 GeV, the very accurate data agree completely with the theoretical prediction, over an impressive seven orders of magnitude. This result has also been obtained by the D0 collaboration at Tevatron. For large transverse energies, the CDF data are tantalizingly above the QCD prediction. It is not yet known whether this slight disagreement is real or is merely a fluctuation.

The second example is from the electron–proton colliding accelerator Hadron Electron Ring Accelerator (HERA) at DESY. HERA is a new type of accelerator; it is the first high-energy colliding beam accelerator which collides particles of different species. Specifically, a 30 GeV/c electron (or positron) beam collides with an 820 GeV/c proton beam, giving a collision energy of 314 GeV. A surprising result is the inflation of soft quarks and gluons in the proton. Figure 11.9 shows data from the H1 collaboration, where the quantity F_2, one of the functions that describes the structure of the proton, is plotted versus the variable x, with the momentum transfer from the electron fixed at 5 GeV/c. x is essentially the fraction of the target proton's momentum carried by the struck quark. $F_2(x)$ describes the density of quarks with this momentum fraction x. Experimentally the region of small x is most difficult to reach and requires very-high-energy electron–proton collisions.

Before HERA, the lowest value of x reached was about 0.01; by contrast, in figure 11.9 showing HERA data, x has attained a value of about 0.0003, a factor of 30 lower. The most important feature is the dramatic increase in the density of low-energy quarks. The curve represents the QCD fit. Furthermore, an analysis of the change in F_2 with

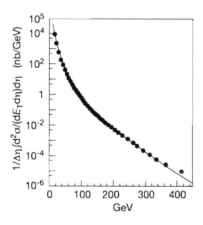

Figure 11.8. Direct comparison of data (full circles) from the CDF experiment at the Fermilab Tevatron proton–antiproton collider with the QCD prediction (full curve). The normalization is absolute. The production of hadron jets decreases rapidly as the jet transverse energy increases.

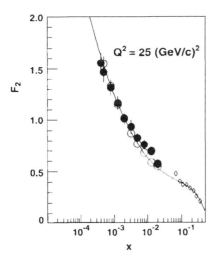

$Q^2 = 25 \ (GeV/c)^2$

Figure 11.9. Measurement of the F_2 structure function (describing the quark density within the proton) by the H1 experiment at the HERA electron–proton collider at DESY. The lozenges at higher x values come from an earlier experiment (BCDMS at CERN). The full curve represents the QCD fit.

the value of the momentum transfer from the electron or positron has shown that the density of gluons in the proton also increases as x becomes smaller. Similar results have also been obtained by the ZEUS collaboration at HERA.

CONCLUSION

The clear experimental observation of three-jet events in electron–positron annihilation, first accomplished by the TASSO Collaboration in June 1979 and confirmed by the other collaborations at PETRA two months later, implies the discovery of a new particle, a boson. Similar to the quarks, this new particle turns into a jet of hadrons. These three-jet events are most naturally explained by hard non-collinear gluon bremsstrahlung. One year later, the spin of the gluon was determined experimentally to be indeed 1, as befits a gauge boson. Since then, with the advent of higher-energy machines in recent years, agreement between experimental data and QCD predictions is impressive.

The 1979 discovery of the second gauge particle, the gluon, occurred three quarters of a century after the prediction by Einstein of the first, the photon. This second gauge particle is also the first experimentally observed

Yang–Mills non-Abelian gauge particle, one with self-interaction. Four years later, in 1983, the second and third non-Abelian gauge particles, the W and the Z of the electroweak interaction, were discovered at the CERN proton–antiproton collider (see the chapter by Rubbia).

The five years from 1979 to 1983 were a most exciting period for particle physics: the discovery of the carrier particles for both strong and weak interactions (gluons and W, Z respectively). With these discoveries, physicists can claim to have a good understanding of the dynamics of quarks.

ACKNOWLEDGMENT

This work was supported by the US Department of Energy contract DE-AC02-76ER00881.

FURTHER READING

Riordan M 1987 *The Hunting of the Quark* (New York: Simon and Schuster)

Hoddeson L, Brown L, Riordan M and Dresden M 1997 *The Rise of the Standard Model* (Cambridge: Cambridge University Press)

ABOUT THE AUTHOR

Born in Hong Kong, Sau Lan Wu graduated *summa cum laude* from Vassar College and received her PhD from Harvard in 1970. She went on to a research associateship at MIT and was a member of Samuel Ting's team that discovered the J/ψ particle at the Brookhaven National Laboratory in 1974. In 1977 she became a faculty member at the University of Wisconsin, Madison, at the same time joining the TASSO collaboration for PETRA at DESY, Hamburg, where her three-jet analysis led to the discovery of the gluon in 1979. For this discovery, she shared the 1995 European Physical Society High Energy and Particle Physics Prize with TASSO colleagues Paul Söding, Björn Wiik and Günter Wolf. In 1980, while continuing her work with TASSO, she joined the ALEPH collaboration at CERN to build a detector for LEP, which has been operational since 1989.

Sau Lan Wu and her group have taken part in many of the new experimental results of LEP, such as the determination of the number of light-neutrino species, and were especially prominent in several major findings of

ALEPH, including the discovery of two new particles, the strange bottom meson B_s and the bottom baryon Λ_b.

In 1990 Sau Lan Wu became the Enrico Fermi Distinguished Professor of Physics at the University of Wisconsin. She is a member of the American Academy of Arts and Sciences and a fellow of the American Physical Society. She received the Outstanding Junior Investigator Award from the US Department of Energy in 1980. Her committee work has included serving as a general councillor of the American Physical Society and as a member of the High Energy Physics Advisory Panel of the Department of Energy. At present she continues to search at LEP for the elusive Higgs boson and for possible new phenomena beyond the Standard Model. For the future, she and her group are also members of the BaBar Collaboration at SLAC's B-factory and the ATLAS collaboration at CERN's LHC.

12 HOW THE STANDARD MODEL WORKS

Guido Altarelli

Editor's Introduction: The Standard Model (SM) is the culmination of a hundred years of patient experimental work, paralleled by theoretical developments which have radically changed our understanding of how Nature operates on the microscopic scale. The SM describes all physical processes in terms of strongly and weakly interacting quarks and weakly interacting leptons, with six types of quark and six leptons grouped pairwise into three generations, successive copies of each other on higher rungs of the mass ladder. Why there are three, and only three, such generations remains a mystery. The discovery of the W and Z carriers of the weak force in 1983 (see the chapter by Rubbia) opened up the electroweak sector of the SM. After the initial discoveries, this new physics had to be accurately charted, and precision determinations of electroweak effects have been one of the main features of particle physics at the close of the twentieth century.

The quark sector of the SM is more intransigent. Quarks do not exist as free particles, and remain firmly locked inside their hadronic prisons. Extracting quark information from hadronic effects requires considerable experimental and theoretical ingenuity. Nevertheless, the precision SM data now available provides a coherent picture of the basic processes of physics. Precision measurements of the Standard Model, as summarized in this chapter, are the cutting edge of physics in the 1990s. The only missing link is the Higgs sector, the spontaneous symmetry breaking mechanism which underpins the electroweak sector of the SM. Unravelling this mechanism remains the major physics goal for the twenty-first century.

THE STANDARD MODEL ARCHITECTURE

The Standard Model (SM) is the ultimate result of a long period of progress in elementary-particle physics. It is a consistent, finite and, within the limitations of our present technical ability, computable theory of fundamental microscopic interactions that successfully explains all known phenomena in elementary-particle physics, describing strong, electromagnetic and weak interactions. All so far observed microscopic phenomena can be attributed to one or the other of these interactions. For example, the forces that hold together the protons and the neutrons in the atomic nuclei are due to strong interactions, the binding of electrons to nuclei in atoms or of atoms in molecules is caused by electromagnetism, and the energy production in the Sun and the other stars occurs through nuclear reactions induced by weak interactions. In principle, gravitational forces should also be included in the list of fundamental interactions but their impact on fundamental particle processes at accessible energies is totally negligible.

The structure of the SM is a generalization of that of quantum electrodynamics (QED) in the sense that it is a renormalizable field theory based on a local symmetry (i.e. separately valid at each space–time point) that extends the gauge invariance of electrodynamics to a larger set of conserved currents and charges. There are eight strong charges, called 'colour' charges and, independently, four electroweak charges (which in particular include the electric charge). The colour local symmetry gives rise to the strong interaction theory which is called quantum chromodynamics (QCD). QCD and the electroweak theory are separately generated by their respective charges but are related in that the quarks possess both colour and electroweak charges, so that they interact with all the corresponding interactions.

In QED the interaction between two matter particles

Figure 12.1. The interaction between two electric charges in QED is propagated by the exchange of one (or more) photons.

Table 12.1. The known leptons and quarks and their gauge quantum numbers. The masses of the confined quarks are only indicative.

	Mass (GeV/c^2)	Electric charge	Weak isospin, I_z	Colour
Leptons				
ν_e	$< 3 \times 10^{-9}$	0	$+\frac{1}{2}$	—
e	5.11×10^{-4}	-1	$-\frac{1}{2}$	—
ν_μ	$< 2 \times 10^{-4}$	0	$+\frac{1}{2}$	—
μ	0.106	-1	$-\frac{1}{2}$	—
ν_τ	< 0.02	0	$+\frac{1}{2}$	—
τ	1.777	-1	$-\frac{1}{2}$	—
Quarks				
u	5×10^{-3}	$\frac{2}{3}$	$+\frac{1}{2}$	3
d	10×10^{-3}	$-\frac{1}{3}$	$-\frac{1}{2}$	3
c	1.5	$\frac{2}{3}$	$+\frac{1}{2}$	3
s	0.2	$-\frac{1}{3}$	$-\frac{1}{2}$	3
t	174	$\frac{2}{3}$	$+\frac{1}{2}$	3
b	4.7	$-\frac{1}{3}$	$-\frac{1}{2}$	3

with electric charges, e.g. two electrons, is mediated by the exchange of one (or more) photons (figure 12.1) emitted by one electron and reabsorbed by the second. In the SM the matter fields, all of spin $\frac{1}{2}$, are the quarks, the constituents of protons, neutrons and all hadrons, endowed with both colour and electroweak charges, and the leptons (the electron e^-, the muon μ^-, the tauon τ^- plus the three associated neutrinos ν_e, ν_μ and ν_τ) with no colour but with electroweak charges. The matter fermions come in three generations, or families, with identical quantum numbers but different masses. Each family contains a weakly charged doublet of quarks, in three colour replicas, and a colourless weakly charged doublet with a neutrino and a charged lepton (table 12.1). At present there is no explanation for this triple repetition of fermion families. The force carriers, of spin 1, are the photon γ, the weak interaction gauge bosons W^+, W^- and Z_0 and the eight gluons g that mediate the strong interactions. The photon and the gluons have zero masses (i.e. they move at the speed of light in all inertial frames) as a consequence of the exact conservation of the corresponding symmetry generators, the electric charge and the eight colour charges. The weak bosons W^+, W^- and Z_0 have large masses ($m_W \simeq 80.4$ GeV, $m_Z \sim 91.2$ GeV) signalling that the corresponding symmetries are badly broken. In the SM the spontaneous breaking of the electroweak gauge symmetry is induced by the Higgs mechanism which predicts the presence of one (or more) spin-0 particles, the Higgs boson(s), not yet experimentally observed. A tremendous experimental effort is under way or planned to reveal the Higgs sector as the last crucial missing link in the SM verification.

We now look in more detail at the basic SM interactions. In QED the interaction vertices describe the quantum-mechanical amplitudes for the emission or absorption of a photon by any electrically charged matter field (figure 12.2(a)): either a quark (which have charges $\frac{2}{3}$ and $-\frac{1}{3}$), or a charged lepton (with charges -1), or their antiparticles (the antiquarks and the antileptons with opposite charges with respect to the corresponding particles). The gauge symmetry implies that the photon is coupled to all charged particles with an amplitude proportional to the particle's electric charge. The photon, being neutral, is not coupled to itself: in the absence of charges the electromagnetic field propagates freely.

QCD, the theory of quark and gluon interactions based on colour SU(3), was put in its final form in 1972 mainly by H Fritzsch, M Gell-Mann and H Leutwyler, D J Gross and F Wilczek, and H D Politzer. In QCD the vertices where a gluon is emitted or absorbed by a quark are quite analogous to the QED vertices where the photon is replaced by one of the gluons (figure 12.2(b)). Neither the photon nor the gluons change the particle flavour; a u quark remains a u quark after emission of a photon or a gluon (the gluons are not coupled to leptons because the latter have no colour

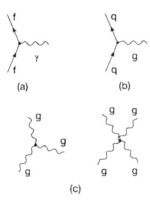

Figure 12.2. (a) The QED interaction vertex involving a charged fermion f (e.g. an electron) and the photon. (b) The QCD interaction vertices involving quarks and gluons. (c) The QCD interaction vertices involving gluons only.

charges). The only difference is that the photon does not affect the colour while each gluon does change the quark colour. This is because the gluons are themselves coloured. As a consequence, in QCD there are additional vertices with no analogue in QED. The gluons interact with all coloured particles but, being themselves coloured, they also interact among each other; there are vertices where a gluon emits a different gluon and changes its own colour (figure 12.2(c)). Thus, even in the absence of quarks, QCD is an interacting theory. The couplings at quark and gluon vertices are the same: they are all proportional to the strong coupling g_s. There is also a vertex with four gluons with a coupling proportional to g_s^2. This is because the gluons have spin 1 so that they are bosons and not fermions like quarks and leptons. (In QED with bosonic matter fields there would also be vertices with two photons and two matter fields at order e_B^2 where e_B is the electric charge of the boson.)

The weak interactions introduce several new basic features. First of all, while the photon and gluon vertices conserve parity (P) and charge conjugation (C), this is not the case in weak interactions. The P and C violation is introduced in the standard electroweak theory of S L Glashow, A Salam, and S Weinberg (1968) because left-handed and right-handed fermion fields are endowed with different electroweak charges. Left-handed fermion fields are doublets (they describe fermions with negative

helicity, the spin component along the direction of flight, or antifermions with positive helicity), while right-handed fermions are singlets (positive helicity for fermions and negative helicity for antifermions). In the symmetric limit such chiral fermions cannot have a rest mass. In fact in the SM all fermions, together with the weak gauge bosons, will get a mass only when the symmetry is broken by the Higgs mechanism. P transforms right-handed into left-handed fields and C transforms fermions into antifermions with the same helicity and are both broken by the different transformations of the right-handed and left-handed fields. P and C are conserved in the photon interactions. In fact γ and Z_0, the states of definite mass after symmetry breaking, are orthogonal superpositions of W_0, the neutral component of the weak isospin triplet W^+, W^- and W_0, and the singlet gauge boson B. As conservation of electric charge is preserved, the photon remains massless after symmetry breaking with a mixing angle θ_W ensuring that the photon couples to a purely vector current in proportion to the electric charge of each fermion:

$$g_W \sin \theta_W = g'_W \cos \theta_W = e \qquad (12.1)$$

where g_W and g'_W are the W^1 and B gauge couplings respectively and e is the positron electric charge. Thus the electromagnetic and colour currents are pure vector currents, while the weak currents are a linear combination of vector and axial vector currents with opposite P and C properties. For charged currents, those coupled to W^+ or W^-, the combination is simply vector minus axial, with a single coupling constant g_W. This is because only left-handed fields have non-trivial transformation properties under the corresponding charges. The combination is more complicated for the neutral current coupled to the Z_0. The axial coupling is specified by $\pm g_W/(4 \cos \theta_W)$ for all quarks and leptons, while the ratio g_V/g_A of the vector to axial vector couplings is different for the quarks and leptons within each family. In general, as a consequence of mixing, $g_V/g_A = 1 - 4|Q|\sin^2 \theta_W$ where Q is the electric charge ($\frac{2}{3}$, $-\frac{1}{3}$, -1 and 0 for up quarks, down quarks, charged leptons and neutrinos respectively).

Another important property of weak interactions is that the charged currents coupled to W^+ and W^- are the only observed interactions that change the flavour of fermions.

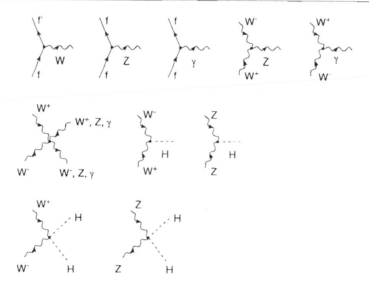

Figure 12.3. The interaction vertices of the electroweak theory involving fermions, gauge bosons (W^+, W^-, Z and γ) and the Higgs boson.

By emitting a W^+ an up-type quark becomes a down-type quark, a neutrino becomes an electron, etc. The weak vertices are shown in figure 12.3. Note that, since the W^+ and W^- carry electroweak charges, they are coupled to γ and Z_0 in triple gauge vertices of order g_w, and quartic gauge vertices proportional to g_w^2.

The very interesting phenomenon of quark mixing is a consequence of the charged weak currents' ability to switch flavour. Consider the quark u that emits a W^+ and becomes a down-type quark. Since the three down-type quarks d, s and b have different masses, in general the down-type quark d' produced in the transition $u \rightarrow W^+ d'$ will not coincide with precisely one state of definite mass but will be a well defined superposition of all three of them. Thus the states with definite charged current couplings do not coincide with those of definite mass. Similarly the c quark is turned into a combination s' and the t quark into b'. The states $q' \equiv (d', s', b')$ are mutually orthogonal and are connected by a unitary transformation to $q \equiv (d, s, b)$: $q' = V_{CKM}q$. The 3×3 unitary matrix V_{CKM} is the Cabibbo–Kobayashi–Maskawa mixing matrix, determined by three mixing angles and one phase. The latter is the unique source of CP violation in the SM. The physics of mixing and CP violation is described in the chapter by Ali and Kayser. Note that if

all neutrinos are massless leptons do not mix. Thus the mass matrix has no preference and we simply call, say, the electron-neutrino ν_e the one which is connected to the electron by the weak transition $e^- \rightarrow W^- \nu_e$. Contrary to the charged currents, the neutral weak current does not change flavour. This is a consequence of the unitarity of the V_{CKM} matrix (this requirement led S L Glashow, J Iliopoulos and L Maiani to predict the existence of a fourth quark, charm, in 1970, when only the u, d and s quarks were known; see the chapter by Schwitters).

We now turn to the Higgs sector of the electroweak theory. As explained in the chapter by Ross the breaking of the electroweak symmetry arises because the Higgs field is non-zero in the vacuum state (like the magnetization of a permanent magnet in the absence of an external magnetic field). One starts from the most general renormalizable symmetric field theory written in terms of the Higgs field with a non-zero value v in the vacuum state. Then one shifts the Higgs field by v in such a way that the new field is zero in the vacuum. It is this shift that makes mass terms appear for fermions and for gauge bosons other than the photon. The photon automatically stays massless if the Higgs component with non-zero value in the vacuum state is electrically neutral. For example, from a trilinear coupling

$g_Y H \bar{\psi} \psi$ (g_Y is a coupling constant), after the Higgs field redefinition, the mass term $g_Y v \bar{\psi} \psi$ is generated for the fermion ψ which corresponds to a mass $m = g_Y v$. Mass terms for fermions involve one left-handed and one right-handed field. Since the left-handed fermions are doublets and the right-handed fermions are singlets, only a doublet Higgs can generate masses for fermions. Instead every non-trivial representation gives mass to the weak gauge bosons (because the gauge particles are coupled to all particles with a non-vanishing corresponding charge). Thus the most economic choice is to assume the existence of one (or more) doublet Higgs field(s). Then, for every fermion, one obtains a mass $m = g_Y v$ but one needs a specific g_Y coupling and a specific interaction vertex for each mass, which admittedly is not very elegant. On the contrary the W and Z masses are specified by the gauge coupling g_W. Precisely $m_W^2 = \frac{1}{2} g_W^2 v^2$ and $m_Z^2 = \frac{1}{2} g_W^2 v^2 \cos^2 \theta_W$. In particular for doublet Higgs field(s) the following important relation holds:

$$m_W^2 = \rho m_Z^2 \cos^2 \theta_W \qquad (12.2)$$

with $\rho = 1$ (apart from small radiative corrections). Precision experiments show that the ρ parameter is close to 1. This is a strong evidence that the Higgs field (either fundamental or effective) is an isospin doublet.

A complex Higgs doublet has four degrees of freedom. Three of these become the longitudinally polarized components of W^+, W^- and Z_0 that become massive (a massless spin-1 particle has two degrees of freedom while a massive spin-1 particle has three degrees of freedom). The remaining degree of freedom corresponds to the physical neutral Higgs particle of spin 0 which remains to be found. Its mass is not predicted by the theory. Experimentally we know from LEP that $m_H > 85$ GeV. The Higgs couplings to fermions are proportional to masses, because we have seen that $g_Y = m/v$. The Higgs particle is also coupled to $W^+ W^-$ and $Z_0 Z_0$ as shown in detail in figure 12.3.

The gauge theories with spontaneous symmetry breaking were proven to be renormalizable in the early 1970s by G 't Hooft and M Veltman, apart from possible chiral anomalies, but the electric charges, the colour and the weak isospin of the fundamental leptons and quarks

in one family are such that these chiral anomalies exactly cancel as pointed out by C Bouchiat, J Iliopoulos and Ph Meyer in 1972. These crucial theoretical breakthroughs were followed in 1973–74 by the discovery of the weak processes induced by the neutral current at CERN and of charm at Brookhaven and Stanford Linear Accelerator Center (SLAC). These results firmly established the SM as the framework of particle physics.

EXPERIMENTAL TESTS OF QUANTUM CHROMODYNAMICS

The QCD theory is rather simple in its basic ingredients. It is the theory of quark and gluon interactions based on the colour unbroken gauge symmetry. While the theory is relatively simple in its formulation, it possesses a surprisingly rich physical content, including the vast menagerie of hadron spectroscopy. The distinguishing features of QCD are 'asymptotic freedom' and 'confinement'.

In field theory the effective coupling of a given interaction vertex is modified by the interaction. As a result the measured intensity of the force depends on the transferred (four-)momentum squared Q^2 among the participants. In QCD the relevant coupling parameter that appears in physical processes is g_s^2. Defining $\alpha_s = g_s^2/4\pi$ (in natural units $\hbar = c = 1$, always used throughout this chapter), the effective coupling is a function of Q^2: $\alpha_s(Q^2)$. Asymptotic freedom means that the effective coupling $\alpha_s(Q^2)$ decreases for increasing Q^2 and vanishes asymptotically. That is, the QCD interaction becomes very weak in processes with large Q^2, called 'hard' or 'deep inelastic' processes (i.e. with a final state distribution of momenta and a particle content very different than those in the initial state). One can prove that in four space–time dimensions all (and only) gauge theories based on a non-commuting symmetry group are asymptotically free. The effective coupling decreases very slowly at large momenta with the inverse logarithm of Q^2: $\alpha_s(Q^2) = 1/b \log(Q^2/\Lambda^2)$ where b is a known constant and Λ is an energy of the order of a few hundred MeV.

Since in quantum mechanics large momenta imply short wavelengths, the result is that at short distances the potential between two colour charges is similar to the electromagnetic Coulomb potential, i.e. proportional to $\alpha_s(r)/r$, with an effective colour charge which is small at short distances. Conversely the interaction strength becomes large at large distances or small transferred momenta, of order $Q \lesssim \Lambda$. In fact the observed hadrons are tightly bound composite states of quarks, with compensating colour charges so that the hadrons are overall neutral in colour.

The property of confinement is the impossibility of separating colour charges, such as individual quarks and gluons. This is because in QCD the interaction between colour charges increases at long distances linearly in r. When we try to separate the quark and the antiquark that form a colour-neutral meson, the interaction energy grows until pairs of quarks and antiquarks are created from the vacuum, giving new neutral mesons. For example, consider the process $e^+e^- \rightarrow q\bar{q}$ at high energies. The final-state quark and antiquark have large energies, and so they separate in opposite directions very fast, but the colour confinement forces create new quark pairs in between. What is actually observed is two back-to-back fast-moving 'jets' of colourless hadrons, accompanied by a central cloud of slow pions that make the exact separation of the two jets impossible. In some cases a third well separated jet of hadrons is also observed; these events correspond to the radiation of a high-energy gluon from the parent quark–antiquark pair (see the chapter by Wu).

In spite of a rich list of QCD phenomena, its experimental verification is difficult because the available methods of solution are not very accurate. Most of the observed phenomena are non-perturbative because the QCD interactions are 'strong'. The most fundamental approach, based on discretization of the theory on a fictitious but convenient space–time lattice, suffers from computer limitations on small lattice sizes and force the use of drastic approximations (e.g. most applications are in the so-called 'quenched' limit where contributions from quark virtual pairs are neglected). In spite of these limitations the simulation of QCD on such lattices has led to important results both of conceptual interest and of practical importance. For example lattice computations have given support to the idea of QCD confinement by proving the linear behaviour at long distances of the $q\bar{q}$ potential and the existence of a deconfinement phase transition. Also

lattice calculation of hadron mass ratios and of hadronic matrix elements have led to encouraging results. With the rapid growth of computer power, this lattice approach is continuously improving.

At present the main connection to experiment is perturbative QCD, founded on the basic property of asymptotic freedom. When all momentum transfers are large, the effective coupling is small and a perturbative approach to hard processes is possible. Rates and distributions for many such processes can be computed and compared with experiment. As a particularly important consequence the perturbative expansion parameter, i.e. the running coupling $\alpha_s(Q^2)$, within a precisely specified definition, can be measured. Q being of the order of the typical energy scale of the hard process selected for the experiment. The set of measurements of $\alpha_s(Q^2)$ presently available provides in itself a quite remarkable quantitative test of the theory.

Since the dependence of $\alpha_s(Q^2)$ on Q^2 is well known in perturbation theory, measurements at different values of Q can be translated into the corresponding value of $\alpha_s(Q^2)$ at $Q = m_Z$ for an easy comparison. A list of the most reliable determinations of $\alpha_s(m_Z^2)$ is shown in table 12.2. Alternatively, for each experiment, we can plot the resulting value of $\alpha_s(Q^2)$ at the scale Q which is typical of the experiment (figure 12.4). This shows the span of energy where the theory has been tested and gives a clear perception of the decreasing trend of $\alpha_s(Q^2)$ at increasing Q^2. Also visible in figure 12.4 is the agreement with the predicted QCD behaviour for the best-fit value of $\alpha_s(m_Z^2)$.

Among the processes listed in table 12.2 that have been used to measure $\alpha_s(m_Z^2)$, a few are particularly important and dominate the average. Starting from smaller values of Q^2, the first of these processes is the hadronic disintegration of the τ lepton. One obtains the value of $\alpha_s(Q^2)$ at $Q = m_\tau$ from the measured ratio of the hadronic yield to that into an electron and neutrinos. As the τ mass is relatively small, $m_\tau = 1.777$ GeV, this method is especially vulnerable to theoretical uncertainties. The resulting value quoted in table 12.2, translated in terms of $\alpha_s(m_Z^2)$, is given by $\alpha_s(m_Z^2) = 0.122 \pm 0.003$. The error quoted there could be more conservatively increased to ± 0.005.

The study of deep inelastic lepto-production is another main source of information on α_s. The general process

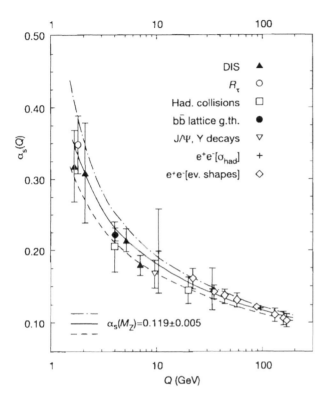

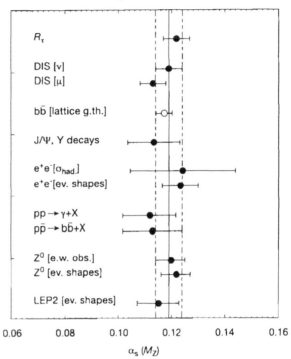

Table 12.2. The measured values of the effective strong interaction coupling, all evaluated at the Z mass $\alpha_s(m_Z)$ as obtained from different experiments.

Figure 12.4. The measured values of the effective strong interaction coupling α_s are plotted against the momentum transfer Q of the process used for the measurement. One clearly sees the decrease in α_s with increasing Q which corresponds to the property of asymptotic freedom. The curves are the predicted running for the indicated value of $\alpha_s(m_Z)$ where m_Z is the mass of the weak neutral gauge boson.

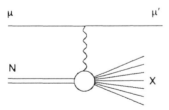

Figure 12.5. The process $\mu + N \rightarrow \mu' + X$ where N is a nucleon or a nucleus and X represents an arbitrary set of particles.

is $\ell + N \rightarrow \ell' +$ anything, where ℓ can be an electron or a muon or a neutrino, ℓ' is some corresponding final lepton and N is a nucleon or a nucleus. For example, if both ℓ and ℓ' are muons and N is a proton, the process is depicted in figure 12.5. The 'deep inelastic' specification means that the transferred momentum squared, i.e. $-Q^2$, between the initial and the final muon is large. This corresponds to a large virtual mass of the exchanged photon and consequently to a large total mass of the final hadronic system, which in general will contain many high-energy particles. The virtual photon acts as a microscope that probes deep inside the target and sees the quarks and the gluons that are inside the proton. From the study of deep inelastic scattering one can measure the 'parton' densities, i.e. the quark and the gluon densities inside the proton $q_a(x, Q^2)$ and $g(x, Q^2)$. The variable x, introduced by Bjorken, is the fraction of the proton longitudinal momentum carried by the parton. There is

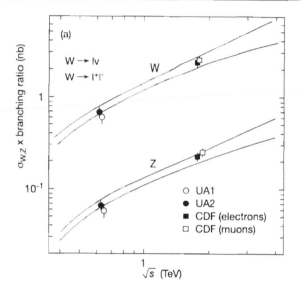

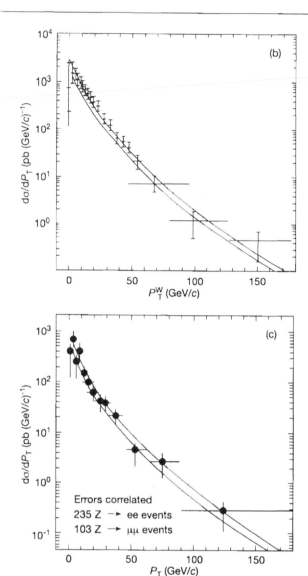

Figure 12.6. (a) Predicted (from QCD) and measured production cross sections (times the branching ratio into $e\nu$ or e^+e^-) for the weak gauge bosons W and Z as measured at proton–antiproton colliders. (b), (c) Predicted (from QCD) and measured distributions in transverse momentum p_T (transverse with respect to the direction of the colliding beams). In both cases the bandwidth indicates the theoretical uncertainty.

a q_a for each type (or 'flavour') of quarks and antiquarks: $q_a = u, d, s, \ldots, \bar{u}, \bar{d}, \bar{s}, \ldots$. The parton densities are the number per unit x of partons seen by the virtual photon with squared mass $-Q^2$ inside the proton. While the initial values of the parton densities at some given Q_0^2 cannot be computed in perturbation theory, the rate of change in Q^2 can be obtained from perturbative QCD by a set of evolution equations written down by V N Gribov and L N Lipatov, and G Altarelli and G Parisi in the 1970s. Thus the departures from flatness in Q^2 of the parton densities can give $\alpha_s(Q^2)$ at an average value of Q. The value reported in table 12.2 and in figure 12.4 is obtained from combining results with muon and neutrino beams. In terms of $\alpha_s(m_Z^2)$, the resulting value is given by $\alpha_s(m_Z^2) = 0.116 \pm 0.005$.

Precise determinations of α_s have been recently obtained by experiments on e^+e^- annihilation at the Z_0 resonance mainly performed at LEP. One method, by far the most reliable and precise, is from inclusive hadronic decays of the Z_0. In particular one measures the ratio $R_h = \Gamma_h/\Gamma_l$

of the hadronic versus the leptonic partial widths of the Z_0 (Γ_l is the average width for $l = e, \mu, \tau$). Assuming the validity of the standard electroweak theory, one can fit the QCD correction that depends on $\alpha_s(m_Z^2)$ demanded by the experimental value. By combining information from all hadronic production by the Z_0, one obtains $\alpha_s(m_Z^2) = 0.121 \pm 0.003$, which is the value shown in table 12.2 and figure 12.4.

An additional important input is derived from the study of the jets from hadronic Z_0 decays. For suitably chosen quantities, the jet variables are computable in perturbation theory at the parton level and depend on $\alpha_s(m_Z^2)$. However, the observed jets are made up of ordinary hadrons and not quarks and gluons. Thus, in order to extract α_s, one has in some way to model the non-perturbative effects due to the transition from partons to hadrons. These effects rapidly decrease with a power of $1/Q$ or $1/Q^2$; so they are quite small for $Q = m_Z$ but still large enough that the related ambiguities are a major source of theoretical error on α_s. The result of this measurement of α_s is given by $\alpha_s(m_Z^2) = 0.121 \pm 0.005$ and is given in table 12.2 and figure 12.4. The above four measurements of $\alpha_s(m_Z^2)$ dominate the world average.

The impressive agreement of so many very different determinations of α_s is a very solid quantitative check of QCD. Yet there are many more; the successful prediction of many cross sections and distributions for hard processes in the framework of the so-called 'QCD-improved' parton model. These predictions are obtained starting from measured quark and gluon densities in the nucleon and the value of α_s at the relevant scale, then computing the perturbative cross sections for quark- and/or gluon-initiated processes into the given final state and finally suitably combining parton densities and cross sections to obtain the desired result. Among the processes which provide the most impressive set of QCD successes are several reactions observed at $p\bar{p}$ colliders: lepton pair production (Drell–Yan processes); W and Z production; heavy-quark pair production; distributions of jets at large transverse momenta p_T. For example W and Z production cross sections measured at collision energies of 0.64 TeV (CERN) and 1.8 TeV (Fermilab) and their transverse momentum distributions are in very good agreement with the predictions (figure 12.6). Similarly the production rate of large transverse momentum jets agrees with QCD both in energy and p_T over very wide ranges (figure 12.7).

EXPERIMENTAL TESTS OF THE STANDARD ELECTROWEAK THEORY

At low energies one cannot detect that the interaction is propagated by the heavy W, so that the two currents act at a space separation of about 10^{-16} cm. The effective interaction appears as a local interaction of the two currents,

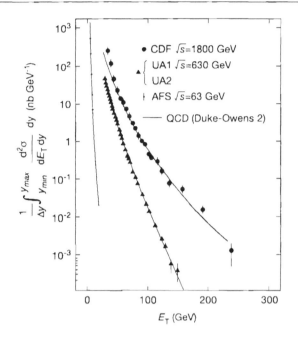

Figure 12.7. Data on the production of one jet with large transverse energy E_T (transverse with respect to the direction of the colliding beams) in proton–antiproton colliders compared with the QCD predictions. This process shows a very rapid energy and E_T dependence, so that it represents a very constraining test for QCD.

each with two fermion fields, in the same point (figure 12.8). The coupling constant G_F of this effective four-fermion theory, called the Fermi constant to recall his pioneer beta decay theory of 1934, is related to the SU(2) gauge coupling g_w and the W mass by the relation

$$\frac{G_F}{\sqrt{2}} = \frac{g_w^2}{8m_W^2} \qquad (12.3)$$

apart from small radiative corrections. The Fermi coupling constant is measured with great precision in muon decay, $\mu^- \to e^- \bar{\nu}_e \nu_\mu$, and is used as an input for the SM tests.

A central role in the experimental verification of the standard electroweak theory has been, and is still, played by CERN. The indirect effects of the Z_0, i.e. the occurrence of weak processes induced by the neutral current, were first observed in 1973 at CERN in the Gargamelle bubble chamber (see the chapter by Sutton). Later, in 1982, still at

Figure 12.8. Feynman diagrams for the decay of a muon: $\mu^- \rightarrow e^- \bar{\nu}_e \nu_\mu$. The W^- propagates the interaction between two points whose distance is very small, inversely proportional to $1/m_W$ and of the order of 10^{-16} cm. Since in muon decay all relevant energies are of order m_μ^2, i.e. extremely small in comparison with the W mass, this small separation can be neglected and the process can be described in terms of a vertex with the four particles drawn from a single point. In this sense the four-fermion theory is an approximation of the more fundamental gauge theory.

CERN, the $W^\pm$ and the Z_0 were for the first time directly produced and observed in proton–antiproton collisions. For this milestone achievement the Nobel Prize was awarded to C Rubbia and S van der Meer in 1984.

In 1989, CERN began to use its new LEP collider, a 27 km ring in which electrons and positrons are accelerated in opposite directions to equal energies in the range 45–100 GeV. The beams collide at four experimental areas where the ALEPH, DELPHI, L3 and OPAL detectors study the final states produced in the collisions. In its first phase, LEP1, from 1989 to 1995, LEP was dedicated at a precise study of the Z_0 properties: mass, lifetime and decay modes, in order to accurately test the predictions of the SM.

For a given luminosity of the machine, i.e. a given intensity of particles, the number of events in a specified final state is proportional to the 'cross section' of the corresponding process, essentially a measure of the reaction probability. One can measure the cross sections of the different possible final states and the associated distributions in energies, angles and so on of the reaction products. By measuring the dependence of the total visible cross section (covering all final detectable states) on the total centre of mass energy, $\sqrt{s} = 2E$, with E the beam energy, one finds a curve which has all the features of a resonant behaviour (figure 12.9). The cross section first decreases when $\sqrt{s}$ is increased. Then it rapidly increases until, in a few GeV, it reaches a sharp peak hundreds of times larger then the underlying continuum level. Then the cross section again decreases, but at a slower rate, with a long radiative tail until it asymptotically returns to the continuum extrapolation. The shape of the resonance peak is determined by its position (mass) and its width. In a first approximation, the probability amplitude A for the

process is the sum of the amplitudes corresponding to the exchange of a photon (A_γ) and to the exchange of a Z_0 (A_Z): $A = A_\gamma + A_Z$. The cross section is proportional to the absolute value squared of the amplitude:

$$\sigma \sim |A|^2 = |A_\gamma|^2 + |A_Z|^2 + 2\,\mathrm{Re}(A_\gamma A_Z^*). \quad (12.4)$$

At small s the photon term $|A_\gamma|^2 = a/s$ dominates but it decreases like $1/s$. For $s \simeq m_Z^2$, near the peak, it is the second term $|A_Z|^2$ that counts, which has a maximum

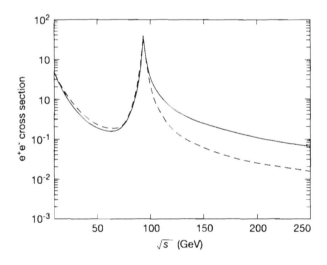

Figure 12.9. Behaviour of the total visible cross section for the process $e^+e^- \rightarrow F$, where F includes all possible final states apart from the invisible neutrinos, as a function of the total centre of mass energy (full curve). The broken curve is from a simple model where only the lowest order diagrams with exchange of a photon or a Z are considered. The difference between the two curves is due to the radiative corrections (see figure 12.10).

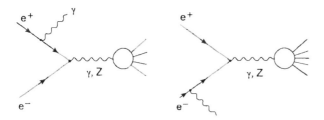

Figure 12.10. Feynman diagrams for initial state radiation in the process $e^+e^- \rightarrow F$. One photon is emitted by the initial electron or positron.

around $s \simeq m_Z^2$ and is sharper the smaller the width Γ_Z. Finally the last term is the interference of A_γ and A_Z, which vanishes at the peak $s = m_Z^2$ but is important on either side. The shape of the resonance is distorted by radiative corrections. The most important effect of this kind is the radiation of photons from the initial electron and/or positron before they interact to form the Z_0 (figure 12.10). This produces the radiative tail above the peak, because electrons and positrons with $\sqrt{s} > m_Z$ can come back on resonance by radiating some energy in the form of photons leading to a displacement of the peak at larger energy by about 100 MeV and to an asymmetry in the resonance profile (because the initial energy of the incoming beams can only be dissipated but not increased by the radiation mechanism).

An accurate reconstruction of the resonance shape, obtained by taking data at different energies in the peak region, allows a precise measurement of the mass m_Z and the width Γ_Z. The combined results from the four LEP experiments, based on the analysis of a total of about 16 million Z events, are

$$m_Z = 91.1867 \pm 0.0020 \, \text{GeV}: \quad \Gamma_Z = 2494.8 \pm 2.5 \, \text{MeV}.$$
$$(12.5)$$

This remarkable precision requires a perfect calibration of LEP's beam energies by a magnetic resonance method. The resulting control of the energy is so precise as to detect tiny effects due to the Moon's gravitation on the Earth that modify the LEP orbit radius by only a fraction of a millimetre. Knowledge of the mass is important for the theoretical prediction of the properties of the Z_0 and is used as an input in the precision tests of the SM.

From the laws of quantum mechanics we know that the lifetime of the resonance is inversely proportional to the total decay width: $\tau_Z = \hbar / \Gamma_Z$. From the measured value of Γ_Z one obtains $\tau_Z = (2.6383 \pm 0.0027) \times 10^{-25}$ s.

In a different series of measurements, the beam energy is kept fixed at the peak and one determines the cross sections at the peak for different final states $e^+e^- \rightarrow$ F. The final state F can be leptonic, i.e. F $= e^+e^-$ or $\mu^+\mu^-$ or $\tau^+\tau^-$, or hadronic, i.e. F $= q\bar{q}$ where q is a quark of given type or flavour with mass $m_q < m_Z/2$, i.e. q $=$ u, d, c, s, b, and $\bar{q}$ is the corresponding antiquark. For the same channels, one can also measure some other important observables in addition to the cross sections, such as certain asymmetries that we shall discuss later. If one adds the invisible channels F $= \nu_i \bar{\nu}_i$, with ν_i being one of the neutrinos (which the detectors cannot reveal), we have almost all the channels that make up the Z_0 decays predicted by the SM. At the peak the Z_0 exchange amplitude is dominant and the peak cross section in the final state F depends on the branching ratio of the decay $Z_0 \rightarrow$ F. More precisely from the rates measured at the peak in the channels F, after correcting for photon exchange and the initial state radiation, one obtains the quantities

$$\sigma_F = \frac{12\pi \Gamma_e \Gamma_F}{m_Z^2 \Gamma_Z^2}. \quad (12.6)$$

As Γ_Z is the sum of Γ_F for all possible final states, both visible and invisible, the branching ratio Γ_F / Γ_Z is the relative probability for the Z_0 to decay into a given channel F.

A very important result which is obtained from the measurements of σ_F is the determination, by difference, of the partial width into invisible channels Γ_{inv}. We have seen that the total width is obtained from the line shape and that the partial widths into visible channels are obtained from the cross sections at the peak. In the SM, Γ_{inv} is due to F $= \nu_i \bar{\nu}_i$ with $i = e, \mu, \tau$, together with possible additional neutrinos from as yet undetected families of fermions. Since the contribution to Γ_{inv} of each type of neutrino can be computed in the SM, from the measured value of Γ_{inv} one can determine the number N_ν of neutrinos. By this method, one finds that $N_\nu = 2.993 \pm 0.011$, in very good agreement with the number of known neutrino types $N_\nu = 3$. This remarkable result implies that the mysterious recurrence of families with the same quantum numbers, with a light neutrino in each family, is terminated with the known three families.

From the measured values of Γ_F, one obtains constraints for the sums $(g_V^F)^2 + (g_A^F)^2$ of the squared vector and axial vector couplings of the Z_0 to all quarks and leptons. In order to disentangle g_V and g_A and to measure the ratio g_V/g_A, a number of asymmetries are also measured at the peak. For example, one measures the forward–backward asymmetry A_{FB}^F for the channel F, defined as the difference over the sum of the events where the fermion goes forwards and the antifermion goes backwards or vice versa, where forwards (backwards) means in the hemisphere of the electron (positron). Alternatively, one measures the polarization asymmetry of a given fermion, defined as the difference over the sum of events where the fermion has a positive or negative helicity. In practice only the polarization asymmetry A_{pol}^τ for the τ lepton can be measured. Another very effective asymmetry is the left–right asymmetry A_{LR}, defined as the difference over the sum of the cross sections from left-handed or right-handed electrons (on unpolarized positrons) into any visible final state. Obviously for this measurement one needs a longitudinally polarized electron beam which is not available at LEP. However, such polarized beams exist at another e^+e^- machine, the Stanford Linear Collider (SLC) at SLAC, which operates at the same energy as LEP1 and produces the Z_0. This linear collider, with its approximately 1.5 km long straight acceleration line, is the prototype for all future high-energy e^+e^- colliders (see the chapter by Wilson). The intensity of SLC is much inferior to that of LEP: while at LEP the four experiments have collected and analysed about 4 million Z_0 events each, only a few hundred thousand Z have been studied by the Stanford Linear Detector (SLD) at the SLC, but an important feature of the SLC is that the electron beam is longitudinally polarized, with the electron spin pointing either forwards or backwards along its direction of motion. Thus A_{LR} has been measured at the SLC and, in spite of the relatively modest sample of collected Z_0 events, the precision attained by the SM test based on A_{LR} is comparable with that of LEP experiments (and can still be improved). This is because A_{LR} is much easier to measure, with all visible final states

contributing and no detailed observations needed on the distributions of the reaction products.

In the SM the Z_0 couplings are identical for the three fermion families. This universality can be experimentally tested by separately measuring the peak cross sections and the forward–backward asymmetries of the charged leptons e, μ and τ (the different flavours of quarks are more difficult to disentangle). Thus one obtains g_V and g_A separately for the three types of charged lepton. The results of this universality test are shown in figure 12.11, where we see that, within the experimental accuracy, the couplings are

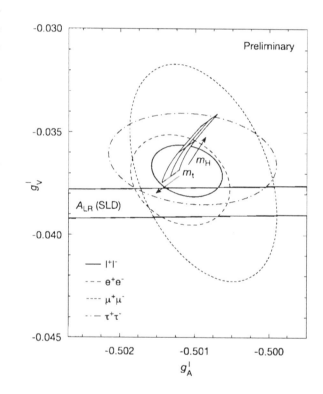

Figure 12.11. The diagram shows the vector and axial vector coupling constants g_V^l and g_A^l for the leptons $l = $ e, μ, τ. The ellipses are the regions selected by experiments at LEP on each separate lepton and the averaged result (full curve). Also shown is the theoretical prediction in the SM with its dependence on m_t and m_H, i.e. the top quark and the Higgs masses. The nearly horizontal band is the region selected by the SLAC measurement of the left–right asymmetry which is in slight disagreement with the LEP results.

indeed the same for all charged leptons. Once universality has been tested the charged lepton results can be combined, to obtain, with increased precision, the values of g_V^l and g_A^l. In the SM we have $g_V^l/g_A^l = 1 - 4 \sin^2 \theta_W$ and $g_A^l \simeq -\frac{1}{2}(1 + \Delta\rho/2)$, where $\Delta\rho$ arises from radiative corrections. Thus one obtains a combined measurement of $\sin^2 \theta_W$ and of $\Delta\rho$ that can be compared with the predictions of the SM.

In summary, the most important independent quantities that, with the assumption of lepton universality, are obtained from the precision measurements at the Z_0 peak are as follows: the total hadronic peak cross section, σ_h, given by equation (12.6) with F any hadronic final state (i.e. one with a $q\bar{q}$ pair of any type plus any number of photons and gluons); the ratio R_h of the hadronic to the leptonic peak cross sections; the ratio R_b of the peak cross sections into a quark–antiquark pair of type b (the heaviest of the five quark types the Z_0 can decay into) to the hadronic cross section; the analogous ratio R_c for charm; finally the value of $\sin^2 \theta_W$ extracted from the asymmetries. The combined LEP and SLC values for these quantities are shown in table 12.3. We see that the accuracies of σ_h, R_h and $\sin^2 \theta_W$ are close to the per mil level, while R_h is measured to a 0.5% precision and the precision on R_c is even less (3%).

Beyond the Z_0 studies, other important experiments for precision tests of the SM are the measurement of the W mass, the study of neutrino–nucleus deep inelastic scattering, of ν_μ–electron elastic scattering and the observations on parity violation in atomic transitions. The crucial measurement of the top quark mass deserves a special role and will be discussed later. The W mass has been measured at $p\bar{p}$ colliders, first at CERN by the UA1 and UA2 collaborations and more recently at Fermilab by the CDF and D0 collaborations. From 1996, the W mass could also be measured at LEP2. The resulting combined value, with an absolute (relative) error of 64 MeV (0.8 per mil), is shown in table 12.3. Additional information on the ratio m_W^2/m_Z^2 is derived from the measured ratio of neutral current versus charged current reactions in deep inelastic neutrino scattering off nuclei. The sensitivity to m_W^2/m_Z^2 arises from an interplay of couplings and propagators. Deep inelastic neutrino scattering has been measured at CERN by the CDHS and the CHARM collaborations and at Fermilab by the CCFR collaboration. The combined result,

Table 12.3. Summary of the high-energy measurements which provide precision tests of the standard electroweak theory. Also shown, in numerical and graphical form, are the so-called pulls, i.e. the deviations, measured in units of the measurement error, from the best fit of the data performed in the SM.

Quantity	Data (May '98)	Pull	Pull graph (-2 -1 0 1 2)
m_Z (GeV)	91.1867(20)	0.1	
Γ_Z (GeV)	2.4948(25)	-0.7	
σ_h (nb)	41.486(53)	0.3	
R_h	20.775(27)	0.7	
R_b	0.2173(9)	1.6	
R_c	0.1731(44)	0.2	
A_{FB}^l	0.0171(10)	0.8	
A_τ	0.1400(63)	-1.2	
A_e	0.1438(71)	-0.5	
A_{FB}^b	0.0998(22)	-1.6	
A_{FB}^c	0.0735(45)	-0.1	
A_b (SLD direct)	0.899(49)	-0.7	
A_c (SLD direct)	0.660(45)	-0.2	
$\sin^2 \theta_W$ (LEP-combined)	0.23185(26)	1.5	
$A_{LR} \rightarrow \sin^2 \theta_W$ (SLD)	0.23084(35)	-1.8	
m_W (GeV) (LEP2+$p\bar{p}$)	80.375(64)	-0.1	
$1 - m_W^2/m_Z^2$ (νN)	0.2253(21)	1.1	
Q_W (Atomic PV in Cs)	-72.11(93)	1.2	
m_t (GeV)	174.1(5.4)	0.6	

expressed in terms of $1 - m_W^2/m_Z^2$, with approximately $\sim 2\%$ accuracy, is shown in table 12.3.

Having described the experimental data, we turn to the corresponding theoretical predictions within the SM and the comparison of theory and experiment. In the SM, as in any renormalizable quantum field theory, a number of input quantities, i.e. masses and coupling constants, must be given in order to make predictions. The commonly adopted set contains the fine structure constant of QED, defined as $\alpha = e^2/4\pi \simeq 1/137.036$ with e the positron electric charge; the Fermi coupling G_F (see equation (12.3)); the Z_0 mass m_Z; the fermion masses; the strong interaction coupling α_s, defined in analogy to the QED fine structure constant, i.e. $\alpha_s = g_s^2/4\pi$; the Higgs mass. The accuracies of the input parameters vary widely among them. The QED coupling α, G_F, m_Z, and the lepton masses are known to a largely sufficient precision. For light quarks, input comes from the experimental data on the hadronic cross section in e^+e^- annihilation, and in practice the light quark masses are replaced by the effective value of the QED coupling at the Z_0 mass, which incorporates the required quark radiative corrections: $\alpha(m_Z) \simeq 1/128.90$. The QCD coupling and the top quark mass are known with a modest precision. The input value used for the effective strong coupling at m_Z is at present $\alpha_s(m_Z) = 0.119 \pm 0.003$. The top quark was observed at Fermilab by the CDF and D0 collaborations in 1994 (see the chapter by Shochet). It is the heaviest fundamental particle known so far; its mass is $m_t = 174.1 \pm 5.4$ GeV, i.e. about the same weight as a rhenium nucleus with 75 protons and 111 neutrons. However the Higgs mass is largely unknown. Searches at LEP1 for rare decay modes of the Z_0 gave the lower limit $m_H > 65$ GeV. This very important achievement of LEP1 improved the previous limit by more than an order of magnitude. This limit has been improved by LEP2 to $m_H > 85$ GeV. Theoretical reasons exclude a Higgs mass above 600–1000 GeV.

While the input parameters are known with widely different precisions, the quantitative impacts of the various input quantities are also very different. Some of the parameters only appear at the level of radiative corrections which are relatively small. The quantity $\alpha(m_Z)$ fixes the largest radiative effects, which are purely electromagnetic. Note that the relative difference between $\alpha \simeq 1/137$ and $\alpha(m_Z) \simeq 1/129$ amounts to an approximately 6% effect, small in absolute terms but very large as far as precision tests of the theory are concerned, because the most accurate measurements reach the level of a few per mil. In fact it turns out that the present uncertainty in $\alpha(m_Z)$ induces an error in $\sin^2\theta_W$ which is comparable to the present experimental error. The masses m_t and m_H only enter in purely weak virtual effects which are much smaller. In these virtual effects, the top quark and the Higgs mass do not appear in the final states, but only as intermediate states in the calculation of quantum fluctuations (in particular in vacuum polarization diagrams and vertex corrections, shown in figure 12.12). The top quark virtual effects are the most important because they increase with m_t^2, while the dependence on m_H of the leading radiative corrections

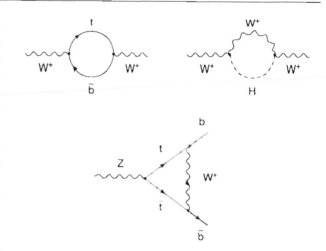

Figure 12.12. Examples of vacuum polarization (above) and vertex (below) diagrams which contribute to the radiative corrections with terms that depend on the top quark and Higgs mass.

is only logarithmic, so that even for $m_H \simeq 1$ TeV the corresponding effects are quite small.

A separate case is that of the strong coupling $\alpha_s(m_Z)$. This is only known to about 3% at best, but essentially it only contributes, with a typical factor $1 + \alpha_s(m_Z)/\pi + \ldots$, to radiative corrections to processes with hadrons in the final state. The theoretical uncertainty in hadronic observables introduced by the error on $\alpha_s(m_Z)$ is of the order of the experimental errors, while the purely leptonic quantities are negligibly affected.

Given the different importance of the input parameters one strategy is to fit the available data in terms of the SM predictions with m_t, m_H and $\alpha_s(m_Z)$ treated as free parameters. From all the available electroweak data, including the direct measurement of m_t, one obtains the best-fit values

$$m_t = 171.1 \pm 5.1 \text{ GeV}$$
$$m_H = 74^{+84}_{-45} \text{ GeV} \qquad (12.7)$$
$$\alpha_s(m_Z) = 0.121 \pm 0.003.$$

The quality of the fit is reasonably good (the χ^2 value divided by the number of degrees of freedom is 13.5/15; a fit is good if this number is around 1 or less). Table 12.3 shows a distribution of the number of standard deviations that each experimental number departs from the corresponding best-fit value. The resulting distribution is remarkably close to what one expects for a normal distribution. A quantity that deviates a little from the prediction is R_b, with a discrepancy now only about 1.4σ, not alarming at all, but was once larger. One looks particularly at R_b, because the b quark is coupled to the top quark and heavy fermions are more likely to be affected by some new physics effect. One could ask what is the value of m_t predicted by electroweak radiative corrections in comparison with the direct measurement? For this one repeats the previous fitting procedure but the CDF–D0 experimental value of m_t is omitted from the input data. The resulting values are

$$m_t = 157.8^{+9}_{-8.5} \text{ GeV}$$
$$m_H = 32^{+41}_{-15} \text{ GeV} \qquad (12.8)$$
$$\alpha_s(m_Z) = 0.121 \pm 0.003.$$

The direct limits on m_H are not taken into account in this fit. The value of m_t is quite close to the experimental value, but slightly on the low side. It is really remarkable that the top quark mass, very different from the other quark masses, can be so accurately predicted in the SM merely from the observed pattern of radiative corrections. However, the Higgs mass is hardly constrained by the fits: the data indicate $m_H \lesssim 300$ GeV. It is important that the Higgs mass is light enough for the perturbative regime to apply safely, lighter than 600–1000 GeV. Thus there is a nice consistency in the sense that the radiative corrections predicted by perturbation theory in the SM agree with experiment and indicate a Higgs mass consistent with the applicability of the employed framework.

CERN has played a large role in testing the SM. Precision electroweak tests have established the relevance of renormalizable, spontaneously broken gauge field theory for physics at the weak-energy scale, but the problem of the experimental unveiling of the Higgs sector is still open. This is CERN's goal in next decade and beyond. From the end of 1995, the total energy of LEP has been increased (LEP2) in steps up to 189 GeV. The energy will be eventually increased to 200 GeV. The main physics goals are the search for the Higgs particle and for possible new particles, the precise measurement of m_W and the experimental study of the triple gauge vertices $WW\gamma$ and WWZ_0. The Higgs particle of the SM could be produced at LEP2 in the reaction $e^+e^- \rightarrow Z_0H$ which proceeds by Z_0 exchange. The Higgs particle can be produced and detected at LEP2 provided that its mass does not exceed 95–100 GeV. In the year 2001, LEP will be dismantled and in its tunnel a new double ring of superconducting magnets will be installed. The new accelerator, the Large Hadron Collider (LHC), will be a proton–proton collider of 14 TeV of total centre of mass energy. Two large experiments ATLAS and CMS will continue the Higgs hunt in the year 2005. The experiments will cover Higgs masses up to about 1 TeV.

FURTHER READING

Halzen F and Martin A D 1984 *Quarks and Leptons: An Introductory Course in Modern Particle Physics* (New York: Wiley)

Huang K 1992 *Quarks, Leptons and Gauge Fields* 2nd edn (Singapore: World Scientific)

Perkins D H 1987 *Introduction to High Energy Physics* 3rd edn (Reading, MA: Addison-Wesley)

ABOUT THE AUTHOR

Guido Altarelli is a senior staff physicist at the Theory Division of CERN, since 1987, and Professor of Theoretical Physics at the Third University of Rome. Before this, he had held various appointments in several universities. These included the University of Florence, New York University, The Rockefeller University, University 'La Sapienza' in Rome, the Ecole Normale Supérieure in Paris, and Boston University. At different times he has been a member of the scientific committees of the Deutsches Elektronen-Synchrotron (DESY), CERN and the Superconducting Super Collider (SSC). His research interests cover a broad range of problems in the phenomenology of particle interactions, within and beyond the SM, in close connection with ongoing and planned experiments at present and future colliders.

13 THE TOP QUARK

Melvyn Shochet

Editor's Introduction: The sixth ('top') quark, with a mass roughly equal to that of a nucleus of tungsten, is by far the heaviest in the Standard Model kit of fundamental particles. The only quark to be systematically hunted, its discovery at Fermilab in 1995 was the culmination of a long and diligent quest. With no indication initially of how heavy the top quark was, physicists had little idea of where and how to look for it. New machines and detectors searched in vain, and each time the physicists pushed the mass limit upwards and reformulated their approach. The problem was that, the higher this limit, the more difficult the top quark would be to find. However, precision measurements on other Standard Model particles began to give a good idea of the top quark mass, and therefore how it could be found (see the chapter by Veltman).

The top quark discovery, by collaborations involving nearly a thousand physicists, is a striking example of the painstaking work needed to unravel the complexities of particle collisions at high energies and to isolate new physics deep inside, a trend which looks likely to continue as we probe still deeper into Nature. With the top quark itself now clearly established, physics priorities switch to exploring this astonishingly heavy particle.

INTRODUCTION

The quest for the top quark began in 1977, just after the discovery of the bottom quark at the Fermi National Accelerator Laboratory (Fermilab). However, to better understand the motivation for the search, we should begin our story three years earlier. In 1974, an unexpected, massive, short-lived resonance was observed simultaneously in electron–positron and hadron collisions. The discovery of the J/ψ, a bound state of the charm quark and its antiquark, marked the beginning of heavy-quark physics (see the chapter by Schwitters).

The charm quark was not merely the fourth in a growing catalogue of quarks. Its importance was the confirmation of an essential feature of the newly developed unified theory of electroweak interactions. In order to explain the absence of a large class of weak decays, Glashow, Iliopoulos and Maiani had hypothesized that Nature created quark types in pairs. The up and down quarks were such a pair, and a partner (charm) was thus needed for the strange quark.

The next important step occurred just a year after the discovery of the J/ψ. In 1975, a third type of charged lepton, the tau τ, was discovered. In the developing picture of pairs of quarks and of leptons (the electron and its neutrino, the muon and its neutrino), the τ lepton was the first indication of a third pair or generation. Three generations fit very nicely with a theoretical observation that had been made two years earlier, before the discovery of the charm quark. Kobayashi and Maskawa had noted that in order to incorporate the observed matter–antimatter asymmetry (CP violation) in a renormalizable theory with no particles other than those in the simplest version of the Standard Model, a minimum of three quark pairs was needed (see the chapter by Ali and Kayser).

The discovery at Fermilab in 1977 of the upsilon Υ, a bound state of the bottom quark and its antiquark, provided evidence for a third generation of quarks to match the third pair of leptons. Immediately, the search began for the partner of the bottom quark b, namely the top quark t.

As we shall see, initial searches for the top quark were unsuccessful. That could have meant that the bottom quark does not fit into the Standard Model picture of quark pairs or doublets but rather is a weak interaction singlet and there is no top quark. This possibility was ruled out at DESY in 1984 with the measurement of the forward–backward asymmetry in the process $e^+e^- \rightarrow b\bar{b}$. If this proceeded only via the electromagnetic interaction, there would be no asymmetry; the bottom quark would be produced in the positron direction as often as in the electron direction. However, if the bottom quark were a member of a quark doublet, there would be a weak interaction effect which would interfere with the electromagnetic production, resulting in parity violation and a substantial forward–backward asymmetry. The observation, a $(22.5 \pm 6.5)\%$ asymmetry, was consistent with the 25.2% prediction of the Standard Model but far from the null result expected if there were no top quark. The same conclusion was drawn from the search for a bottom-quark decay that would be prevalent if the bottom quark were a singlet. Thus either the top quark was awaiting discovery or the Standard Model was wrong.

The discovery of the top quark depended critically on one parameter, its mass. If it were close to the bottom-quark mass, discovery would come quickly. If, on the other hand, the top quark were very massive, higher-energy accelerators would have to be built before the top quark could be seen. A number of mass estimates were made, all relying on a natural progression in the mass of the different quarks. The predictions were approximately 15 GeV. If this were correct, the electron–positron colliders being built at the time would find the top quark.

SEARCHES FOR THE TOP QUARK IN ELECTRON–POSITRON COLLISIONS

A top quark with mass up to 15–20 GeV could be detected at the PETRA e^+e^- collider which was operating at DESY in the late 1970s. Three search techniques were employed. If a bound $t\bar{t}$ state were produced, analogous to the J/ψ ($c\bar{c}$) or Υ ($b\bar{b}$), a narrow resonance would be seen. If a top quark and antiquark were produced without forming a bound state, the rate of e^+e^- collisions that produce hadrons would be 36% larger than in the absence of top quark production. With the third technique, the more spherical angular distribution of particles from top-quark decay would be differentiated from the more planar distribution from light quarks. From

the absence of such signatures, the PETRA experiments ruled out the top quark with a mass below 23 GeV.

In the early 1980s, the TRISTAN electron–positron collider was built in Japan. One of its major goals was the discovery of the top quark. By the end of that decade, three experiments had used techniques similar to those employed at PETRA to increase the lower limit on the top-quark mass to 30 GeV.

The final chapter in the search for the top quark in e^+e^- collisions occurred at the Stanford Linear Collider (SLC) at the Stanford Linear Accelerator Center (SLAC) and at the Large Electron–Positron Collider (LEP) at CERN, which studied the decay of Z^0 bosons produced in e^+e^- collisions. Since they did not observe $Z^0 \rightarrow t\bar{t}$ decays, they set a top mass limit of 45 GeV, half of the mass of the Z^0.

EARLY SEARCHES FOR THE TOP QUARK IN HADRON COLLISIONS

When the top quark was not seen in e^-e^- collisions up to masses well above the early theoretical predictions, attention turned to hadron collisions. The advantage of hadron collisions is that they occur at higher energy and thus reach a higher mass. The discovery in 1983 of the carriers of the weak force, the W and Z bosons, was made possible by the construction at CERN of a collider in which 270 GeV protons struck 270 GeV antiprotons but, as we shall see, sources of background, other processes that can mimic top-quark decay, presented a serious obstacle to top-quark discovery. Moreover, the complexity of top-quark decay made interpretation of candidate events very difficult. The discovery of the other quarks occurred with collisions producing as few as two particles. The top quark, on the other hand, decays into unstable particles that in turn decay into other unstable particles. Top quark events can contain up to 100 particles that have to be carefully disentangled.

The production of the top quark in proton–antiproton collisions is represented schematically in figure 13.1. A quark from the proton and an antiquark from the antiproton annihilate, producing a top quark and a top antiquark. The higher the top quark mass, the more difficult is the search for at least two reasons. First, the production rate for the process shown in figure 13.1 drops with increasing top mass. Second, higher-energy quarks and antiquarks are needed to produce heavier top quarks, since on average the annihilating quark and antiquark must each have an energy equal to the top quark mass. The problem is that a quark carries only a fraction of its parent proton's energy, with a probability that drops as the fraction increases.

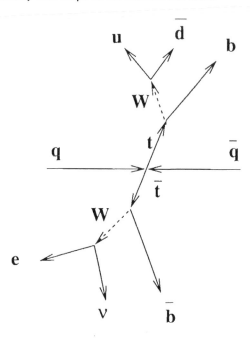

Figure 13.1. Schematic diagram of a high-energy quark (from a proton)–antiquark (from an antiproton) annihilation producing a top quark and a top antiquark, each decaying into a bottom quark (antiquark) and a W boson. Here, one of the W bosons decays into an electron and a neutrino, while the other W decays into a quark–antiquark pair.

In the Standard Model, a top quark always decays into a bottom quark and a W boson. The W decays into a charged lepton and a neutrino or into a quark–antiquark pair. Since both the top quark and the W boson have lifetimes less than 10^{-24} s, the particles observed in the detector depend on how the two W particles decay. Table 13.1 shows the probabilities for different configurations. For example, 14.8% of events would contain an electron, a neutrino, and four quarks ($q\bar{q}$ from a W plus the b and $\bar{b}$ from the t and $\bar{t}$). If the mass of the top quark is less than the sum of the W and bottom-quark masses, 86 GeV, a real W cannot be produced in the top-quark decay. Rather it is a virtual W,

lower in mass than a real W, which can exist for an instant in accord with the uncertainty principle. However, its decay scheme is the same as for real Ws.

Table 13.1. The decay probabilities for the pair of W bosons in a t$\bar{t}$ event. The rows represent the decay modes of one W, while the columns specify the decay modes of the other W. The values below the diagonal have been included in those above the diagonal since for the eν $\mu\nu$ configuration, for example, we do not care which W decayed into eν and which decayed into $\mu\nu$.

	eν	$\mu\nu$	$\tau\nu$	q$\bar{q}$
eν	1.2%	2.5%	2.5%	14.8%
$\mu\nu$		1.2%	2.5%	14.8%
$\tau\nu$			1.2%	14.8%
q$\bar{q}$				44.4%

In 1984, the lower limit on the top-quark mass was 23 GeV. The UA1 experiment at the CERN p̄p collider was searching for the process shown in figure 13.1, with one of the W bosons decaying into either eν or $\mu\nu$. Experimentally, each quark fragments into a collimated jet of hadrons (pions, kaons, protons, etc). The neutrino is not observed because it does not interact with the detector. Rather its presence is inferred because an unseen particle is needed for conservation of momentum to hold in the reaction. The UA1 collaboration required candidate events to contain an electron or muon, a neutrino of moderate energy (to suppress the background from real W bosons which produce higher-energy neutrinos), and two hadron jets. Even though figure 13.1 shows four quarks in the final state (b, $\bar{b}$, u and $\bar{d}$), only two were required so that the analysis would also be sensitive to a second process, $q\bar{q} \to W \to t\bar{b} \to e\nu b\bar{b}$, which would produce only two jets. By 1985, the group had found 12 events, with an expected background of 1.6 events due mostly to hadrons that are misidentified as electrons. For a 40 GeV top quark, they expected approximately 10 signal events, in agreement with their observation. They concluded in a journal article and in presentations at international conferences that their results were consistent with a 30–50 GeV top quark, but they stopped just short of claiming discovery.

By 1988, the group had increased their data sample by a factor of two. More importantly, they had a much better understanding of the numerous sources of background. Other than misidentified electrons, the highest background was due to the direct production of W bosons with large momentum recoiling against two jets from quarks or gluons (figure 13.2). Another important source of background was the production of other heavy quark–antiquark pairs, $b\bar{b}$ and $c\bar{c}$. With an optimized set of selection criteria, they observed 36 events, to be compared with a 35-event background prediction. There was no sign of the 23 additional events expected from a 40 GeV top quark. Consequently they no longer spoke of seeing the top quark but instead set a lower limit of 44 GeV on its mass.

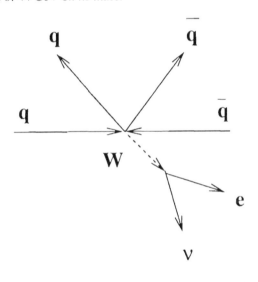

Figure 13.2. The major background to the top-quark signal. A real W boson is directly produced with large momentum in the proton–antiproton collision. It recoils against a quark–antiquark pair.

At this time, a new player entered the quest. A proton–antiproton collider was now in operation at Fermilab. The use of superconducting magnets in the Tevatron accelerator enabled the protons and antiprotons to reach 900 GeV energy, nearly three times the energy at the CERN collider. Higher-energy quarks could create higher-mass top quarks. The Collider Detection at Fermilab (CDF) collaboration was operating a new large detector at Fermilab

and analysing their data rapidly in the hope that the top quark would appear.

The race for the top-quark discovery was now between CDF and UA2, the second large experiment in operation at the CERN collider. During 1988–89, each experiment periodically heard rumours that the other group had seen a signal. The competition was so keen that both the BBC in the UK and public television in the USA produced documentaries on 'The Race for the Top'. However, when both groups had analysed the data, the top quark still had not appeared.

The CDF group carried out two analyses. In the first, they studied approximately 100 events with an electron, a neutrino, and two or more hadron jets. They used the $e\nu$ mass to separate the dominant background, direct W production (figure 13.2), from a top-quark signal. For the background, the $e\nu$ mass would peak at the W mass, 80 GeV, well above the masses expected from the decay of a 40–80 GeV top quark. The data were consistent with background only, resulting in a lower limit on the top-quark mass of 77 GeV.

The second CDF analysis searched for events with a high-energy electron and a high-energy muon. This channel has the advantage of very low background, estimated at 1.5 events. The disadvantage, shown in table 13.1, is the expected number of top events, which would be a factor of six lower than that for the electron–neutrino–jets channel. CDF observed one event, consistent with the estimated background. In this channel, they ruled out a top quark with a mass below 72 GeV.

The UA2 detector at CERN had been significantly upgraded between 1985 and 1987 both to handle the increased beam intensity of the accelerator and to improve physics capability for such difficult analyses as the top-quark search. During 1988–89, UA2 accumulated eight times as much data as in all previous running of the CERN collider. They studied their sample of 137 events with an electron, a neutrino and at least one hadron jet. By allowing events with one jet into their sample, they significantly increased their top-quark detection efficiency for the production process $q\bar{q} \rightarrow W \rightarrow t\bar{b} \rightarrow e\nu b\bar{b}$ which, although insignificant at Fermilab's accelerator energy, is the dominant mechanism for the energy of the CERN collider. In a region of $e\nu$ mass well below the 80 GeV W mass, they observed 17 events compared with 26 expected from background and another 10-15 from a 65 GeV top quark. From these data, they excluded a top-quark mass below 69 GeV.

The CERN–Fermilab race to the top was over. Neither group had observed the elusive object, and future progress could only be made at Fermilab. The energy of the protons and antiprotons at CERN was too low to produce top quarks with mass above 75 GeV.

In a final paper on the data taken in 1988–89, the CDF group combined two analyses to obtain improved sensitivity. Their dilepton analysis (two charged leptons) was expanded from events containing an electron and a muon to include events with either two electrons or two muons. The single-lepton analysis, previously restricted to electron plus neutrino plus jets, was expanded to include events with a muon plus neutrino plus jets. However, the analysis strategy also had to change. The range of top mass now being explored was nearing 85 GeV. Above that, a top quark would decay into a bottom quark and a *real* rather than a virtual W boson. Then the $e\nu$ or $\mu\nu$ mass distribution would be indistinguishable from that of the dominant background, directly produced W bosons. They needed another distinguishing characteristic. The greatest difference between signal and background is the presence of two bottom quarks in each $t\bar{t}$ event; the background process would rarely produce bottom quarks. Since a bottom quark decays into a muon 10% of the time, while light quarks rarely do, CDF searched their 200 events with an e or μ plus neutrino plus jets for the presence of an additional muon from b-quark decay. They observed none, expecting one from the background. When combined with their dilepton search, this result extended the top-quark mass lower limit to 91 GeV. This meant that, if the Standard Model top quark existed, it would decay into a bottom quark and a real W boson.

THE CHALLENGE OF A VERY MASSIVE TOP QUARK

It is useful to summarize the situation as it existed in 1990. The top quark, initially predicted to have a mass of 15 GeV, had to have a mass above 91 GeV, if it existed at all. According to the Standard Model, it would decay 100% of the time into a W boson and a bottom quark, but more than that was known. Experiments at the LEP $e^{+}e^{-}$ collider at CERN had collected their first large samples of Z^{0} boson decays. The Standard Model imposes consistency constraints on various parameters in the theory. From the measured Z^{0} properties (mass, lifetime, production rate, decay rate into various final states, etc) the LEP experimenters deduced that the Standard Model top quark could be no larger than approximately 225 GeV (see the chapter by Veltman). Thus the search would not go on for ever; either the top quark would be found with mass between 91 and 225 GeV, or our theory of the electroweak interaction would be proven wrong!

The only existing accelerator that could produce such massive objects was the Fermilab collider, but even there it would not be easy. The production rate of top quarks would be small, with potential backgrounds high. A few numbers will make this clear. If the Fermilab collider were to run at its design intensity, there would be 2×10^{11} p$\bar{\text{p}}$ collisions per year. The expected number of observed top quark events would be 5 to 10 per year for a top-quark mass between 150 and 175 GeV, less than one for each 10^{10} collisions that occurred in the detector! The dominant background to the top-event signature, directly produced W bosons recoiling against three or more hadron jets, would be seen 2.5–5 times more often than the signal.

It was imperative to increase the signal and to reduce the background. Accelerator performance provided a great help. The accelerator p$\bar{\text{p}}$ collision rate exceeded the design goal in 1992, and by 1995 it had reached 20 times the design value. The experimenters reduced background by identifying bottom quarks or applying other criteria that suppress sources other than top quarks. The signal was enhanced by using many $t\bar{t}$ decay channels, both dilepton (ee, eμ, and $\mu\mu$) and e or μ plus jets.

THE FERMILAB EXPERIMENTS

The first Fermilab collider run in the 1990s saw the commissioning of a new experiment. The D0 collaboration had been constructing their detector for half a decade, and they now joined CDF in the quest for the top quark. These two experiments are typical examples of large modern collider detectors. Each is tens of metres in each

Figure 13.3. The CDF detector at Fermilab's Tevatron proton–antiproton collider. The central cylindrical region of 3 m diameter contains the superconducting magnet coil and tracking chambers. Surrounding the coil are the calorimeters (here rolled back for maintenance) and the muon detectors.

direction, has a mass of many thousands of tons and contains approximately 100 000 individual particle detectors.

The goals of the two experiments are the same, to measure as precisely as possible the energies and directions of all of the particles produced in p̄p collisions. Since there can be as many as 200 particles produced in a collision, it is important to surround the beam with a highly segmented system of particle detectors. The techniques employed by the two groups are somewhat different. CDF stresses measuring particle trajectories in a magnetic field; D0 focuses on measuring particle energies in a uranium–liquid-argon detector.

The CDF detector is shown in figure 13.3. It consists of a superconducting solenoid magnet of 3 metre diameter filled with tracking detectors that measure the curvature of charged particles in the 1.5 T magnetic field. Of particular importance to the top-quark search is the vertex detector with a 15 μm position resolution that allows measurement of the few millimetre distance between a bottom quark's

production and decay points. Surrounding the magnet coil is a set of calorimeters to measure particle energy, followed by muon detectors.

The D0 detector, shown in figure 13.4, has a smaller tracking region and no magnetic field. However, its calorimeters have intrinsically better energy resolution for hadrons and are more finely segmented than those in CDF. A large system of muon detectors built around magnetized iron surrounds the calorimeters.

Each of the two experiments was designed, constructed and operated by a team of over 400 physicists from 30–40 universities and national laboratories around the world. The sociology of these groups is interesting. In the design, construction and maintenance of the apparatus, a university or laboratory group will take responsibility for a detector subsystem. However, once the data are taken, each physicist works on a physics problem of his or her particular interest. Scientists working on similar problems, the electroweak interaction for example, meet regularly as

Figure 13.4. The D0 detector at the Tevatron. The heart of the apparatus is the cylindrical cryostat of 5 m diameter containing the uranium–liquid-argon calorimeters. At the centre of the detector are the tracking chambers. Outside the cryostat are the muon detectors.

a physics group led by one or two people appointed for two or three years by the scientific spokespeople. CDF and D0 of course each had a Top Quark Physics Group. The overall leadership of the collaboration is vested in scientific spokespersons elected by all members of the collaboration and an executive or institutional board with representatives from each collaborating laboratory and university.

FIRST EVIDENCE FOR THE TOP QUARK

The Fermilab collider had a long data run, the first part in 1992–93 and the second in 1994–95. By the beginning of 1994, D0 had analysed their 1992–93 data and published a new top-quark mass limit. They searched in the $e\mu$ and ee dilepton modes and for events with an electron or muon, a neutrino, and at least four jets. In addition to requirements on the numbers of leptons and jets observed, they further discriminated against the background with a requirement on the spherical nature of the event topology, similar to the technique employed in e^+e^- collisions. They observed

three events, while expecting six from the background. They were thus able to rule out the top quark with a mass below 131 GeV.

At the same time, within the CDF collaboration, there was enormous activity under way to understand the source of an enticing signal that had been noticed in August 1993. As is almost inevitable when the first hint of a new phenomenon is seen, statistics were meagre.

As in their previous analysis, CDF searched in both the dilepton and single-lepton modes: $q\bar{q} \rightarrow t\bar{t} \rightarrow W^+bW^-\bar{b}$ with both W bosons decaying into $e\nu$ or $\mu\nu$ for the dilepton case, while one W decayed to $q\bar{q}$ for the single-lepton case. There were changes in the analysis strategy, however, as a result of the higher mass range being explored and new apparatus that had been added to the detector. In the dilepton analysis, they suppressed the background by a factor of 20 more than previously by requiring two hadron jets in addition to the two charged leptons and the signature of neutrinos. This was possible because a top quark of mass above 130 GeV would decay into a bottom-quark jet of energy above 40 GeV, which is easily observed in the detector. In their previous analysis, which covered the 77–91 GeV range, the bottom quark could be produced with little or no kinetic energy, making jet detection almost impossible.

As mentioned previously, events containing a single charged lepton, neutrino and hadron jets were very difficult to distinguish from the dominant W + jets background. The most significant difference was the presence of bottom quarks in top events. As in the past, CDF searched for extra electrons or muons near a jet as a characteristic of bottom-quark decay, but now they had an even more powerful method. Before the 1992–93 run, they had built and installed a vertex detector employing silicon technology. With this device they could distinguish particles produced in the $\bar{p}p$ collision from those coming from the decay of a bottom quark a few millimetres away. This technique for bottom-quark identification produced higher detection efficiency and lower background.

CDF found two dilepton events, six single-lepton events with a jet tagged as a bottom quark using the vertex chamber, and seven with an extra lepton consistent with bottom-quark decay. Three of these single-lepton events had both of the bottom-quark signatures. The expected background in the three channels were 0.6, 2.3 and 3.1 events respectively.

CDF calculated that the probability that these results were consistent with background alone was 0.26%, which corresponds to a 2.8 standard deviation excess. Although odds of 400 to 1 against the null hypothesis may seem quite good, the usual standard is at least 4 standard deviations, odds of 15 000 to 1. Thus it was necessary to look at other features of the data for supporting or contradicting evidence.

There were a few features of the data that were worrisome. The best control sample for the single-lepton channel contained events with hadron jets plus a Z boson (decaying into e^+e^- or $\mu^+\mu^-$) instead of the W boson in the top search. Since top quarks should not decay into Z bosons, the number of bottom-quark jets in this sample should agree with the background estimate. They expected 0.6 events and saw two. Although this is not a large disagreement, there was concern that it could indicate an additional source of background. There were other similar checks with higher statistics that agreed with the background estimates, but this

sample was most similar to top events. The other problem was an insufficient number of events in the single-lepton channel with four jets when the predicted top-quark and background rates were summed.

On the other hand, there were additional features of the data that did support the top-quark hypothesis. The most important of these is the mass distribution. For seven of the ten single-lepton events, the energies and directions of the detected particles allowed the mass of the parent top quark to be calculated. The distribution, shown in figure 13.5, is peaked near 170 GeV and does not look like the expected background shape. Under the assumption that these events are from top-quark production, CDF made the first direct measurement of the top-quark mass at 174 ± 16 GeV. Another piece of evidence in favour of the top-quark interpretation was one of the dilepton events which was tagged as having a bottom-quark jet by both tagging algorithms. This event has a 13 times larger probability of being a top event than background.

In the end, the CDF collaboration had to weigh all the evidence for and against and to decide on the strength of the

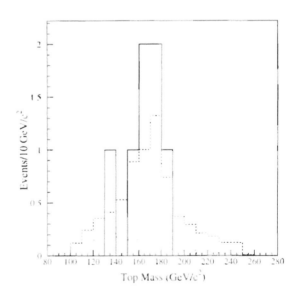

Figure 13.5. The top-quark mass distribution for candidate events (solid line) compared with the expected distributions for background (dotted line) and background plus a 175 GeV mass top quark (dashed line).

case. Since all this information was laid out in excruciating detail in a 60-page paper in *Physical Review*, the issue came down to choosing a title for the paper. It would be 'Search for the top quark', 'Evidence for the top quark', or 'Observation of the top quark'. There was broad consensus within the collaboration that the evidence was strong, but it was felt that the statistics were not quite sufficient to call this the definitive discovery. That would have to await more data. Thus 'Evidence for the top quark' was the title.

When CDF submitted the paper for publication in May 1994, the D0 group released a statement. They had reoptimized their analysis strategy for a higher-mass top quark and increased the size of their data sample. Combining the dilepton and single-lepton modes, they observed seven events when four to six events were expected from background. Although their results were not inconsistent with those of CDF, their small statistics led them to conclude that they had no independent evidence for the existence of the top quark. In November of 1994, D0 submitted a paper with slightly increased statistics and the same conclusion. In this paper, D0 included the use of a bottom-quark tag, in which they searched for muons from bottom-quark decay.

CDF submitted one additional top-quark paper on the 1992–93 data. Here they focused on the kinematic features of the events (particle energies and directions) rather than the presence of bottom quarks. By comparing the jet-energy distributions with those expected for background and top-quark signal, they provided additional independent evidence that top quarks were present in their data sample.

FIRMLY ESTABLISHING THE EXISTENCE OF THE TOP QUARK

Early in 1994, the Fermilab collider had resumed operation, and both experiments were continuing to increase their data samples. In February 1995, the CDF group had sufficient evidence to firmly establish the top quark. Following a previously agreed-upon procedure, they informed the D0 collaboration. Since D0 also had a clear top-quark signal, the two groups made a joint announcement and simultaneously submitted their papers to *Physical Review*. There were now papers from two groups, each reporting odds of a million to one against the observation being pure background. The search for the top quark was over.

In the three months since their November 1994 paper, the D0 team had further optimized their analysis for a very heavy top quark, with mass near 180 GeV. The most significant change was an increase in the requirement on the total energy of the jets in an event. For a 180 GeV top quark, there would be at least 360 GeV energy in the decay products of the $t\bar{t}$ pair. Much lower energy was expected from background sources. The expected signal-to-background ratio was now a factor of 2.5 better and their data sample 3.5 times larger than reported in November. Summed over the dilepton and single-lepton channels, they found 17 events compared with a four-event expected background. This corresponds to a 4.7 standard deviation excess. The production rate that they observed was consistent with the theoretical prediction for the top-quark mass that they measured: 199 ± 30 GeV.

The CDF paper differed in two important ways from their 'evidence' paper of six months earlier. First, they had 3.5 times as much data as before. Second, they improved the efficiency for detecting the bottom-quark decay point by almost a factor of two. Their efficiency for identifying at least one of the bottom quarks in a top event was now over 50%. The sample of bottom-quark-tagged single-lepton events and dilepton events was now up to 43 events, a 4.8 standard deviation excess over background. With larger statistics, the two worrisome features of the previous paper were now in good agreement with expectations. In fact, the relative numbers of single-lepton and dilepton events with 0, 1, and 2 jets tagged as bottom quarks were as expected for events containing two W bosons and two bottom quarks, as in $t\bar{t}$ events. With the increased number of events, CDF improved its mass measurement to 176 ± 13 GeV. A sketch of an event is shown in figure 13.6. It has all the essential characteristics of $t\bar{t}$ production: an electron and a neutrino consistent with the decay of a W boson, two jets identified as bottom-quark jets by the displacement of the jet vertex from the $\bar{p}p$ collision point, and the two remaining jets with a net reconstructed mass as expected for the decay $W^- \rightarrow \bar{u}d$.

STUDYING THE NEWLY DISCOVERED QUARK

Once the existence of the top quark was firmly established, the focus of the investigation changed to the properties of this extremely massive object. Because the top quark is

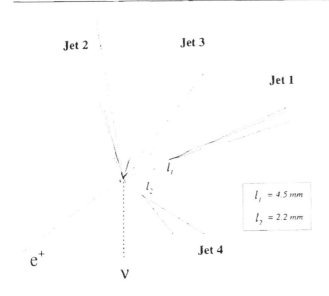

Figure 13.6. Sketch of one of the top events collected by the CDF. The two bottom-quark jets (jets 1 and 4) have vertices displaced from the collision point by a few millimetres.

35 times more massive that the next heaviest quark and is the only quark or lepton with a mass near the scale of the electroweak symmetry breaking (set by the W and Z boson masses), it is possible that the top quark's properties may provide some unexpected surprises. These could help us to understand how the Standard Model fits into a more complete description of the elementary particles and forces. Moreover, the top quark has a very short lifetime, $< 10^{-24}$ s, because it decays into a real W boson and there is a very large energy released in its decay. As a result, the top quark is unique among the quarks, decaying before binding into a meson. Thus the decay products reflect only the properties of the top quark rather than the properties of a two-quark bound state.

Initial studies are often severely limited by small numbers of events, and so it was for the top quark. However one proceeds with what is available. Here we shall consider the search for additional top-quark decay modes, the use of the top-quark mass for a precision test of the Standard Model, and the search for decay modes forbidden in the Standard Model.

Additional decay modes

The discovery of the top quark relied only on the electron and muon single-lepton modes and the ee, $\mu\mu$, and eμ dilepton channels. As seen in table 13.1, this represents only 35% of all top-quark decays. The remaining decay modes contain either τ leptons or no leptons at all. Using the complete 1992–95 data set, the experimenters were able to study the all-jet channel and to make the first measurement of top-quark decay into τ leptons.

The CDF observation of the all-hadronic decay of $t\bar{t}$ pairs used events containing at least five hadron jets with large total energy. Because such events do not contain any leptons, the pure strong-interaction background due to the scattering of quarks and gluons in the $\bar{p}p$ collision is very large, with an expected signal-to-background ratio (S/B) of 1/200. By requiring that at least one of the two bottom quarks in an event be identified, S/B was improved to 1/20. Since this was still not good enough, two additional sets of requirements were applied. In the first, they required that the distribution of energy in the event, such as how spherically it is distributed, be as expected for top-quark production. In the second method, they required that both bottom-quark jets be identified. In either case, S/B was reduced to 1/3–1/4, sufficient to permit extraction of a signal. The observed $t\bar{t}$ production rate was consistent with that obtained in other decay channels. Moreover, they extracted an independent measurement of the top-quark mass, 186 ± 16 GeV.

CDF found evidence for τ leptons in top-quark decay by searching for dilepton events where an electron or muon is accompanied by a τ lepton that decays into hadrons plus a neutrino. The hadron jet from the decay of a high-energy τ is extremely narrow and contains either one or three charged particles. These characteristics were used to separate τ candidates from generic hadron jets. Four events were observed while two were expected from background. The meagre statistics made it difficult to conclude that a signal had been observed. However, the fact that three of the four events contain a jet identified as a bottom quark made the $t\bar{t}$ interpretation much stronger.

The top quark mass and tests of the Standard Model

In the Standard Model, the masses of the charged and neutral carriers of the weak interaction, W and Z, are related. The largest effect is due to the mixing of the weak and electromagnetic interactions, but there is a more subtle effect from the ability of a Z to turn into a b$\bar{\text{b}}$ pair for an instant, whereas a W would in the same way, because it has electric charge, become a t$\bar{\text{b}}$ for a short time, after which the t$\bar{\text{b}}$ would become a W again. If the top and bottom quarks had identical masses, the effect on the W and Z would be the same. The larger the mass difference between the top and bottom quarks, the larger the difference in the W and Z masses. Since the Z mass has been measured with extraordinary precision at the e$^+$e$^-$ collider at CERN, the plot that is usually made to show this effect is W mass versus top-quark mass.

Both CDF and D0 have improved their techniques for measuring the top-quark mass. The D0 group employs a multivariate discriminate to calculate the relative likelihood that the topology of the lepton and jets comes from t$\bar{\text{t}}$ decay or background. Using this discriminate along with the calculated mass for each event provides them with a better separation of the signal and background and thus improved mass resolution. Their 1998 result for the top-quark mass is 172 ± 7 GeV.

CDF has improved its mass measurement by separating data into subsamples based on the number of identified bottom-quark jets. They now take into account the decreased background and improved resolution that come from identifying more of the bottom quarks. Their 1998 mass measurement is 176 ± 7 GeV.

The test of the Standard Model is shown in figure 13.7. The top-quark mass from CDF and D0 and the W boson mass from CDF, D0 and the LEP e$^+$e$^-$ experiments at CERN are compared with the theoretical predictions for a number of possible Higgs boson masses. The data are consistent with the Standard Model and favour a relatively low-mass Higgs boson H.

Rare decay modes of the top quark

In the Standard Model, the top quark always decays to a bottom quark and a W boson. If there were other decay

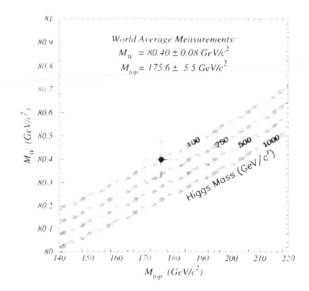

Figure 13.7. Top-quark and W-boson mass measurements compared with the predictions of the Standard Model for different Higgs boson masses.

modes with substantial branching fractions, it could be the first sign of physics beyond the Standard Model. We shall consider the possibility of t → Wd or t → Ws decay. Then we shall see whether a W boson is produced in each top decay. Finally we shall look at a top-quark decay channel containing neither a W boson nor a bottom quark.

CDF measured the ratio of decay fractions of t → Wb relative to t → Ws and t → Wd by comparing the numbers of observed t$\bar{\text{t}}$ events with 0, 1 or 2 identified bottom-quark jets. Their result is 100%, but with a large statistical uncertainty of 30%. This is one of many measurements that will greatly benefit from a large increase in the data sample.

In some extensions of the Standard Model, such as supersymmetry, there is more than one Higgs boson. If one is electrically charged and its mass is below the top-quark mass, then the decay t → Hb is possible and could in fact dominate in top-quark decay. A charged Higgs boson can decay into either a pair of quarks or a τ lepton plus a neutrino. CDF considered this possibility by searching for events containing either two τ leptons and neutrinos or one τ lepton, a neutrino and jets, one of which is identified as a bottom quark. They observed seven events, consistent with the expected background. The absence of a signal enabled them to exclude a charged Higgs boson with mass as high as 150 GeV for versions of the theory in which the Higgs decays into $\tau\nu$.

The final search was for top-quark decay in which neither a W boson nor a bottom quark is produced. Such decays, termed flavour-changing neutral current decays, are $t \rightarrow u\gamma$, $t \rightarrow uZ$, $t \rightarrow c\gamma$ and $t \rightarrow cZ$. CDF searched for $t\bar{t}$ events in which one top quark decayed in the usual mode, $t \rightarrow Wb$, and the other decayed into a quark and either a photon or a Z^0 that subsequently decayed into e^+e^- or $\mu^+\mu^-$. They required a high-energy γ or Z^0 plus jets or leptons from the W, the bottom quark, and the up quark u or charm quark c. The absence of a signal above background led to the conclusion that top-quark decays into $q\gamma$ less than 3% of the time and t decays into qZ^0 less than 33% of the time. The latter mode was severely limited in statistics because Z^0 bosons decay into e^+e^- or $\mu^+\mu^-$ only 7% of the time.

Thus, with the limited data available to date, no decay mode of the top quark has been observed other than the Standard Model $t \rightarrow Wb$ mode.

PROSPECTS FOR TOP-QUARK INVESTIGATION

The top-quark discovery period ended in the beginning of 1996 when the 3.5 year long Fermilab Tevatron collider run ended. Further significant progress in understanding the properties of Nature's heaviest building block requires a large increase in the number of top-quark events that can be studied. This will occur in the year 2000, after Fermilab completes its new Main Injector accelerator. When this is able to deliver protons and antiprotons to the Tevatron, the $t\bar{t}$ production rate will increase by a factor of 10, and perhaps ultimately by a factor of 50. When combined with the major CDF and D0 detector upgrades now under construction, Fermilab experimenters should collect in the first two years of Main Injector operation 40 times the data that they collected during the 1992–95 period. Both CDF and D0 are completely rebuilding their tracking detectors to handle the higher accelerator beam intensities and to increase the fraction of $t\bar{t}$ events that they observe. Most significant is the addition of a magnet to the D0 detector. This will allow that group to identify bottom-quark jets by

their flight path of a few millimetres, as CDF has been able to do.

By the end of 2001, CDF and D0 should each have collected a few hundred dilepton and 1000–2000 single-lepton $t\bar{t}$ events. With that much data, the uncertainty in the top-quark mass could be reduced to 2 GeV or less. Since the W mass should be measured to ±25 MeV by that time, the data point in figure 13.7 will have error bars reduced by approximately a factor of three in both directions. This will provide an extremely precise test of the Standard Model and an indirect determination of the mass of the Higgs boson.

The large data sample will allow important studies of top-quark production and decay. The lifetime of the top quark, estimated to be approximately 10^{-25} s, is too small to be measured directly. However, it can be deduced by measuring the production rate of single top quarks via the weak interaction, rather than the usual $t\bar{t}$ production via the strong interaction. Each experiment should collect a few hundred such events and thus determine the top-quark lifetime. In addition, there will be sensitivity to rare top-quark decay modes at the few tenths of a per cent level as well as the ability to see anomalous production properties such as $t\bar{t}$ as decay products of new very massive particles.

In 2005, there will be another major improvement in our ability to study the top quark. When the 14 TeV Large Hadron Collider (LHC) begins operation at CERN, experimenters will collect 200–1000 $t\bar{t}$ events per day! This will provide high sensitivity for measuring the properties of the top quark as well as searching for unexpected effects in top-quark production and/or decay.

CONCLUSION

The search for the top quark, under way since the discovery of the bottom b quark in 1977, culminated in its discovery in 1994–95. It is by far the most massive of Nature's elementary building blocks so far observed. Since it alone has mass on the scale at which the electroweak symmetry is broken, its properties could provide important clues on the mechanism for that symmetry breaking. The measured mass of the top quark is in good agreement with the prediction of the Standard Model. With the expected increase in the number of top-quark events of two to three orders of magnitude in the next 5–10 years, we shall learn a

great deal about the properties of the top quark and perhaps, if we are fortunate, of physics beyond the Standard Model.

FURTHER READING

Campagnari C and Franklin M 1997 The discovery of the top quark *Rev. Mod. Phys.* **69** 137

Liss T M and Tipton P L 1997 The discovery of the top quark *Sci. Am.* **277** 54

Quigg C 1997 Top-ology *Phys. Today* **50** 20

Wimpenny S J and Winer B L 1996 The top quark *Ann. Rev. Nucl. Particle Sci.* **46** 149

ABOUT THE AUTHOR

Melvyn Shochet is Elaine and Samuel Kersten Professor of Physical Sciences at the University of Chicago. He received his undergraduate education at the University of Pennsylvania and a PhD from Princeton University. His early experiments were on the rare decays of K mesons. When the Fermilab accelerator began operation in 1972, he worked on experiments that helped elucidate the quark composition of hadrons.

Shochet was a member of the small group formed in 1976 to study the possibility of operating the Fermilab accelerator as a collider. From this project, the CDF collaboration was formed. Shochet was scientific co-spokesman of CDF from 1988 to 1995.

14 THE STANDARD MODEL AND BEYOND

Graham G Ross

Editor's Introduction: This chapter reappraises the Standard Model (SM), the cornerstone of our current understanding of particle physics, before going on to point out why the SM cannot be the ultimate theory. The deficiencies of the SM hint at deeper symmetries, while radically new ideas such as supersymmetry, incorporating the SM, have far-reaching implications for our interpretation of the world around us. Dispensing with conventional ideas of points in space–time, new fundamental descriptions in terms of 'strings' in multi-dimensional spaces hold out the exciting possibility of incorporating gravity into the picture as well, providing an ultimate unification of all Nature's forces—the long-sought 'Theory of Everything'.

INTRODUCTION

The second half of the twentieth century has seen the remarkable development of theories of the strong and weak interactions to match quantum electrodynamics (QED), the theory for the electromagnetic interactions. These theories are all quantum field theories based on the principle of local gauge invariance. Somewhat prosaically christened the 'Standard Model' (SM), this development represents one of the triumphs of modern physics. It describes essentially all the observed phenomena of the fundamental interactions as proceeding through the exchange of the elementary quanta of the force carriers: the photon for the electromagnetic interaction, the gluons for the strong interaction and the $W^{\pm}$ and Z bosons for the weak interactions. However many physicists think that the Standard Model is incomplete and are pursuing the elusive 'Theory of Everything' which they think represents the ultimate unification of the fundamental forces, including gravity. In this chapter I shall try to explain why the SM is thought to be inadequate and describe some of the exciting ideas that have been proposed to extend it.

The Standard Model

The Standard Model describes all the matter that we observe in the universe in terms of a small number of elementary particles with interactions described by simple laws. The protons and neutrons which make up the atomic nucleus are made from elementary constituents, the (charge $\frac{2}{3}$) up

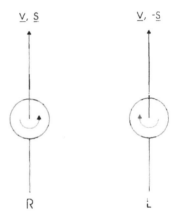

Figure 14.1. Spin and helicity. At the quantum level, elementary particles possess a property called spin which is an intrinsic angular momentum, such as is possessed by a spinning top. Massless spinning states may be distinguished by whether they are right handed (R) with their spin angular momentum s aligned along the velocity v, or left handed (L) with their spin angular momentum anti-aligned along the velocity. Helicity is defined as the projection of the spin along the direction of velocity. Two classes of spinning particles are found. Those with integer spin are called bosons. In the SM these include the gauge bosons with spin 1 and the Higgs boson with spin 0. Gravity is mediated by the graviton with spin 2. There are also particles with half-integer spin known as fermions. In the SM the fermions are the quarks and leptons.

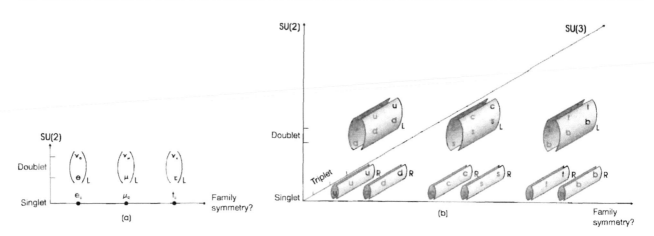

Figure 14.2. The fundamental states of matter. (a) The symmetry patterns of the lepton states in the SM. The left-handed (L) charged lepton and neutral neutrino states transform as a doublet and are connected via $W^{\pm}$ emission, e.g. $e_L^- \rightarrow W^- + \nu_{eL}$. The right-handed (R) charged lepton states are singlets, i.e. they do not couple to the W bosons. This asymmetry means that the weak interactions violate parity, a symmetry which interchanges L and R states. (b) The symmetry patterns of the quark states. They have the same pattern under the SU(2) symmetry but extended into a new dimension, the SU(3) colour symmetry. All quarks are coloured, with red, green or blue colour charge. (In the shaded regions of the figure, the quarks are as follows: leftmost, red; middle, green; rightmost, blue.) These states are connected via gluon emission. Since both LH and RH states carry the same colour the strong gluon interactions respect parity.

and the (charge $-\frac{1}{3}$) down types of quark and the 'gluons', the carriers of the strong force which holds them together. In addition the SM includes the electron (charge -1) as an elementary state and its neutral partner, the neutrino. These 'leptons', like the quarks, feel the electromagnetic and weak forces but, unlike the quarks, do not feel the strong force. To complete the SM, three families, or copies, of the quarks and leptons are needed to describe the states currently observed. These states also possess a property called 'spin' (see figure 14.1).

The laws that govern the interactions are derived using the same principle which leads to the extremely successful theory of electromagnetism, QED, the quantum version of Maxwell's theory of electromagnetism. Like QED, they follow from the principle of local gauge invariance applied to a relativistic field theory. The underlying ingredient is the recognition of a symmetry relating the states of the theory. Such a symmetry is based on patterns and the patterns of the SM are illustrated in figure 14.2. Consider first the pattern of the electron and its neutrino shown in figure 14.2(a). The experimental observation of the weak interactions of

the electrons and its neutrino suggest that these states are intimately related and have similar properties. The charged weak interaction arises through the exchange of a W boson which changes a left-handed electron into a left-handed electron neutrino (see figure 14.1 for the meaning of handedness). The neutral weak interaction arises through the exchange of a Z boson which couples an electron to an electron or a neutrino to a neutrino with similar coupling strengths. Given these similarities it is tempting to suggest that the electron and its neutrino are different aspects of the *same* state. This is what is done in the Standard Model, assigning them to a 'doublet' as is shown in figure 14.2. This implies that physics is unchanged if one replaces the electron by its neutrino and vice versa. Indeed the symmetry is much richer, saying that the physics is unchanged by the replacement of the electron by a combination of the electron and the neutrino. This is known as an SU(2) 'gauge' invariance, the 2 referring to the (complex) rotation of the two members of the doublet. The right-handed components of the electron, muon and tau are singlets under SU(2) and hence do not change under an SU(2) rotation. In turn this

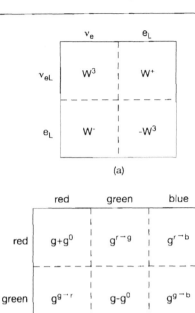

	v_e	e_L
v_{eL}	W^3	W^+
e_L	W^-	$-W^3$

(a)

	red	green	blue
red	$g+g^0$	g^{r-g}	g^{r-b}
green	g^{g-r}	$g-g^0$	g^{g-b}
blue	g^{b-r}	g^{b-g}	$-2g$

(b)

Figure 14.3. Gauge boson couplings. (a) The coupling of the W bosons to the leptons. For SU(2) there are $2^2 - 1 = 3$ gauge bosons. The minus sign in the diagram indicates that the couplings of the neutral boson W^3 to the neutrino and to the electron have equal magnitudes but opposite signs. (b) The couplings of the gluons to the quarks. For SU(3) there are $3^2 - 1 = 8$ gauge bosons. Note that for the diagonal entries the charges describing the coupling of the colour neutral gluons, g and g^0, sum to zero.

means that they carry no weak charge and do not couple to the weak gauge bosons and do not participate in the weak nuclear force. Note that we have not observed right-handed components of neutrinos and they are not included in the SM particle content.

The invariance principle is further extended to allow for independent SU(2) rotations at different space–time points. This is known as a 'local gauge' symmetry and is motivated by the idea that in a relativistic theory it is impossible for observers at different space–time points to

communicate instantaneously because this would violate causality through the propagation of information faster than the speed of light. However in time the information about the precise 'definition' of an electron must be transmitted from one space–time point to another and so one may expect that such a local symmetry should be accompanied by a means of 'standardizing' the definition. Implementing local invariance in a gauge theory shows how this is achieved through the appearance of 'gauge bosons' which carry this definition from point to point in a manner that does not violate causality. For SU(2) the gauge bosons are the $W^\pm$ and Z bosons mediating the weak interactions. Figure 14.3(a) shows how a gauge boson couples to each pairing of the fundamental doublets. The two 'weak charges' needed to distinguish the electron and the neutrino lead to three weak bosons carrying three different weak charges. The remarkable thing is that the observed structure of the weak interactions corresponds to just this structure which follows from the initial recognition that there is a local SU(2) gauge symmetry.

Our modern view of the origin of the fundamental forces rests on the realization of the importance of symmetry in determining the dynamics of the system. Our theory of the strong interactions is based on a new symmetry, SU(3), with each quark coming in three 'colours' and belonging to a triplet of states related by SU(3) rotations as shown in figure 14.2(b). There are eight gauge bosons, 'gluons', associated with SU(3) corresponding to the different ways that they can couple to the three colours of quarks as is illustrated in figure 14.3(b). The gauge theory of the strong interactions is called quantum chromodynamics (QCD).

In order to include electromagnetism in the Standard Model, it is necessary to add a 'U(1)' gauge symmetry corresponding to the local conservation of electric charge. This gives rise to a single gauge boson B (since there is only one type of electric charge). The photon is made up of a mixture of B and the W^3 gauge boson of figure 14.3(a). The orthogonal combination of the B and W^3 gauge bosons is the neutral weak boson, the Z. This completes the gauge symmetry structure of the Standard Model. The full gauge symmetry is SU(3) ⊗ SU(2) ⊗ U(1).

The SM has been found to be amazingly successful in describing all phenomena associated with the strong and electroweak interactions, the success culminating with the

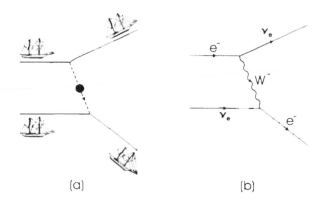

(a)   (b)

Figure 14.4. Exchange forces. Forces in a quantum field theory arise owing to the transmission, or exchange, of 'virtual' states, states which cannot exist classically because their production would violate the conservation of energy and momentum. At the quantum level they can exist for a short time Δt owing to the uncertainty principle which allows a fluctuation ΔE in the energy provided that $\Delta t \Delta E \le \hbar$. The quantum 'exchange force' may be illustrated through an analogy with the classical case. For example a cannon ball 'exchanged' between ships carries momentum and induces a force on both ships as shown in (a). Similarly the exchange of a gauge boson generates 'exchange' forces. For example the W boson in (b) induces a force between the electron and the neutrino. Unlike the classical case the force may either be attractive or repulsive. For a virtual W exchange $\Delta E \simeq M_W$ where M_W is the mass of the W boson and from the uncertainty principle we see that the force acts only over a very short time $\Delta t \le \hbar/M_W$. Hence the virtual W can 'travel' only a short distance and thus the force produced acts only over a short range. Since the net force is proportional to the time that it acts, the strength of the W exchange force is g_w/M_W, where the constant of proportionality, g_w, is the weak charge describing the coupling of the W boson to the doublet containing the electron and its neutrino.

discovery of the W and Z bosons in 1983 at CERN (see the chapter by Rubbia) and the gluons at DESY (see the chapter by Wu). The exchange forces associated with these gauge bosons provides us with a unified quantum description of the strong, the weak and the electromagnetic interactions (figure 14.4) and, to date, essentially all observed phenomena are consistent with the predictions of the SM.

An incomplete theory?

Given this success why is it that many physicists think that the Standard Model is only a partial step towards what has been called the 'Theory of Everything'? Because the SM leaves so many questions unanswered, it seems that there must be some some more basic underlying theory if we are to claim a detailed understanding of the fundamental forces of nature.

- Question 1: Why do the weak interactions violate parity, i.e. why do the quarks and leptons only

interact with the charged weak current in a left-handed manner?

- Question 2: Where are the missing (right-handed) neutrino states? Why are only the neutrinos massless?

- Question 3: Why are the charges of the quarks quantized in fractions a third of that of the charged leptons?

- Question 4: Why are there three families of quarks and leptons?

- Question 5: What determines the relative strengths of the various gauge interactions?

- Question 6: What determines the masses and mixing angles of the various states in the SM?

There are still more ways in which the SM seems incomplete. Although it unifies the strong, weak and electromagnetic interactions, describing all by local gauge field theories, the SM does not complete this unification for the strengths of these interactions are not related and no explanation is given of why they should be so different when measured in the laboratory. Moreover there is no explanation given for why the weak interactions should be short range, mediated by massive gauge bosons, while the electromagnetic and strong forces are mediated by massless gauge bosons. (The short range of the strong interactions may be understood by the fact that at long distances the strong interactions become very strong (see the chapter by Wu), preventing the quarks from moving far from each other.) The Standard Model is able to parametrize this difference by the spontaneous breaking of the weak gauge symmetry through the Higgs mechanism, postulating the existence of a further fundamental particle, the Higgs boson, which determines the vacuum state of the system (see the boxed text on page 160). The SM employs a special choice of Higgs boson to give the weak gauge bosons a mass. However, it provides no explanation of why it is the weak and not the strong or electromagnetic interactions that must be altered in this way.

Finally, and perhaps most importantly, the unification is of only three of the four fundamental interactions and there is no connection with the gravitational force; so we have another fundamental question to answer.

- Question 7: Does unification of the fundamental forces *including* gravity occur?

Attempts to unify with gravity have focused attention on a further question:

- Question 8: What generates the hierarchy of mass scales needed to describe the fundamental interactions?

The gravitational coupling is proportional to Newton's constant with inverse mass dimensions:

$$G_N \propto M_{\text{Planck}}^{-2}.$$

M_{Planck} is the Planck mass, approximately 10^{19} GeV/c^2 (the unit of mass used in particle physics is 1 GeV/c^2, approximately the mass of the proton). The weak coupling is given at low energies by the Fermi constant $G_F \propto M_W^{-2}$, where the W boson mass M_W is approximately 10^2 GeV/c^2, some 17 orders of magnitude smaller than the Planck mass. The quark and charged lepton masses range from $\frac{1}{2} 10^{-3}$ GeV/c^2 for the electron to approximately

175 GeV/c^2 for the top quark while the neutrinos in the SM are massless. For many the origin of this hierarchy is the most pressing of all the questions raised by the SM.

TOWARDS A THEORY OF EVERYTHING

History has been very unsympathetic to attempts to construct a 'Theory of Everything'. From the Greeks' description of the physical world in terms of just four indivisible elements to the discovery of the nucleon and the suggestion that it was the fundamental building block of the nucleus, the idea that we have at last found the ultimate elementary structure has been proved wrong as further substructure has been discovered. Is there any reason to think that the Standard Model is different or is there a new set of fundamental constituents, building blocks of the SM states, just waiting in the wings to be discovered? Some support for the former possibility is provided by the fact that no structure has been found in the quarks and leptons down to distance scales of 10^{-16} cm; for the electron this corresponds to a scale 10^5 times smaller than its Compton wavelength and the ratio is not much less for the light quarks. No elementary-particle candidate has survived such detailed scrutiny in the past.

The attempts to go beyond the SM broadly divide into two classes, characterized by whether the states of the SM are composite or elementary. In both classes the need to understand the hierarchy of mass scales imposes the most stringent constraint and requires new structure close to the energy scale that we have experimentally explored and probably accessible to the new accelerator, the Large Hadron Collider (LHC) currently being constructed at CERN.

Explaining the hierarchy of mass scales poses a problem, the 'hierarchy problem', because the point-like nature of elementary particles means that there may be large radiative corrections to their masses coming from virtual processes in which, as a result of the fundamental interactions, the particle splits into more than one elementary state. Although classically forbidden, such processes may proceed provided that, in accordance with the uncertainty principle, the virtual states exist only for a short time (see figure 14.4(b)). These radiative corrections are an integral part of the quantum theory and, in the

SPONTANEOUS SYMMETRY BREAKING

In the SM the strong, electromagnetic and weak interactions are all generated by local gauge field theories. While this explains the similarities between these interactions (they are all mediated by vector spin-1 bosons), it does not by itself explain the gross differences observed between the interactions. In particular the weak interaction gauge bosons, the W^+, W^- and Z, are massive with mass of order 100 GeV, while the photon and gluon are massless. In order to explain this difference the SM employs a mechanism known as 'spontaneous symmetry breakdown'. In this, although the laws describing the various interactions are the same, the symmetry between them is broken by the choice of ground state.

Perhaps the most familiar example in nature of this phenomenon is the spontaneous magnetization of a ferromagnet. As the ferromagnet cools below its critical temperature, the interaction between the magnetic moments at the atomic level causes them to align in domains of macroscopic dimension. Within a domain the rotational symmetry is obviously lost, although the underlying laws governing the magnetic moment interactions are clearly rotationally invariant. What has happened is that the aligned ground state breaks this symmetry so that the properties of the system are no longer manifestly symmetric. Only if the cooling is done many times is the rotational invariance obvious for the spontaneous breakdown occurs in a random direction and the magnetization averaged over many samples will be zero.

In the SM the symmetry breakdown giving the W and Z bosons a mass is also spontaneous. In this case the role of the order parameter, corresponding to the magnetization in our ferromagnetic example, is played by a new scalar (spin-0) field known as the Higgs boson. In the ground state of the theory the field acquires a non-zero expectation value spontaneously breaking the $SU(2) \otimes U(1)$ gauge symmetry and generating mass for quarks and leptons as well as the W and Z bosons. By demanding that the Higgs scalar should carry no electric or colour charge the strong and electromagnetic interactions are left unbroken, leaving the gluons and photon massless. A prediction of this mechanism is that there should be a new scalar state with an uncertain mass of the order of the Z boson mass. So far this state has not been found and a considerable part of the future experimental effort will be devoted to its discovery. Given its central role in the SM for the generation of mass it is clear its discovery is a crucial test of the theory. In Grand Unified Theories (GUTs) the process of spontaneous symmetry breaking is repeated but at a scale 10^{14} times higher than in the SM.

case of the SM, an estimate of such processes involving virtual momentum states up to the order of the Planck mass suggests they give a mass to the Higgs particles of the order of the Planck mass. This in turn would mean that the W and Z masses are similarly of the order of the Planck mass, quite unacceptably large. There are two ways to avoid the hierarchy problem, one in which some or all of the states of the SM are composite, and one in which the states of the SM are fundamental.

If the Higgs particle is not elementary but has structure at a scale characterized by the mass scale Λ (mass and distance are inversely proportional; 1 $(GeV/c^2)^{-1}$ corresponds to $\frac{1}{5} 10^{-13}$ cm), then virtual processes involving pointlike couplings to the Higgs particles cannot involve virtual momentum states above Λ; above that scale the Higgs particle must be described by its constituent parts. As a result the virtual processes only give a mass of order Λ and provided that $\Lambda \leq O(10^3 \, GeV/c^2 \equiv 1 \, TeV/c^2)$ these corrections are consistent with the measured masses of the W and Z bosons. Thus one is led to the conclusion that the scale of structure of the Higgs particle (and the associated massive W and Z particles) must be close to the scale that is currently being probed in the laboratory and should be accessible to discovery at the LHC.

If the states of the SM are elementary and we are to evade the hierarchy problem, there must be a reason why the radiative corrections driving the Higgs mass to the Planck scale are absent. This would follow if the theory possessed an approximate symmetry which, in the limit that the symmetry is exact, would require the electroweak group to be unbroken. Then the W and Z masses would naturally be small (small on the Planck scale even though the W and Z are among the heaviest states known!) if the breaking of the symmetry is also small. The possible symmetries consistent with relativity have been classified and it is known that only one, supersymmetry (SUSY), can play this role. An exciting possibility is raised by this new symmetry because, if it is a local symmetry, it requires the introduction of the graviton and the gravitational interaction. Thus SUSY raises the possibility of unifying gravity with the strong, electromagnetic and weak forces. Indeed string theories (see later), which are the only known candidates for providing a quantum theory of gravity, are consistent only if they are made supersymmetric and it is quite plausible that a relic of this SUSY may survive to low energies. For these reasons much work has been devoted to constructing the minimal supersymmetric version of the SM as the paradigm of a non-composite theory which avoids the troublesome hierarchy problem.

SUPERSYMMETRIC EXTENSION OF THE STANDARD MODEL

Like SU(2), SUSY relates two states of matter which are paired in new doublets. However, there is a fundamental difference between the symmetries. SUSY relates integer-spin bosonic states to half-integer-spin fermionic states.

This is in contrast with the local gauge symmetries of the SM which only relate states with the same spin. For this reason the extension of the SM to include SUSY requires the addition of many new states (see figure 14.5). These include the 'squarks' and 'sleptons', spin-0 partners of the spin-$\frac{1}{2}$ quarks and leptons respectively. Also needed are the 'gluinos', the 'Wino', the 'Zino' and the 'photino', spin-$\frac{1}{2}$ partners of the SM gauge bosons, the gluon, the W, the Z and the photon respectively. Although these new states

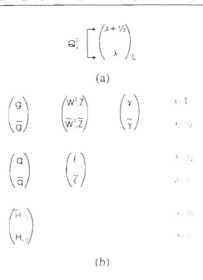

(a)

(b)

Figure 14.5. Supersymmetric patterns. (a) The basic symmetry pattern of supersymmetry (SUSY) involves *different* helicity states λ and $\lambda + \frac{1}{2}$ with the SUSY generator $Q^\pm$ relating integer-spin bosons to half-integer-spin fermions. These make up the basic 'supermultiplet'. (b) The supersymmetric assignments of the minimal supersymmetric standard model require the addition of new states. The gauge bosons are assigned to 'vector' supermultiplets with the helicity-1 gauge bosons partnered by new states, the helicity-$\frac{1}{2}$ 'gauginos'. The helicity-$\frac{1}{2}$ quarks and leptons are assigned to 'chiral' supermultiplets with helicity-0 partners, the 'squarks' and 'sleptons'. Finally the helicity-0 Higgs scalars are assigned to chiral supermultiplets with helicity-$\frac{1}{2}$ partners, the 'Higgsinos'. Note that in supersymmetric theories there must be two doublets of Higgs particles, one more than in the Standard Model. The local version of SUSY is supergravity and includes a theory of gravity. In this the helicity-2 graviton is partnered by the helicity-$\frac{3}{2}$ fermion, the 'gravitino'.

are thought to be considerably heavier than their partners and hence should not yet have been found directly in laboratory experiments, they will contribute as virtual states to the radiative corrections discussed above. The exciting observation that solves the hierarchy problem is that the new SUSY radiative corrections have the same magnitude as the SM contributions but *opposite* sign. Thus, when SUSY is exact, there is *no* correction to the Higgs mass. When SUSY is broken at a scale M_{SUSY} the new supersymmetric states acquire mass of $O(M_{SUSY})$ and this cancellation is spoilt. However the resultant contribution is of $O(M_{SUSY})$ and provided that this is not too large (i.e. $O(1\,\mathrm{TeV})$ or less) it is consistent with the observed electroweak breaking. As was the case for composite models, the supersymmetric solution to the hierarchy problem requires new physics beyond the SM at a scale accessible to the new accelerators. In this case the new physics is the spectrum of new SUSY states. Thus we see the SUSY SM answers one of the crucial questions posed earlier.

- Answer to Question 8: Supersymmetry solves the hierarchy problem.

However, the price paid is substantial since it more than doubles the number of states needed in the SM without offering any explanation for its multiplet structure and couplings. This defect may be remedied in 'Grand Unified' extensions of the theory and, as we shall now discuss, these extensions *do* offer explanations for all the unanswered questions raised above and can even generate the hierarchical difference on the Planck scale and the electroweak breaking scale. Indeed there are hints that the new SUSY states are needed to bring the predictions of GUTs for the gauge couplings of the SM into agreement with the measured values.

GRAND UNIFICATION

Grand Unified Theories (GUTs) seek to embed the Standard Model in a unified structure which can relate its multiplet structure and interactions by extending the patterns that led to the SM. The archetypical GUT is based on the symmetry group SU(5), chosen because it is the smallest group which can accommodate the $SU(3) \otimes SU(2) \otimes U(1)$ gauge group of the SM. In this the strong, the weak and the electromagnetic forces are just different facets of the one underlying (SU(5)) local gauge interaction. The quarks and leptons are also related in SU(5) because they belong to the common multiplets as shown in figure 14.6(a).

Just as described above, the symmetry now dictates how the gauge bosons couple. In particular the left- and right-handed states interact with the gauge bosons of the SM as shown in figure 14.6(b).

- Answer to Question 1: The charged weak interactions couple only to the left-handed states.

In addition, the coupling of the neutral gauge bosons must have charges that sum to zero for each of the multiplets. The coupling to the states of the five-dimensional representation contains three down quarks, the electron and its neutrino is given in figure 14.6(b). This explains one more puzzle.

- Answer to Question 3: The down quark charges must be quantized in units of $\frac{1}{3}$ so that the sum of three down quark charges can cancel the charge of $+1$ on the positron, just as required. Clearly the factor 3 is just the number of quark colours.

Of course there must be an explanation for the non-observation of the additional gauge bosons implied by SU(5). This must be due to another stage of spontaneous symmetry breaking with a new Higgs boson and giving a large mass, M_X, to the new X bosons (see figure 14.6). Provided that M_X is larger than the current energy of particle accelerators we shall not have been able to find the X bosons directly. However, a much stronger bound is available through the virtual effects of the X bosons since they mediate new processes which have not, so far, been observed. In particular the new interactions of SU(5) mediate proton decay as is shown in figure 14.7. The current experimental limit on the decay lifetime of the proton is an impressive 10^{32} years, giving a limit on the X boson mass of $M_X > 10^{16}\,\mathrm{GeV}/c^2$!

Note also that in SU(5) there is no room for a right-handed neutrino component. This offers an explanation for the different properties of the neutrino for the absence of this state forces the neutrino to be massless. In extensions of SU(5) to more unified GUTs such as SO(10) the missing right-handed neutrino appears. However it is expected to acquire a mass at the stage of spontaneous symmetry breaking of SO(10) to SU(5) which we expect to be very large and greater than M_X. This means that GUTs offer an answer to another of our questions.

- Answer to Question 2: Even if right-handed neutrinos exist they are superheavy. As a result the left-handed neutrino states remain very light with mass

$$\begin{pmatrix} d_r \\ d_y \\ d_b \\ e^+ \\ -\nu^c \end{pmatrix}_R \qquad \begin{pmatrix} 0 & u_b^c & -u_y^c & u_r & d_r \\ -u_b^c & 0 & u_r^c & u_y & d_y \\ u_y^c & -u_r^c & 0 & u_b & d_b \\ -u_r & -u_y & -u_b & 0 & e^+ \\ -d_r & -d_y & -d_b & -e^+ & 0 \end{pmatrix}_L$$

(a)

	d_q^{red}	d_q^{green}	d_q^{blue}	e_n^+	$\bar{\nu}_e$
d_R^{red}	g^0, γ, Z^0	g^{r-g}	g^{r-b}	$X_{-4/3}^{red}$	$X_{-1/3}^{red}$
d_R^{green}	g^{g-r}	g^0, γ, Z	g^{g-b}	$X_{-4/3}^{green}$	$X_{-1/3}^{green}$
d_R^{blue}	g^{b-r}	g^{b-g}	g, γ, Z^0	$X_{-4/3}^{blue}$	$X_{-1/3}^{blue}$
e_R^+	$X_{4/3}^{red}$	$X_{4/3}^{green}$	$X_{4/3}^{blue}$	γ, Z^0	W^+
$\bar{\nu}_e$	$X_{1/3}^{red}$	$X_{1/3}^{green}$	$X_{1/3}^{blue}$	W^-	Z^0

(b)

Figure 14.6. (a) The matter multiplet structure in SU(5). (b) The coupling of the gauge bosons of SU(5). The quarks and leptons of a single family may be accommodated in a 5 + 10 representation of SU(5) as shown where the superscript c refers to the antiparticle and L (R) refers to the left-handed (right-handed) helicity component. The weak gauge group SU(2) acts only on the last two rows and columns showing that the weak interactions involve only the left-handed quarks and leptons (in agreement with observation). There are $5^2 - 1 = 24$ gauge bosons which include the gluons, the $W^\pm$, Z and the photon. The remaining 12 'X' gauge bosons couple leptons to quarks. These acquire a very large mass on the GUT scale, M_X, explaining why they have not been observed. They mediate new interactions between the quarks and the leptons and lead to proton decay. In larger GUTs the multiplet structure may be further simplified; for example in SO(10) a single 16-dimensional representation contains the 5 + 10 of SU(5) together with a (right-handed) neutrino state which is also expected to acquire a mass of $O(M_X)$ or greater.

of $O(M_W^2/M_X)$ with the result that the left-handed neutrino states can only acquire a very small mass[1].

There remains, of course, the question of determining the couplings and masses in the SM. In a GUT, such as SU(5), with a single group factor there is a single gauge coupling constant describing the strong, weak and electromagnetic couplings. At first sight this seems disastrous, for the couplings of the various interactions as determined in the laboratory are quite different. The

[1] There is some evidence in experiments studying extraterrestrial neutrinos that the neutrino does have a very small mass of $O(10^{-11}$ GeV/$c^2)$, a value consistent with this explanation.

situation is drastically changed, however, by the inclusion of radiative corrections which take account of the effect on the couplings of virtual off-mass-shell states. These non-trivial corrections, inherent to the theory, may be calculated in any given GUT and, as we shall now discuss, offer a quantitative test of the theory.

Gauge coupling predictions

The success of the SM in predicting the properties of the strong force rests largely on a property of the SU(3) gauge theory known as 'asymptotic freedom'. Because of the ubiquitous radiative corrections the strength of the gauge

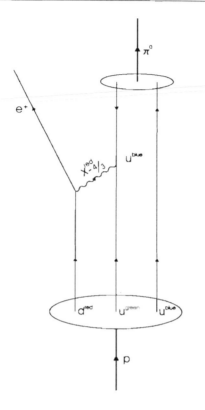

Figure 14.7. Proton decay. The process in SU(5) responsible for the decay of a proton into a positron and a pion proceeding via the exchange of a virtual superheavy X boson. The rate is proportional to g_X^4/M_X^4.

interaction depends on the length scale being probed. This is analogous to the screening of the nuclear electric charge in the atom by its electrons. Close to the nucleus, at nuclear length scales, the full nuclear charge is seen, but this reduces at longer length scales owing to the screening effects of the electrons. In the case of the radiative corrections to gauge coupling strengths the screening is due to the cloud of virtual states which can couple to the gauge bosons. In QCD it turns out that the special nature of the gauge group produces *anti*screening due to self-coupling effects amongst the gluons. As a result the strength of the strong interaction becomes smaller at short length scales; at progressively shorter separations, quarks behave more and more like free particles. This is the property known as 'asymptotic freedom'. Such variation in or running of the strong

coupling is essential to explain the observed properties of quarks as measured at particle accelerators. When probed at very small distances, less than 10^{-13} cm, the quarks in the nucleus appear to be almost free, corresponding to a small value of the strong coupling, while at distances larger than this they have large coupling and are strongly bound to form the nucleons: so-called 'quark confinement'.

The relevance of the running of the couplings to GUT predictions follows because the different couplings of the SM evolve differently. This is because the magnitude of the radiative corrections due to virtual states is dependent on the number of such virtual states. Since the number of such states is different for the various gauge group factors of the SM, the radiative corrections cause their couplings to evolve in different ways, with the SU(3) coupling falling faster than the SU(2) coupling and the U(1) coupling increasing (it has no antiscreening) as the length scale at which they are computed decreases. Remarkably the qualitative behaviour is in the correct direction for unification of the coupling constants, as may be seen in figure 14.8. The strong coupling is driven smaller than the other couplings for large values of the energy scale μ ($\equiv$ small distance scale $\propto 1/\mu$). However, the radiative corrections depend only logarithmically on the ratio μ and the couplings come together only at a very large value, $\mu = M_X$, $M_X = O(10^{16} \text{ GeV}/c^2)$. Such a large scale of Grand Unification proved to be essential for the viability of the theory for, owing to the unification of quarks and leptons, new processes become possible and GUTs typically violate baryon and lepton number because of the new gauge bosons associated with the enlarged gauge group (see figure 14.6(b)).

Although this qualitative picture is encouraging, the detailed quantitative comparison with SU(5) fails. Using the recent precision measurements for the weak couplings from CERN's Large Electron–Positron Collider (LEP) (see the chapter by Altarelli) and from neutrino scattering experiments together with the electromagnetic coupling constant which is accurately known, and the somewhat less accurately measured value for the strong coupling, one finds that the radiatively corrected couplings fail to meet at a point. Moreover the best value for M_X, although very large, still leads to an unacceptably fast rate of proton decay.

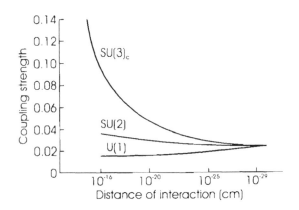

Figure 14.8. The evolution of the gauge couplings. In the supersymmetric version of the SM the couplings meet at a point in agreement with simple GUT predictions such as supersymmetric SU(5) which break in a single step to the SM. In this case M_X is large enough adequately to inhibit proton decay. Agreement with the GUT predictions is only maintained if the masses of the new supersymmetric states lie in the range from 100 GeV/c^2 to 10 TeV/c^2, the first indirect evidence for supersymmetry.

The situation changes dramatically in the supersymmetric versions of GUTS because, in calculating the evolution of the couplings, the new supersymmetric states must be included. Initially the non-supersymmetric SU(5) predictions were favoured, but the recent experimental improvement in the precision measurements of the gauge couplings has led to the conclusion that the supersymmetric prediction is quantitatively better than the non-supersymmetric result, thus answering Question 5. This only works if the new supersymmetric states are 'light' on the Grand Unified scale with masses less than 1 TeV/c^2 just as is required for a solution of the hierarchy problem and comfortably within the reach of the LHC.

Thus one may argue that there is already evidence for new forms of (relatively) light supersymmetric matter *and* an underlying GUT. The result is so dramatic that it needs to be evaluated with some caution. The most obvious reservation is that it relies on the theoretical extrapolation of the SM, albeit with the inclusion of the new supersymmetric states, twelve orders of magnitude beyond the energy scale at which it has been tested. No theory has proved to be so robust in the past; so

it is understandable if one views this extrapolation with some caution. Even given the framework of Grand Unification there is considerable uncertainty because the relative normalization of the coupling constants at M_X may differ in different unification schemes. Within the (large) class of theories which give the SU(5) predictions shown in figure 14.8 there are corrections coming from virtual states with mass of O(M_X) or greater which are not included in the analysis and which can affect the results substantially. Nonetheless it is remarkable that the simplest possible extension of the SM to include supersymmetry, coupled with the simplest assumption about Grand Unification, yield predictions in detailed agreement with experiment and require a very low scale of mass for the new supersymmetric states.

STRING THEORY: THE ULTIMATE UNIFICATION?

We have seen that many of the questions about the structure of the Standard Model have an answer in the context of Grand Unification. However even Grand Unification has fundamental problems which suggests there is a still more basic level of unification. The most obvious omission is that it does not provide a unification with gravity. Attempts to build a quantum field theory description of gravity have failed because of the appearance of uncontrolled infinities when calculating radiative corrections due to the exchange of virtual states. Such infinities are also present in any field theory such as the SM itself. While we have learned to live with the divergences encountered in the SM by a process called renormalization (see the chapter by Veltman), many see their existence as a signal of an underlying problem with the field theory approach. The origin of the infinities follows from the fact that the fundamental interactions occur at a point (see figure 14.9(a)). Thus in the case of a virtual exchange correction, such as is shown in figure 14.9(b), the virtual W^- quantum will be produced at the point x and absorbed at the scale y. In quantum theory, one must sum over all possible configurations consistent with the uncertainty principle. This includes the case where $x = y$ corresponding to the weak charges at the two interaction points coinciding. The same applies to electromagnetic and colour charge through equivalent graphs involving virtual photons or gluons. Classically Coulomb's law tells us there

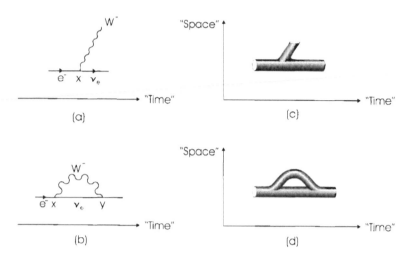

Figure 14.9. (a) Interaction at a point in field theory. (b) A radiative graph giving rise to an infinite correction in field theory. (c) An extended string interaction. (d) A radiative string correction.

is an infinite energy associated with coincident charges and it is the quantum analogue of this problem that leads to the infinities in the quantum field theory description of virtual states.

String theories provide a method of controlling the infinities of field theory and quantum gravity. The underlying reason for this is illustrated in figures 14.9(c) and 14.9(d). The string theory description of fundamental states involves an extension in a new 'space' dimension. Thus the basic pointlike interaction of field theory is replaced by an interaction extended in this new dimension, as shown in figure 14.9(c), and cannot be associated with a point. As a result the virtual corrections such as shown in figure 14.9(d) do not require configurations with coincident charges, smearing the would-be overlapping interactions in the new dimension. Because of this, string theories ameliorate the problem of divergences and provide us with the only known quantum theory capable of describing gravity and answering Question 7.

String theories have many novel features. In them our space–time is a derived quantity and often involves more than four space–time dimensions. To avoid conflict with observation these additional dimensions must be visible only at very short distances. In the 1920s, Kaluza and Klein showed how this may occur through a process known as

'compactification' in which 'space' in the extra dimensions is curved and will only be seen when probed at very short distances (of the order of the inverse Planck mass, a scale of $O(10^{-32}$ cm) at which the gravitational interactions become as important as the other fundamental interactions and cannot be ignored!).

String theory offers a method of achieving the ultimate unification of the four fundamental interactions including gravity. Does it answer any of the remaining questions (Questions 4 and 6), left unanswered by Grand Unification, which are related to the origin of the three families of quarks and leptons? A potential explanation comes from the Kaluza–Klein compactification. Since the states of the theory are confined in these extra dimensions, the level structure is quantized and one finds identical copies of light states distinguished only by a new quantum number associated with the compact dimensions. (The situation is similar to the structure found in atomic or nuclear physics where identical orbital and spin angular momentum states are found distinguished by different principal quantum numbers.) In string theories, where the idea of compactification has been most extensively studied, examples have been found with just the three families needed to describe the quarks and leptons.

As we have discussed, GUTs offer an explanation for

much of the multiplet structure of the SM and even make quantitative predictions for the gauge couplings. Since the quarks and leptons are related there may also be relations between their masses and couplings following from the GUT and some of these prove to be quite successful. However the original GUTs failed to reduce the overall number of parameters needed, largely because they require a much more complicated Higgs sector with its associated couplings in order to explain the pattern of symmetry breaking needed to reduce the GUT gauge group to that of the SM at the scale M_X. However, the modern versions of GUTs, based on compactified string theories, do offer the prospect of predicting *all* such couplings and masses. This follows because the underlying string theory has only one parameter, the string tension, related to the Planck mass. All other quantities are, in principle, predicted by the theory.

In practice these predictions have proved to be elusive because, although the underlying dynamics of the theory are uniquely determined, the vacuum structure is very complicated and determined by non-perturbative (at present incalculable) effects. As a result the work so far has analysed only a few promising candidates for the string vacua, chosen because they generate the three-generation structure of the SM. The qualitative structure is encouraging, specific cases giving multiplet patterns very like that of the SM. The one quantitative prediction that has been obtained so far is a prediction for the unification scale, M_X. While this varies slightly in different string compactifications, the general prediction is that it should be close to the only fundamental scale in the theory, set by the string tension which is quite close to the Planck scale. Remarkably this prediction, of order 5×10^{17} GeV/c^2, is not so far from the gauge unification scale which is found to be $(1-3) \times 10^{16}$ GeV/c^2. Considering that this is obtained from an extrapolation from the energy scale of only 10^2 GeV/c^2 of the present experiments, the agreement is not bad at all. It is difficult to overemphasize the importance of this result. Our belief that there is a stage of Grand Unification of the strong, weak and electromagnetic interactions rests largely on the quantitative success of the unification of the associated couplings. The successful prediction of the unification scale would be the first indication of unification with gravity. Given the lack in our present understanding of the choice of string vacua, the prediction can only be

viewed as exploratory, but it does suggest that compactified string theories may indeed provide the ultimate Theory of Everything.

CONCLUSIONS AND OUTLOOK

Our attempts to look beyond the Standard Model have been largely guided by the need to understand the observed hierarchy of mass scales and in particular to evade the hierarchy problem. The two explanations offered rely either on a further stage of compositeness or on a new symmetry, supersymmetry. In either case, new structure is predicted close to the electroweak breaking scale and accessible to the next generation of accelerators.

In the case of compositeness, the need to calculate the effects of the new strong binding force has prevented detailed predictions, although simple schemes have been constructed giving much of the structure of the SM below the composite scale. Attempts to build composite models have suggested ways to understand the pattern of quark and lepton masses, usually through the introduction of additional gauge interactions. It seems likely that to explain the structure of these interactions, some stage of Grand Unification with an enlarged gauge group will be necessary but, until the composite structures are identified, attempts at such unification seem premature.

By contrast, it is quite remarkable how predictive are the attempts to unify using SUSY and Grand Unification. There is even some indication, albeit very circumstantial, that the new supersymmetric states are needed to bring the predictions of Grand Unification into agreement with experiment. The extension of Grand Unified ideas to include compactified string theories offers answers to all the questions posed by the SM, and in principle determines all its parameters in terms of a single constant. A first comparison of the predictions is encouraging for the unification scale needed for gauge coupling unification is quite close to the expectation in string theory.

While such indirect hints are encouraging, the reader may justifiably ask what are the direct observable implications of these theories? The remarkable phenomenological success of the SM does not at present suggest the need for anything beyond it. Perhaps the only hint of new structure comes from astrophysics and cosmology. Astrophysical

1

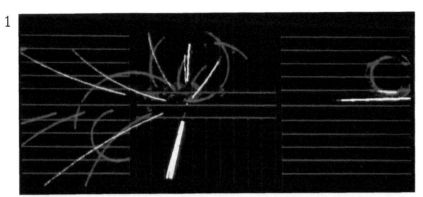

Plate 1: One of the proton–antiproton collisions recorded by the UA1 detector, showing the decay of a W particle into a tau and a tau neutrino. The subsquent tau decay is seen as a pencil-like jet (see the chapter by Rubbia). Photo CERN.

2

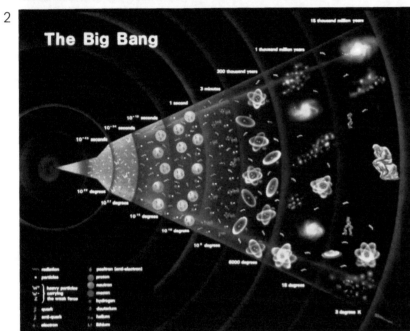

The Big Bang

Plate 2: Milestones in the evolution of the Universe from the Big Bang (see the chapters by Ellis and Shafi). Photo CERN.

Plate 3: An electron–positron annihilation event reconstructed in the ALEPH detector (see the chapter by Saxon). Photo CERN.

Plate 4: The ALEPH detector at LEP (see the chapter by Saxon). Photo CERN.

4

3

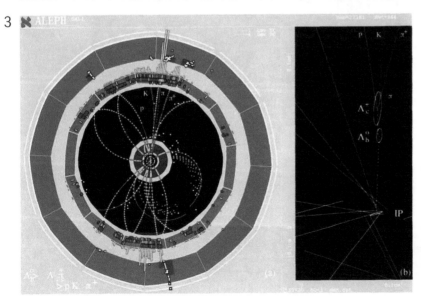

Plate 5: A simulated event in the central part of the ATLAS inner detector at CERN's LHC (see the chapter by Saxon).

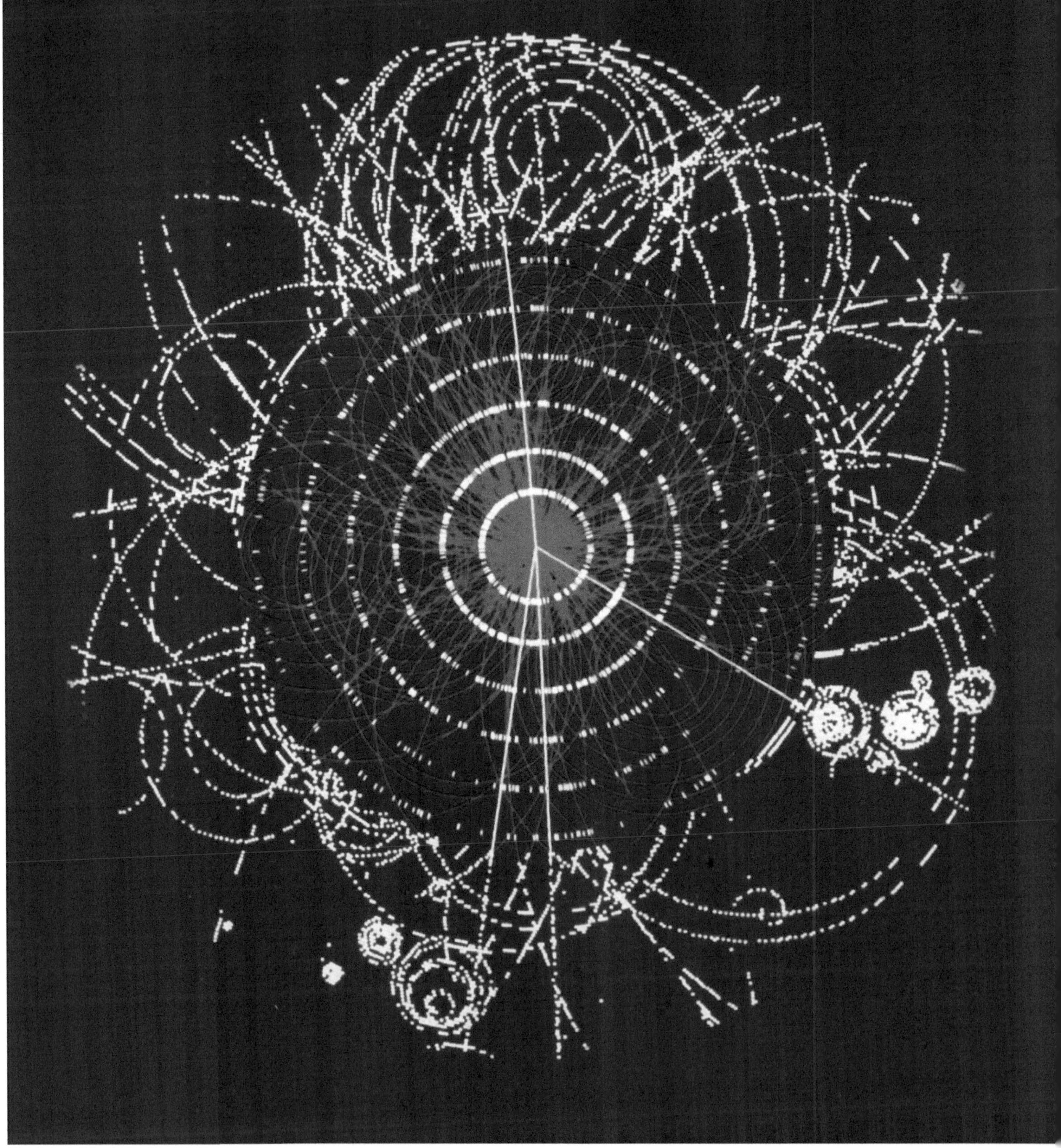

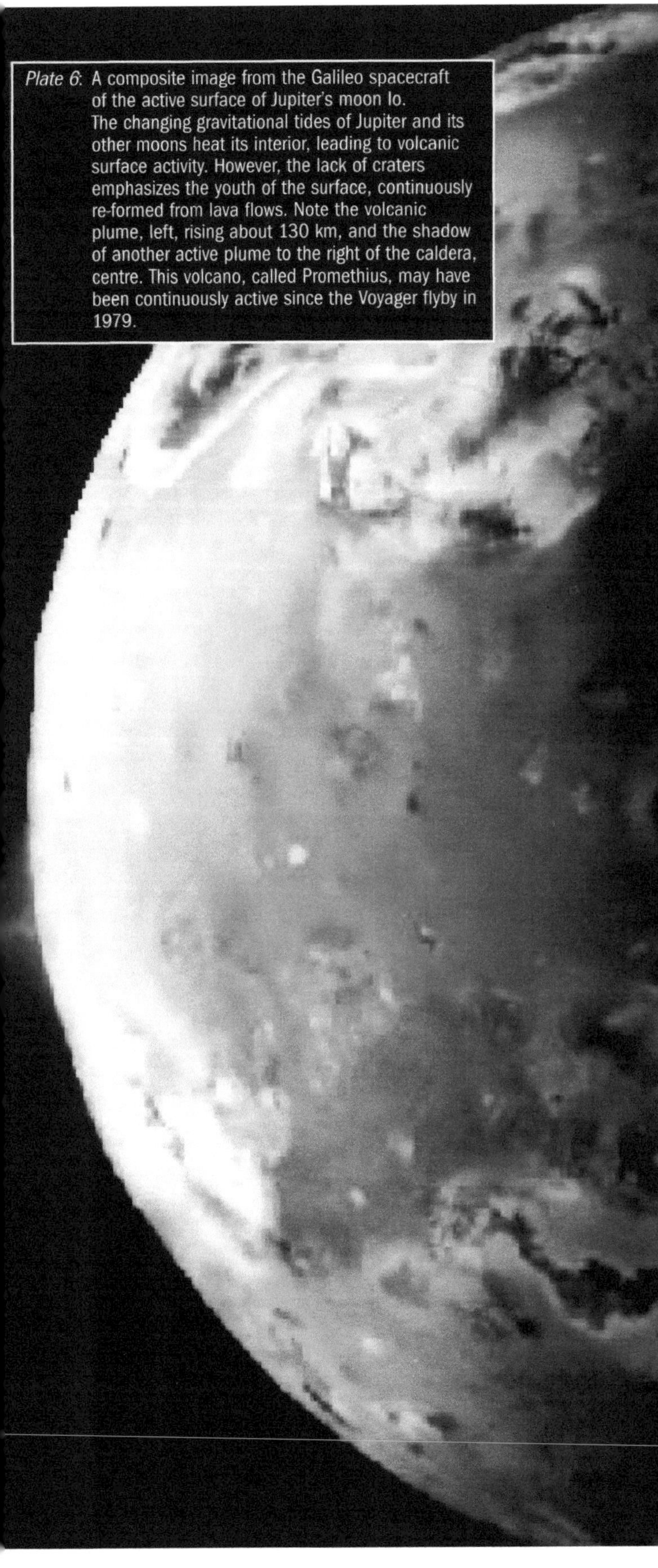

Plate 6: A composite image from the Galileo spacecraft of the active surface of Jupiter's moon Io. The changing gravitational tides of Jupiter and its other moons heat its interior, leading to volcanic surface activity. However, the lack of craters emphasizes the youth of the surface, continuously re-formed from lava flows. Note the volcanic plume, left, rising about 130 km, and the shadow of another active plume to the right of the caldera, centre. This volcano, called Promethius, may have been continuously active since the Voyager flyby in 1979.

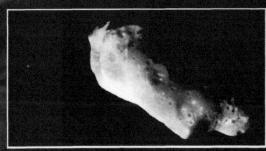

Plate 7: The 50 km long asteroid Ida photographed by the Galileo spacecraft on its way to Jupiter. Asteroids are sufficiently small that self-gravity does not overwhelm internal forces, so they need not be spherical. Also visible is the 2 km diameter moon, Dactyl, orbiting Ida.

Plate 8: The Sojourner rover examining a rock on the surface of Mars as part of the Pathfinder mission. The rover, equipped with an alpha-particle source and spectrometers to detect scattered alphas, protons and x-rays, is able to probe the composition of Martian rocks and soil.

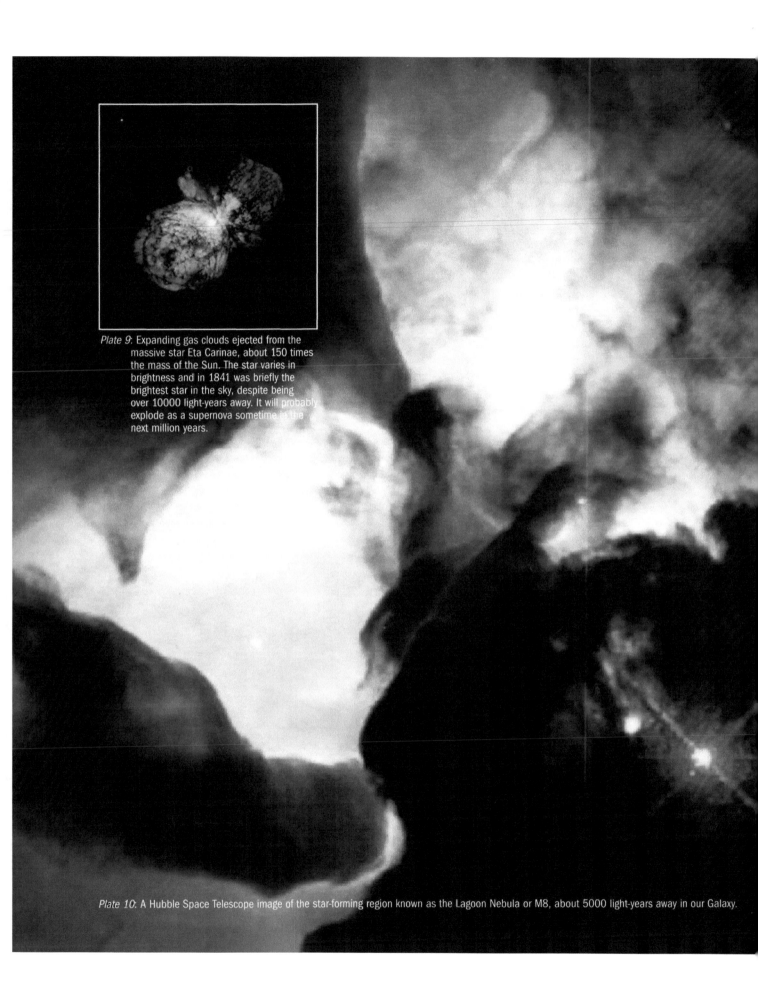

Plate 9: Expanding gas clouds ejected from the massive star Eta Carinae, about 150 times the mass of the Sun. The star varies in brightness and in 1841 was briefly the brightest star in the sky, despite being over 10000 light-years away. It will probably explode as a supernova sometime in the next million years.

Plate 10: A Hubble Space Telescope image of the star-forming region known as the Lagoon Nebula or M8, about 5000 light-years away in our Galaxy.

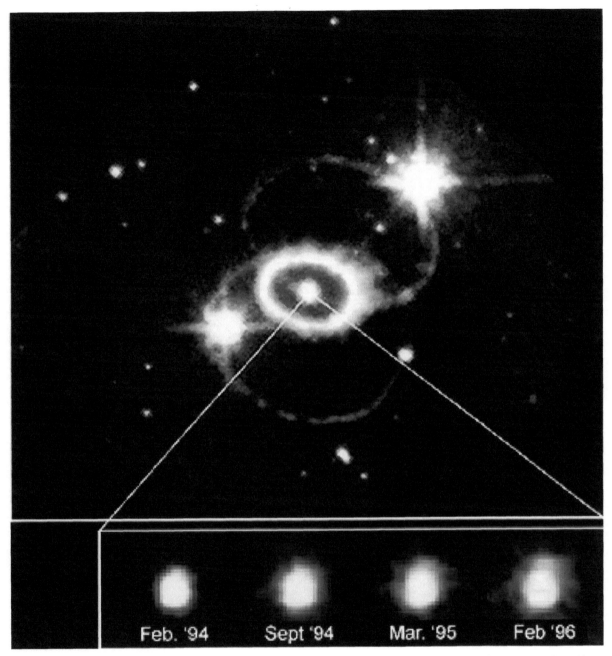

Plate 11: A Hubble Space Telescope image of the aftermath of supernova 1987A. The supernova remnant can be seen expanding with time in the inset pictures. SN 1987A in our satellite galaxy, the Large Magellanic Cloud, was the nearest and brightest supernova since the invention of the telescope. Note the bright ring around the remnant and the two unexplained outer rings.

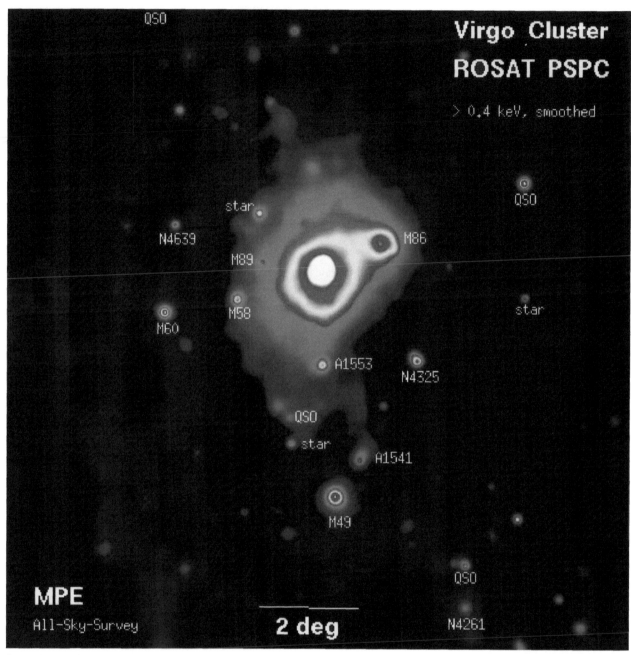

Plate 12: X-rays from the Virgo cluster of galaxies imaged by the position-sensitive proportional counter on ROSAT. The x-rays are emitted by the hot (about 20×10^6 K) intracluster gas. The main emission is centred on the giant elliptical galaxy M87 (see plate 14). Other points are mostly due to the hot (10×10^6 K) interstellar medium of individual elliptical galaxy members of the cluster (e.g. M49 and M60), some foreground stars in our Galaxy, some background quasars and a cluster (A1553). (Courtesy Dr H Böhringer.)

Plate 13a: Gravitational lensing of a single distant blue galaxy by the core of the massive cluster CL0024+1654. Most of the yellow galaxies seen in this Hubble Space Telescope image are in the cluster; the blue objects are multiple images of the same background galaxy. It is blue because it is actively forming massive stars; the cluster galaxies are red because they are not and mostly consist of old stars. Unlike a spectacle or telescope lens, a gravitational lens is non-linear and commonly produces multiple distorted images, which may be considerably amplified. In a sense, the cluster core is acting as the largest (highly distorting) telescope possible (ignoring the effects of the whole Universe).

Plate 13b: One of the most distant objects known is the galaxy at a red shift of 4.92, which has been lensed by the foreground cluster. This time the lensed galaxy appears orange because of the large red shift. The distant galaxy is shown larger to the top right and a reconstruction of its true shape is shown below. A few galaxies have since been seen at red shifts beyond 5.

Plate 14:
M87 is the nearest giant elliptical galaxy, in the Virgo cluster at a distance of about 50 million light-years. Doppler shifts seen in optical spectra of gas in a disc surrounding its nucleus indicate that it is orbiting a three-billion solar mass black hole. Top right, a large jet is visible. Radio studies in the innermost parts of the jet reveal relativistic motion.

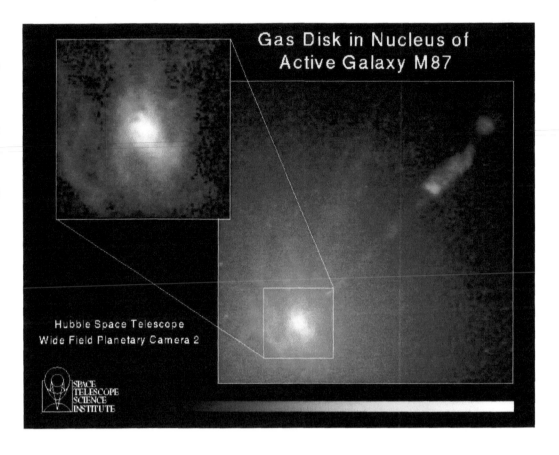

Plate 15:
Hubble Space Telescope optical spectra of gas motions in the heart of another elliptical galaxy in the Virgo cluster, M84, showing the presence of a black hole of about 300 million solar masses. The position of the spectrograph slit is shown in the image at the left, while the spectrum of a bright emission line is shown on the right. Note the large wavelength shifts across the nucleus, characteristic of gas orbiting a massive central object. (The total wavelength shift is only about 1 nm; the colours shown greatly exaggerate it.)

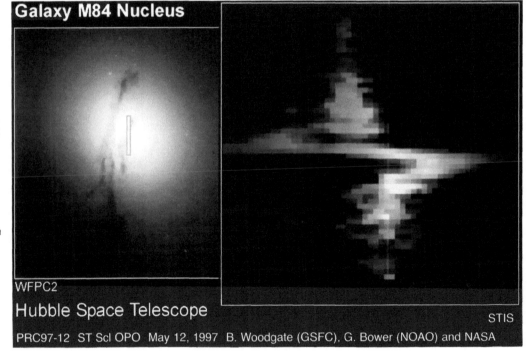

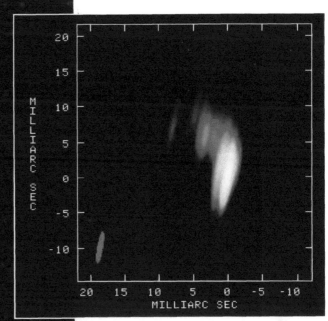

Plate 16b: The very long baseline interferometric image of the quasar 1156+295 at a red-shift of 0.73 in which the baseline has been enlarged using an 8 m diameter antenna on the HALCA satellite, whose orbit has an apogee of 21000 km.

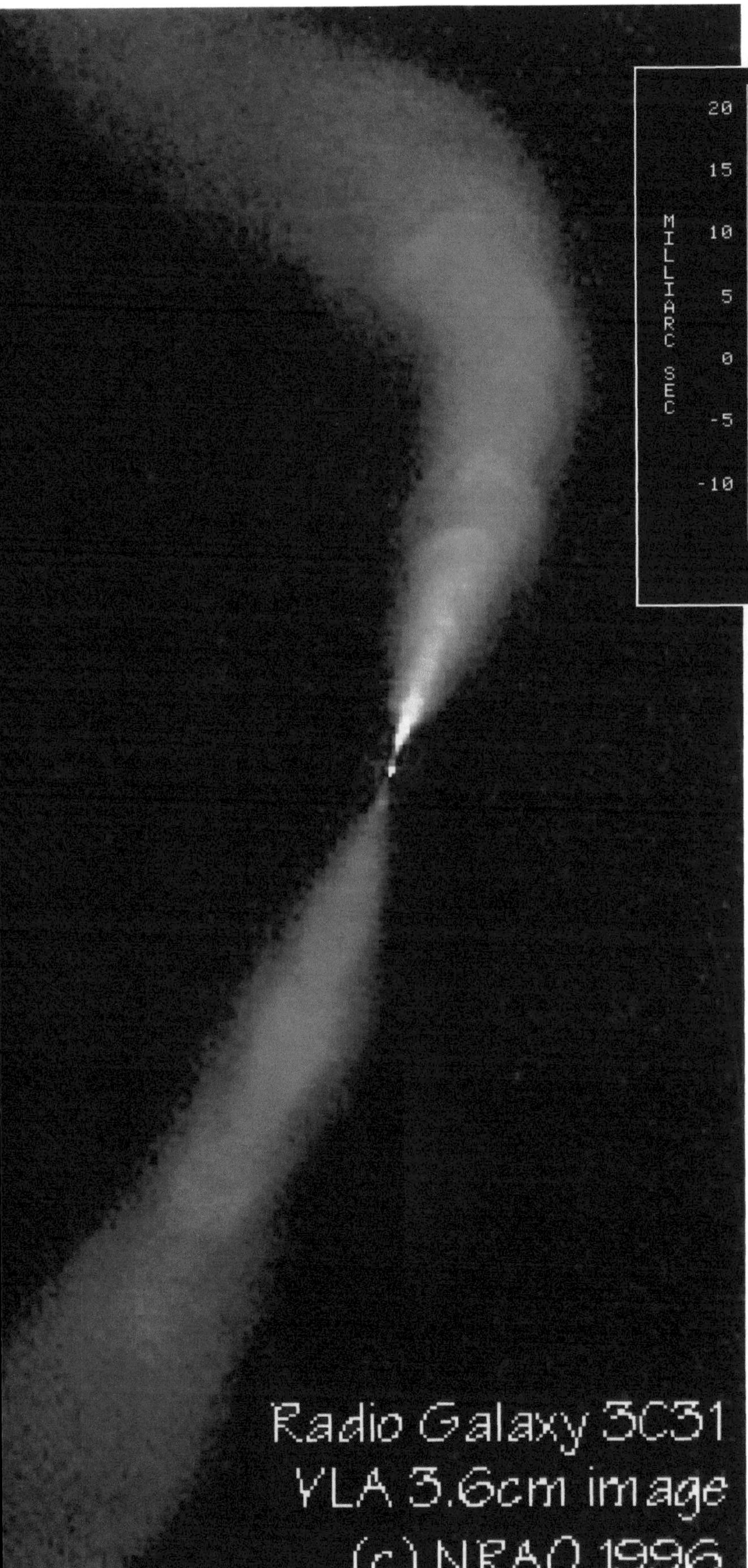

Plate 16a: Radio image of the jet and counterjet in the radio galaxy 3C31, made with the Very Large Array at 8 GHz. The host galaxy, NGC 383, is some 300 million light-years away. Other observations made with a very long (intercontinental) baseline on a milliarcsecond (few light-years) scale show only the northern (brighter) jet. This is presumably approaching at relativistic speeds and Doppler boosting its radiation towards us.

Radio Galaxy 3C31
VLA 3.6cm image
(c) NRAO 1996

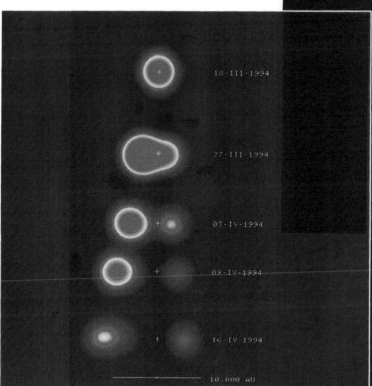

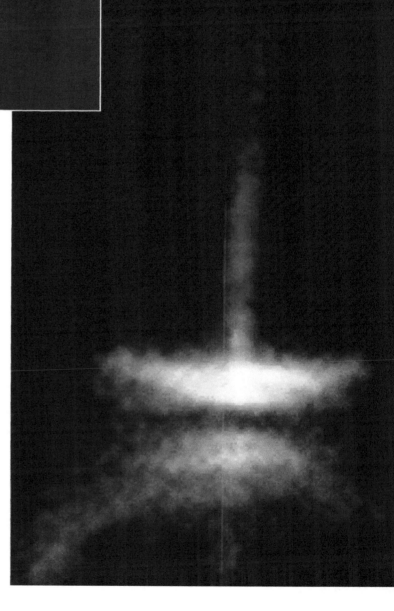

Plate 17b: A sequence of radio images of the radio–x-ray source GRS 1915+105 near the centre of our Galaxy, probably a stellar mass black hole fed by a companion star. Emitted blobs are ejected on either side of the centre (indicated by white crosses). They appear to move about 10000 times the Sun–Earth distance in about a month, a velocity which exceeds that of light. This is a relativistic illusion; the true speed is about 92% of the velocity of light at an angle of about 70° to the line of sight.

Plate 17a: A Hubble Space Telescope image of the proto star HH30 at 450 light-years. It is surrounded by a disc of gas and dust which we see edge on. A red jet emerges.

Plate 18: The Hubble Deep Field, taken over ten days of imaging through four different colour filters onto charge-coupled devices. The width of the field is about 4'. Most of the objects are galaxies at red-shifts approaching 1, with some above 3. The faintest objects are about 4 billion times fainter than the faintest stars visible with the naked eye.

Plate 19: The microwave background sky as seen by the COBE satellite. The sky is brighter in one half and fainter in the other because we are moving at 380 km s^{-1} with respect to this background (when our motion round our Galaxy is accounted for, our local group of galaxies is moving at about 600 km s^{-1}). The variation is a relativistic effect, analogous to those seen in the arrival of cosmic rays. The sky is plotted so that the plane of our Galaxy runs horizontally down the middle; emission from some galactic objects can be seen there.

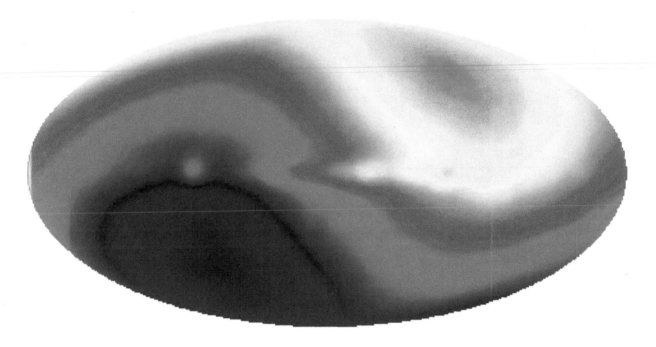

Plate 20: The microwave background sky as seen after the accumulation of 4 years of data from COBE after removal of the effects of known objects and our motion (see plate 19). The temperature fluctuations seen are in the surface of the Big Bang fireball when the photons were last scattered, when the Universe was about a million years old. Much smaller-scale versions of these inhomogeneities have produced individual galaxies. Left image; North Galactic Hemisphere; Right image; South Galactic Hemisphere.

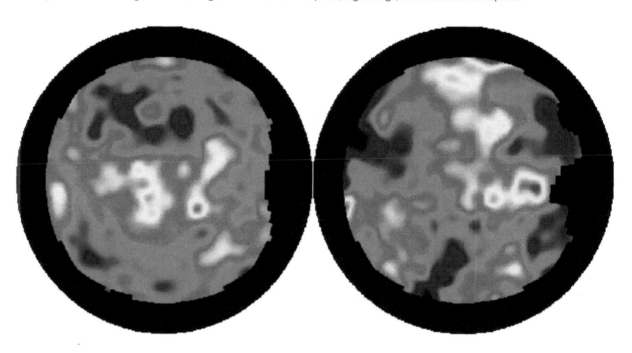

15 THE TOOLS: ACCELERATORS

E J N Wilson

Editor's Introduction: In the first half of the twentieth century, cosmic rays and radioactive emission were the main 'probes' of particle physics. However, the parallel development of new techniques to take beams of particles to high energy led to the construction of huge new particle accelerators to explore new physics territory and to make possible the discoveries described elsewhere in this book.

In this chapter, accelerator specialist E J N Wilson describes the ingenious technology developed to accelerate particle beams, and the series of obstacles which have had to be overcome in the continual quest for higher energy and/or more particles. In parallel he also points out how the large resources needed to build these accelerators have led to international scientific collaboration on an ever-larger scale.

HISTORY OF ACCELERATORS

To illuminate the deep interior of matter and reveal its underlying structure requires beams of particles. The machines which produce and accelerate these beams are the microscopes of the particle physicist. A microscope, whatever its magnification, cannot see objects smaller than the wavelength of light. Waves will merge owing to diffraction around a smaller object and leave no shadow. Quantum mechanics teaches us that particles too have a wavelength and, when we direct a stream of particles at an object in the hope of reconstructing its detail from the pattern of the scattered particles, we cannot hope to discern details smaller than this de Broglie wavelength. This wavelength decreases inversely with increasing momentum or energy of the particles and, since each discovery in particle physics stimulates a search for even finer detail, the tools must provide streams of particles at ever-increasing energies.

The other principle of modern physics which motivates us to higher energies is the discovery that mass and energy are interconvertible according to the equation $E = mc^2$. Much of particle physics consists in hunting for new particles and the top energy of an accelerator represents a threshold in the mass of the new particles that one may hope to produce. The energy of the accelerated particles, measured in electron volts, i.e. the voltage that one would have to apply between anode and cathode of some huge high-voltage column to give a single electron the same energy, has risen from a few hundred electronvolts to 10^{12} eV or 1 TeV.

Just after the First World War, Rutherford in his pioneering scattering studies bombarded nuclei with α particles from the decay of a radioactive source. Such decay products have energies of several MeV, higher than could be produced at that time by electrical machines. Various high-voltage electrostatic generators had been invented in the closing years of the nineteenth century but, even if the acceleration of particles between two high-voltage terminals had been considered, it was clearly not judged to be competitive at that time. Later, in the 1930s, electrostatic acceleration was used by Cockcroft and Walton in Cambridge for their fission experiments. The early 1930s also saw the invention by R J Van de Graaff at Massachusetts Institute of Technology (MIT) of an electrostatic generator which used a moving belt to carry charge into the high-voltage terminal and could reach several million volts (MV). Van de Graaff accelerators are still a useful source of low-energy particles but are inevitably limited by problems of high voltage.

The accelerators used today overcome such breakdown limits by repeated acceleration between terminals carrying a few million volts. The terminals are usually in the form of 'drift tubes', or diaphragms, which form the end faces of a resonant cylinder which we call a radio-frequency (RF) cavity.

An alternative to harnessing an electrical field is to make use of the fact that particles are accelerated in a time-varying magnetic field as if they were the current in the secondary of a transformer. There are two such possibilities: the circular accelerator, conceived by the fertile imagination of a Norwegian electrical engineer, Rolf Wideröe in 1923–25, and the linear accelerator, also developed by Wideröe, from a concept published by Gustav Ising in Sweden in 1924.

THE RAY TRANSFORMER

In 1919, Wideröe, still at school in Oslo, read that Rutherford had observed the disintegration of nitrogen nuclei. Fascinated to hear that the alchemists dream had been realized, he wrote, 'It was clear to me even then that natural alpha rays were not really the best tools for the task; many more particles with far higher energy were required to obtain a greater number of nuclear fissions.'

Later Wideröe began his studies at Karlsruhe Technical University, still looking for the alchemist's tool. He quickly appreciated that the voltage limitations of electrostatic devices meant they could never compete with natural α particles. In 1923, wondering whether electrons in an evacuated ring would flow in the same way as the electrons in the secondary winding of a transformer, he hit upon the novel idea of the first circular accelerator. His notebooks contain sketches of a device he called a 'ray transformer' which later became known as the 'betatron'.

In his first preserved sketches the beam tube, in the form of an annulus, of radius R, is placed in the gap between the parallel poles or faces of an electromagnet, whose yoke in the form of a C (on the left in figure 15.1).

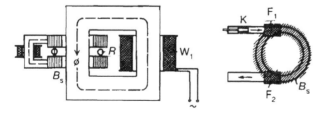

Figure 15.1. Wideröe's sketch of the ray transformer, the idea which developed into the betatron.

The field B_s between the poles guides particles in a circular orbit in the midplane between the poles. At the same time the beam tube forms the secondary of a transformer 'wound' on an iron yoke independent of the C magnet but which threads through holes in the centre of the two pole pieces. The rate of change in flux, ϕ, in this yoke provides the acceleration. The two windings, that of the C magnet and the primary W_1 of the transformer, give independent control of the guide field and accelerating flux. Both windings are powered with alternating current (AC) from the mains.

Wideröe calculated that electrons circulating in a ring of only 10 or 20 cm diameter would reach several MeV within one quarter wave of the AC excitation of the transformer. The guide field had to be matched to the rate of change in flux in the transformer or the orbit would expand or contract. Imagine that the particle circulates in a uniform vertical field and that this field rises at a certain rate. He found that the flux linking the orbit would then be exactly half that necessary to accelerate the particle and to maintain an orbit of constant radius. The flux linking the orbit must then be doubled. To arrive at this conclusion he used the recently discovered theory of special relativity to correctly describe the motion of particles close to the speed of light. This was perhaps its first application in the field of electrical engineering. Wideröe was dissuaded from building the ray transformer by difficulties with surface fields and by his professor, who wrongly assumed that the beam would be lost because of gas scattering. Nevertheless this 2 : 1 ratio, now known as the Wideröe principle, was an important discovery.

LINEAR ACCELERATORS

In 1924, while Wideröe was thinking about the acceleration of particles in a circular accelerator, G Ising in Sweden had the idea of overcoming voltage breakdown in a single stage of acceleration by placing a series of hollow cylindrical electrodes in a straight line. Today we would call such a device a drift-tube linear accelerator (linac). In Ising's scheme a pulsed waveform is applied to each drift tube in turn. This sets up an accelerating field in each gap while the particles are shielded inside the drift tubes.

In 1927, Wideröe, discouraged from pursuing the precursor betatron, was still searching for a PhD subject and, happening upon Ising's paper, switched his attention to linear devices. His crucial contribution was to realize that an oscillating potential applied to one drift tube, flanked by two others which are earthed, can accelerate at both gaps. One must ensure that the oscillator's phase changes by 180° in the flight time between gaps. He also realized that one might extend such a series of tubes indefinitely.

His three-tube model accelerated sodium ions. This was accepted as a thesis but Wideröe did not take the idea any further since he then started professional employment in circuit breaker research for high-voltage transmission lines. It was left to D Sloan and E O Lawrence at Berkeley to continue linac development. Between 1931 and 1934 they constructed mercury ion linacs with 30 drift tubes but these were never used for research.

It was W Alvarez at the Radiation Laboratory of the University of California who started to build the first serious proton linac in the mid-1940s. He proposed mounting a series of drift tubes inside a resonant cavity in the form of a cylinder. The cylindrical cavity was excited with RF waves at one of its natural resonant modes. He realized that, if the phase shift between drift tube gaps was 360° rather than 180°, there would be no need for alternate tubes to be earthed and each gap would appear as an identical accelerating gradient to the particle.

Figure 15.2 shows an Alvarez linac: a copper-lined cylinder excited by a radio transmitter. The resulting potential difference between the ends of the drift tubes accelerates particles from left to right. The fields oscillate but the particles inside the metallic drift tube are protected from the decelerating phase.

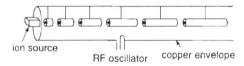

Figure 15.2. The concept of the Alvarez linac.

The distance between gaps increases as the particle is accelerated since it travels a longer distance in one swing of the radio frequency oscillation. However, when the energy is large, the length of the drift tubes and their spacing become identical—a direct consequence of special relativity.

The wartime development of 200 MHz radar power amplifiers made such a structure sufficiently compact to be practical as an accelerator. Transverse fields between the accelerating gaps have a defocusing effect which was compensated first with wire grids over the ends of the drift tubes and later with alternating focusing quadrupoles within the tubes. The Alvarez structure is still widely used, especially for non-relativistic proton and ion beams.

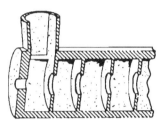

Figure 15.3. Iris-loaded structure from P M Lapostolle, 1986, Proton linear accelerators, *Los Alamos Report LA-11601-MS*. The 'chimney' is the input waveguide.

Others also had the idea of accelerating with radio frequency waves in simple waveguides, but the phase velocity of a wave in a simple pipe is faster than that of light and particles cannot remain perched on the rising crest of an accelerating wave. However, the phase velocity can be reduced by a series of iris diaphragms in the pipe. Such a structure (figure 15.3) is very popular in electron linacs and storage rings for accelerating particles close to the velocity of light.

THE CYCLOTRON

The cyclotron idea is universally attributed to Ernest Lawrence, who was musing on the possibility of using a magnetic field to recirculate the beam through two of Wideröe's drift tubes. Writing down the equations of motion of a particle describing a circular orbit in a uniform magnetic field, he noticed that the circulation frequency remained constant as the particle was accelerated to a higher energy and a larger radius. The idea was published in 1930 and Stanley Livingston was given the job of making a working model as his doctoral thesis. Figure 15.4 shows a schematic diagram of that model, designed to slip between the poles (diameter, 4 in) of an electromagnet.

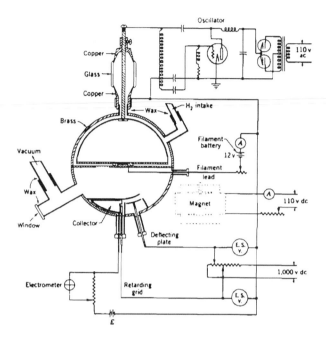

Figure 15.4. Diagram of Lawrence's early cyclotron.

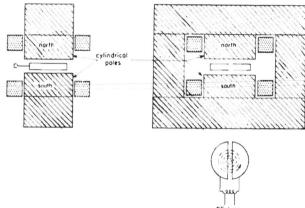

Figure 15.5. Concept of the cyclotron.

The constant circulation frequency enables a continuous beam of particles from a central ion source to spiral outwards in a vertical magnetic field synchronous with an accelerating field applied between two D-shaped resonators called dees (figure 15.5). It was also found, almost by accident, that a field which decreased towards the edge of the poles provided a vertical focusing force to prevent beam loss to the walls of the dees. Such a negative field gradient has an adverse defocusing effect horizontally but, provided that it is not too strong, there is sufficient natural centrifugal focusing to ensure horizontal stability.

In 1939, Lawrence had already started on a 184 in diameter cyclotron. The cost and bulk of the poles and the return yoke had reached what then seemed to be the practical limit. Smaller cyclotrons had already experienced difficulty in surpassing the energy of 30 MeV at which the proton begins to become relativistic. As their velocity approaches that of light, the revolution frequency drops and particles are no longer synchronous with the accelerating potential. In theory the proton's circulation frequency might be restored by a positive radial field gradient but this would destroy the vertical focusing. The alternative was to change the RF as

particles accelerate but, for this to work, acceleration must be as a sequence of bursts. This solution was applied later to the synchrocyclotron and to the synchrotron once the principle of phase stability was understood, but in the early 1940s there was a reluctance to abandon the cyclotron's continuous beam. Hundreds of cyclotrons are still used throughout the world, mainly for nuclear physics, industrial and medical applications.

THE BETATRON

During the war years, Lawrence's cyclotron programme was diverted towards the production of fissile material by electromagnetic separation of uranium isotopes. Accelerator development continued as D W Kerst and R Serber took up Wideröe's beam transformer idea, calling it the betatron. Accelerating electrons, it complemented the cyclotron, used for protons and ions. Its topology was similar to the cyclotron but it was pulsed and the beam did not spiral out but stayed at the same orbit radius. A short batch of electrons was injected and accelerated by the rate of change in the magnetic flux linking the orbit. Kerst found that, by shaping the poles, a single magnet yoke could provide both guide field and an accelerating flux which obeyed Wideröe's 2:1 ratio. Working at the University of Illinois and later at General Electric Corporation Laboratories in the USA, he constructed machines which surpassed the energies of

Lawrence's cyclotrons. By the mid-1940s, betatrons had begun to become as bulky as cyclotrons and the magnet of a 300 MeV machine at the University of Illinois had a mass of 275 tons.

THE SYNCHROTRON

New predictions for fundamental particles requiring an energy just above that of existing machines has been a recurring challenge to the accelerator builder. In the 1940s the largest cyclotrons and betatrons were just too small to make physics discoveries. The stage was set for the discovery of the synchrotron principle which opened the way to an ever-increasing series of circular accelerators and storage rings. The ideas of varying the frequency, focusing the beam with a field gradient and pulsing the magnet had been applied in diverse contexts; it remained to combine them. It was the Australian physicist Mark Oliphant, then supervising uranium separation at Oak Ridge, who invented the synchrotron in 1943 describing his invention in a memo to the UK Atomic Energy Directorate:

'Particles should be constrained to move in a circle of constant radius thus enabling the use of an annular ring of magnetic field...which would be varied in such a way that the radius of curvature remains constant as the particles gain energy through successive accelerations by an alternating electric field applied between coaxial hollow electrodes.'

A short pulse is injected and the field rises as the particles are accelerated (figure 15.6). Unlike the betatron, the acceleration is not by induction, but by fields within a hollow cylindrical resonator excited by a radio transmitter. Particles pass through apertures in the resonator, receiving another increment in energy at revolution. The fact that the machine is pulsed and the frequency must be controlled to track the increasing speed of particles solves the difficulty that isochronous cyclotron builders had in accelerating relativistic particles. Unlike cyclotrons and betatrons, the synchrotron needs no magnetic field within the circular orbit of the high-energy particles. The guide field is instead provided by a slender ring of individual magnets (see figure 15.8).

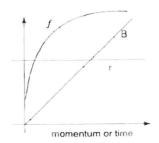

Figure 15.6. Field and frequency rise together in a synchrotron.

PHASE STABILITY

It was not clear to Oliphant whether particles would remain synchronous with the applied accelerating fields, although he was convinced that a means would be found. In any case, V I Veksler in Moscow in 1944 and Edwin McMillan in Berkeley in 1945 independently discovered the key principle of phase stability. Particles orbiting the synchrotron are timed to ride on the rising edge of the voltage wave in the accelerating cavity (see A in figure 15.7). They receive more energy if they are late (B) and less energy if they are early, so that they oscillate about the stable synchronous phase. For all particles, the time average of their energy gain matches the rising magnetic field.

VARIABLE-ENERGY CYCLOTRONS

The first attempts to overcome the relativistic barrier of the cyclotron followed the discovery of phase stability. A slight radial decrease in focusing field must be preserved, but by changing the RF driving the dees it is possible to follow the variation in revolution frequency as the circulating proton approaches the speed of light. Such pulsed 'synchrocyclotrons' reached many hundred MeV. Lawrence's pre-war 184 in cyclotron was adapted as a synchrocyclotron and in 1946 accelerated deuterons to 190 MeV and He^{2+} to 380 MeV. Other machines followed in the USA, in Canada at McGill, in Europe at Harwell (1949) and in Russia at Dubna (1954).

THE FIRST SYNCHROTRONS

A betatron, from Kerst's stable, had been imported into the UK to produce x-rays for wartime applications. At

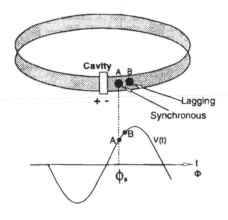

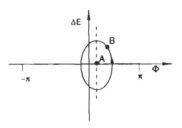

Figure 15.7. The principle of phase stability. The cylindrical coordinate system rotates with the beam, demonstrating the meaning of the RF phase angle in longitudinal phase space.

the Telecommunications Research Laboratory in Malvern, Frank Goward, hearing of McMillan's work on the synchrotron and phase stability, realized that the betatron could be turned into an electron synchrotron, doubling its energy. The idea was first demonstrated in August 1946 by Goward and his colleague D E Barnes.

Close runners-up were a team at the General Electric Co. at Schenectady who were constructing a purpose-built 70 MeV electron synchrotron. They just failed to beat Malvern to the post by a month or two. However, they had the consolation that this machine, with a glass vacuum chamber, was the first to produce visible synchrotron radiation.

These first synchrotrons were electron machines but projects for proton synchrotrons aiming at energies above 1 GeV were not far behind. Oliphant, now back at the University of Birmingham, had made his bid early to construct a 1 GeV proton machine but was becoming

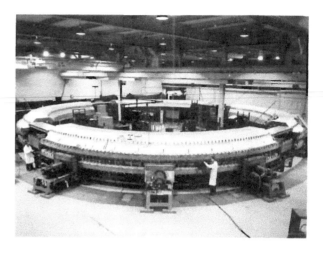

Figure 15.8. The Cosmotron: the first large proton synchrotron.

entangled in the red tape and lack of imagination which abounded in post-war Britain.

Meanwhile in 1948 two huge proton synchrotrons had been authorized in the USA. The larger of these, the Bevatron, aimed at 5–6 GeV, or a billion (B) electronvolts in US parlance, was to be constructed at the University of California, Berkeley while on the East Coast, Brookhaven built the 2.5–3 GeV Cosmotron.

These synchrotrons, like the cyclotrons and betatrons before them, had pole pieces shaped to produce a gentle radial field gradient. The focusing was weak and the beam was large. Estimates of the size of the beam tube, which needed to allow for orbit distortion due to field errors, were rather primitive. Calculations for the Bevatron varied between a low-energy version with a 4 ft (1.2 m) vertical aperture and one of only 1 ft (0.3 m). A quarter-scale model was built which showed that the largest aperture was not required. The model also showed that splitting the magnet into quadrants with intervening space for injection and acceleration equipment did no harm. The first beam at 5.7 BeV was in October 1954.

Their rivals, the Cosmotron team including Stan Livingston, John and Hildred Blewett, Ernest Courant, Ken Green and N Blackburn, had refined their aperture to a mere 1.2 m by 0.22 m (figure 15.8). When they switched on in early 1952 they began to doubt the optimism of their calculations, but after two worrying months found that they had been fooled by a burnt-out voltage divider in the RF system. Once this was fixed they made what the *New York Times* in May 1952 headlined their first 'Billion volt shot'.

The pulsed power to its magnet was provided by a generator attached to a huge motor-driven flywheel. Synchronizing the frequency of the accelerating voltage with the rise in magnetic field to a precision of 0.1% was a tricky problem. Saturable ferrite components in the low-level oscillator were controlled by a magnetic field pick-up whose output was trimmed with 28 bias circuits. Signals from an electrode measuring the radial position of the beam completed the fine tuning of the acceleration process.

These two machines placed the USA in the forefront of high-energy physics, a position that it was to maintain for three decades. The age of the large synchrotrons had commenced but there was still one major improvement to come.

STRONG FOCUSING AND ALTERNATING GRADIENTS

A much more effective form of focusing was invented at the Cosmotron. This had been designed as a typical weak focusing synchrotron with a C-shaped magnet open to the outside. The energy of the Cosmotron was limited by instabilities. Livingston and Courant wanted to compensate this by reinstalling some of the C magnets open to the inside and were surprised to calculate that this seemed to improve the focusing. Courant and Snyder explained this with an optical analogy of alternating focusing by equal convex and concave lenses, representing magnets with strong positive and negative radial field gradients, but were soon to find that their idea had already been patented by Nicholas Christofilos in Athens.

The pole pieces of an alternating gradient (AG) synchrotron, flared into a hyperbola, produce a strong gradient, and consecutive bending magnets have the wider side of their gap towards the inside and outside of the ring successively (figure 15.9). This focusing, since it acts in both planes, is no longer weaker than the horizontal centrifugal focusing. Stronger focusing makes for a more compact beam and the magnet gap can be made much smaller—comparable with a human hand rather than a whole body.

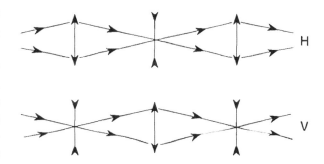

Figure 15.9. An alternating gradient optical system. An alternating pattern of lenses convex in one plane and concave in the other will carry rays through the centres of defocusing lenses. The upper diagram shows the horizontal motion, and the lower diagram shows the vertical motion.

USA AND EUROPE: NECK AND NECK

The strong focusing principle was discovered just after the Europeans had created a central European laboratory, CERN, for a 10 GeV synchrotron, too big for any one European nation to construct. Within weeks, CERN visitors O Dahl, F Goward and R Wideröe visited the Cosmotron and, hearing of the new idea, immediately modified their plans for a 10 GeV weak focusing machine to a 25 GeV proton synchrotron (PS) which they could build for the same price. Brookhaven had already planned such a machine, the Alternating Gradient Synchrotron (AGS), as their next step.

Although in a sense rivals, the AGS and CERN teams shared their expertise. John and Hildred Blewett joined O Dahl and K Johnsen in Bergen to build the first prototype of CERN's high gradient synchrotron magnet before moving to Geneva in the fall of 1953. John Blewett reports that a youthful Kjell Johnsen showed there was little to fear from the phenomenon of transition, one of the headaches stemming from the alternating focusing system.

Transition depends on the fact that a tardy particle, given an extra increment of energy, speeds up and overtakes its synchronous and well behaved neighbour. However, the extra speed must be balanced against the extra circumference that it must travel on a larger-radius circle in the bending field. At low energies the extra speed predominates but, as the particle approaches the velocity of light, the path length argument wins and the RF wave

Figure 15.10. CERN's Proton Synchrotron in 1959 showing three alternating gradient bending magnets (one pointing inwards).

has to be repositioned on a falling slope to *decelerate* tardy particles rather than to speed them up so that they take a shorter inside radius. Between is a region of transition and in strong focusing this usually occurs midway in the acceleration cycle.

Another headache was that, in their enthusiasm for the new principle, designers proposed large numbers of lenses, and particles oscillated many times as they went round the ring. (The number of these oscillations is called Q.) This would make the focused beam very compact and the transition energy (roughly equal to Q in units of GeV) would not need to be crossed. John Adams, Mervyn Hine and John Lawson at Harwell pointed out that AG machines with a high Q were particularly susceptible to field errors of all types. Johnsen convinced the designers that a more modest Q was better and that transition could be crossed. This did not completely solve the problem of field errors and Adams and Hine, by now at CERN, ensured that the necessary precision of alignment and magnet performance was respected.

CERN recruited a team of painstaking European engineers. This team became a legendary asset and tackled a series of 'difficult' synchrotrons.

By 1959 CERN's PS (figure 15.10) was ready for testing, ahead of Brookhaven's AGS. It did indeed falter at transition until Wolfgang Schnell produced an ingenious circuit built in a coffee tin to change the phase at the moment of transition.

The advent of strong focusing stimulated the imagination of accelerator builders. Synchrotron projects were redesigned and cyclotron builders remembered that AG focusing resembled a scheme proposed just before the war by L H Thomas with ridged pole pieces for a cyclotron. These focused the beam vertically as an alternative to a negative radial field gradient. Kerst and others, working in the Midwestern Universities Research Association (MURA) centred in Chicago, realized that, if Thomas' idea had been heeded, there would have been no need to resort to frequency modulation and pulsing to overcome relativity effects. A gradually increasing radial field might be combined with strong focusing. By making the radial increase in field very strong and sweeping the ridges into a spiral this fixed-frequency alternating gradient (FFAG) machine would simultaneously support beam from injection to top energy within a relatively small range of radii.

STORAGE RINGS

Although these ideas eventually gave birth to the modern sector-focused cyclotron, they came rather too late in view of the plans to build powerful synchrotrons in the USA and at CERN. However, when Symon and Kerst presented this idea in 1956 they also pointed out that by joining two such machines to appear as a figure of eight in plan view, beams of particles might collide head-on—the modern collider concept.

It had been realized for some time that an accelerator colliding with a stationary nucleon has an effective collision energy which grows only with the square root of the accelerator energy. It is this available energy that interests the experimenter. The remaining energy is 'wasted' in recoil. On the other hand, it had been realized in the 1930s, and in fact was a common physics examination question, that, if only one could collide two accelerated particles head on, all their energy would become available for collision. This had been in Wideröe's mind working in isolation in wartime Europe and, when he met B Touschek in Hamburg in the closing years of the war encouraged him to construct such a collider. Touschek was initially dismissive but much later in Rome in 1960 built the first storage ring based on this idea. It was called AdA (Annello d'Accumulazione) and had the added innovation that, rather than using two separate rings of electrons, electrons and positrons (antielectrons) could circulate in opposite directions in a single ring. The machine first worked as an electron storage ring and in 1961 was transported to Orsay where there was a positron injector, and where it became the first electron–positron collider.

Similar ideas were being investigated at Stanford and at Novosibirsk. The Stanford–Princeton electron–electron ring was proposed as early as 1956 but completed only in 1966 while the Novosibirsk VEP-I electron–electron collider was started in 1962 and began colliding in 1965.

The study of the non-linear fields in FFAG machines laid the foundations of the theory of field errors and stimulated particle tracking by computer simulation. MURA also studied tracking in the longitudinal direction as the beam is accelerated. With this came the important realization that beams could be accumulated side by side, 'stacked' by phase displacement of the RF and MURA workers. Those at Novosibirsk had also to wrestle with

understanding the innumerable instabilities which threaten intense beams.

To achieve useful 'event rates', beam current densities had to be very high and the builders of these machines and those at the Cambridge (MA) Electron Accelerator (CEA) made many notable contributions to our understanding of beam behaviour.

This knowledge became essential when CERN built the first really large proton–proton Intersecting Storage Rings (ISR), 30 GeV on 30 GeV, completed in 1969. Even then there was more to be learned. A 'brick wall' barring the way towards higher intensities had to be scaled by understanding the interaction of the beam with the ionization in the vacuum chamber and by improving vacuum technology by several orders of magnitude.

COOLING

For particle–antiparticle storage rings, antiparticles were hard to create and antiprotons particularly were in short supply. A well known theorem suggested that, once a beam was created, it could not be manipulated into a 'brighter beam'. However, ways of beating this Liouville conservation theorem with electron cooling and stochastic cooling were invented respectively by Budker at Novosibirsk in 1966 and by van der Meer at CERN in 1968, published in 1972. These aimed at reducing the longitudinal and transverse oscillations of the beam and made it possible for antiproton beams to be concentrated, slimmed, tapered and shaped before stacking them in the vacuum chamber of the ring, so accumulating antiprotons harvested over many hours of production. It takes a pulse of 10^{13} protons to produce 10^6 antiprotons and it takes tens of thousands of pulses of protons to produce enough antiprotons for a proton–antiproton collider.

As such storage rings proved the key to higher-energy collision processes, both the pulsed Super Proton Synchrotron (SPS) at CERN and the Fermilab Tevatron were operated to collide protons and antiprotons. These were followed by LEP, the 8.5 km diameter electron–positron collider at CERN coming on stream in 1989, and by HERA, a 2 km diameter collider for 26 GeV electrons and 820 GeV protons at DESY, Hamburg, which first operated in 1991. The Large Hadron Collider (LHC), a 7 on 7 TeV proton collider for the LEP tunnel is being constructed, while smaller but very high luminosity rings as ϕ, beauty and charm factories are coming into operation. The storage ring collider idea has extended the energy available for production by many orders of magnitude and stimulated most of our present understanding of beam behaviour.

ELECTRONS VERSUS PROTONS

Protons are complex objects consisting of three quarks held together by gluons. In collision with other hadrons (protons or antiprotons) the interaction is dominated by this strong force. However, only one quark in each of the colliding hadrons is involved in the interaction and it is difficult to identify which quarks or gluons have taken part in any given interaction. In addition, the quarks which interact carry only a fraction of the total energy imparted by the accelerator so that much of the energy is wasted. Nevertheless hadrons are unique probes for the study of the strong interaction.

Electrons, on the other hand, are ideal probes for electromagnetic interactions but are unable to distinguish between quark flavours. The pointlike nature of the electron ensures that all its energy is put to good use in producing particles of interest.

LIMITS AND HOW TO OVERCOME THEM

LEP is likely to be the last large circular electron machine. All charged particles radiate electromagnetic radiation as they accelerate, decelerate or change direction. The prime example of course is the emission of radio waves by electrons oscillating in a radio antenna. This radiation varies with the square of the acceleration (the lowest power that nature can allow if energy is not to be absorbed as well as emitted). The compression of space and dilation of time in special relativity means that such 'synchrotron radiation' is emitted in a narrow cone tangential to the path of the particle.

The parameter which determines how close a particle is to the velocity of light is γ, the ratio of its total energy to its rest mass $m_0 c^2$. Electrons with a mass m_0 almost 2000 times smaller than that of a proton will, at the same energy, have a γ value 2000 times larger. The power emitted by a particle circulating in a synchrotron is proportional to γ^4/ρ^2 where r is the radius of its path. Thus, electron and proton machines of the same radius and energy differ by many orders of magnitude in the power radiated. Electron machines soon encounter a steeply rising barrier once the power radiated, sometimes amounting to megawatts, can no longer be restored by radio frequency acceleration. One of the reasons for LEP's huge radius is an attempt to take advantage of the ρ^2 in the denominator. Any higher-energy machine must be a linear machine where opposing electron and positron beams are accelerated in linacs to clash head on and to avoid synchrotron radiation losses.

One of the challenges of experiments with proton storage rings is the need to increase the probability of collision. This is necessary in order to supply the production rate of the rare interactions of interest to high-energy physicists, which falls as $1/M^2$, where M is the mass. Likening the accelerator to a microscope using particles as de Broglie waves of shorter wavelength $\lambda = hc/E$ to discern finer details of structure, a higher energy produces a shorter wavelength to reveal smaller objects, but these smaller objects have lower production rates.

LUMINOSITY

The figure of merit of a collider is its luminosity. To understand this let us look at a probe particle which encounters an oncoming beam—a cloud of N particles each of which presents a cross section σ for the production of an event of interest (see figure 15.11). The probability that the event occurs is just the fraction of the beam's area occluded:

$$P = \frac{N}{A}\sigma = L\sigma.$$

In this trivial case of a single encounter of one particle passing through a beam, the luminosity $L = N/A$. It is

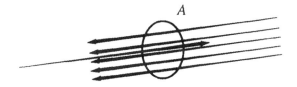

Figure 15.11. A probe particle encounters a target—a beam of particles with cross-sectional area A travelling in the opposite sense.

independent of the cross section under study and depends only on the beam geometry. We may think of it as a probability of producing an event normalized to unit interaction cross section.

In practice the probe beam encounters many target particles and such encounters occur as often as the many bunches in the circulating beams in a storage ring meet each other. The luminosity for two such equal beams colliding is

$$L = \frac{N^2 f_b}{A}$$

where N is the number of particles per bunch, f_b is the bunch frequency and A is the transverse beam area at the crossing.

A luminosity of 10^{33} cm^{-2} s^{-1} will produce one event per second from a process for which the cross section is 10^{-33} cm^2.

The choice of storage ring parameters must fall within the constraints of a quantity known as the beam–beam tune shift, a measure of the perturbation to beam focusing due to electromagnetic fields set up by one beam and experienced by the other. This must be no more than a few per cent for colliding electron beams and a factor of 10 smaller for proton beams:

$$dQ = \frac{r_p \beta^*}{\gamma} \frac{N}{A}$$

where γ is the usual relativistic factor, r_p the classical proton radius and β^* is a measure of how tightly focused the beams are at the waist where they interact; a small β represents a narrow waist.

One can see from the above rather similar expressions for luminosity and tune shift that the only route to improving luminosity without increasing dQ is to squeeze the beams to make β^* as small as possible, to increase N to the limit imposed by beam instability and to increase the number of bunches to have more frequent crossings.

The radius of a synchrotron is proportional its top energy and inversely proportional to its magnetic field. As synchrotrons and proton colliders have grown, their builders have sought stronger magnetic fields to reduce the real estate that they occupy and the cost of tunnelling. Room-temperature magnets have steel pole pieces which define the field shape. These and the return yoke usually saturate

at a field of about 2 T. One can of course imagine air-cored windings precisely shaped around a cylindrical pipe to produce a uniform field without the need for iron. The ideal coil shape should mimic a pair of intersecting ellipses as closely as possible. However the field in such a magnet is severely limited by the problem of cooling the coils which must carry a considerable current density. Superconducting coils, which in theory do not dissipate heat, seem to offer a means to increase the current density and to allow modern synchrotron designers to increase the guide field to as high as 9 T.

THE SUPERCONDUCTING SUPER COLLIDER

In the early 1980s CERN was congratulating itself on the discovery of the W and Z, and Fermilab had just completed and commissioned the Tevatron. To continue the study of the subinfinite into the next century, the USA launched the Superconducting Super Collider (SSC), a double proton synchrotron interleaved in the way that the ISR had been but aimed at 20 TeV per beam. With the help of superconducting magnet technology developed for the Tevatron, the circumference was contained within 87 km, about three times that of LEP.

The design had been thoroughly prepared by a Central Design Group at Berkeley and prototype magnets had been built at Berkeley, Brookhaven and Fermilab. The machine was to cost US$6000 million and with so much construction work at stake different states vied to host the new laboratory. In the end Texas won and, with no existing accelerator infrastructure, the injectors, the staff, even the offices and cafeterias, all had to be built from scratch. The whole complex including its linac, a chain of three booster rings and the collider ring itself, not to mention the tunnel and the detectors, were on a scale that matched the traditions of that state.

As more precise designs and cost estimates were made, the undertaking seemed to become considerably more expensive. At a time when President and Congress were arguing about how to balance the ballooning US budgetary deficit, the cost was questioned and in the annual funding review Congress decided in 1993 to cancel the plan. Three years into the construction and with more than US$2000 million spent, the SSC was cancelled and the 2000 or more

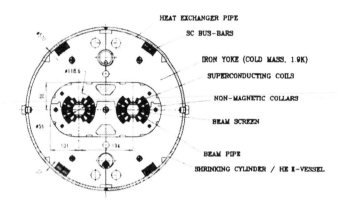

Figure 15.12. Cross section of the LHC twin-bore superconducting dipole magnet in its cryostat.

staff asked to return to their old jobs or to seek employment elsewhere.

For the first time the steady growth pattern in the tools for particle physics of the last six decades had faltered and with it the initial hopes of American particle physics for the twenty-first century.

THE LARGE HADRON COLLIDER

The sad demise of the SSC left Europe as the sole contender for the high-energy race with its 7 TeV hadron collider, the LHC (figure 15.12). With only about one third of the energy of the SSC it was much cheaper and arguably quicker to build since it made use of the existing injector synchrotrons and the existing LEP tunnel. With the demise of the SSC the LHC became the machine on which the particle physics world pinned its hopes. At the end of 1994 CERN embarked on a 10 year programme to construct the machine which may be the last large synchrotron to be built, but hopefully not the 'last great machine'.

LINEAR COLLIDERS

So steep is the rise of synchrotron energy loss for electron accelerators that a circular electron ring for more than LEP's 2×100 GeV is not a practical proposition. Linear accelerators will provide an electron–positron collider to complement the discoveries made with LHC.

The first essay in this direction was made at the Stanford Linear Accelerator Center (SLAC) in the late

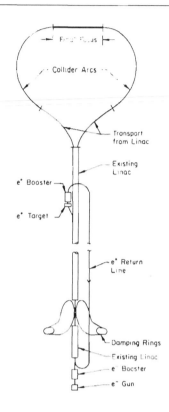

Figure 15.13. The SLC.

1980s where the famous 'SLAC 2 mile linac' was used to accelerate 50 GeV electron and positron bunches. In this Stanford Linear Collider (SLC) project the bunches were guided around opposite semicircular arcs to collide at a single interaction point (figure 15.13). Higher energies must accelerate electrons and positrons in two opposing linacs since synchrotron radiation in the arcs will rule out the use of the SLC topology.

Higher energy imposes a higher luminosity. While the particles in a circular collider may be re-used, producing a new encounter each time that they circulate, in a linear collider a fresh batch must be accelerated for each encounter. The beam power (the product of the current and the final energy) is impressive. The frequency of encounter can be only as fast as the linac can cycle and very intense beams must be accelerated. Not only must the efficiency of the RF system which converts power from the wall plug into beam power be high, but the voltage gradients in the accelerating cavities must be exceptionally high if the linacs are to be only 10 km or so in length. Higher-frequency (smaller) linac structures help since the energy stored in their fields is smaller and they can be run at higher fields. However, instability problems arise because the walls are closer to the beam. We must also perfect RF power sources at frequencies far beyond those normally used for telecommunication use. A world-wide collaboration is studying the design of a future linear collider, which must provide at least 2×500 GeV to be a tangible step.

APPLICATIONS

As well as large machines for high-energy physics, many thousands of accelerators have been put to more practical use in other branches of scientific research as well as in industry and medicine.

For medical applications, cyclotrons are at work producing isotopes to supply hospitals with radiochemicals and modified biological agents whose location can be traced in the body by the particles that they emit. Some, by their biochemical nature, can even select certain sites in the body for diagnosis or treatment.

Recently there has been a great interest in Europe, notably in Italy, and particularly in Japan, to build proton machines of a few hundred MeV to irradiate deep tumours. Protons deposit most of their energy near the end of their path, doing minimum damage to surface tissue and sparing delicate organs just beyond the target zone. Ions such as carbon should be even more effective than protons.

In industry, heavy-ion beams, such as those accelerated by GSI at Darmstadt, are widely used to implant atoms in the surfaces of semiconductors to 'print' the circuits of modern computer chips. Other industrial uses are hardening metal surfaces for bearings and etching in silicon microcircuitry.

Synchrotron radiation sources have mushroomed all over the world and their highly collimated and tunable radiation is put to a huge variety of applications. In research, x-ray diffraction techniques reveal the structure of proteins and enzymes and the crystal lattices of promising new materials such as high-temperature superconductors.

They are complemented by intense proton machines (the so-called spallation sources) such as ISIS at the UK Rutherford Appleton Laboratory to produce neutrons.

Neutron diffraction extends many of the research techniques of synchrotron light sources and adds a new dimension in that the source may be pulsed to allow time-of-flight identification techniques.

Even more impressive accelerators of only a GeV or so, but very high intensity, are under study to bring pellets of deuterium or tritium into the self-sustained thermonuclear reaction regime. Intense linacs would also allow transmutation of long-lived nuclear waste into isotopes which rapidly decay to become harmless or alternatively provide the beam which 'fans the flames' of the 'energy amplifier', a fail-safe form of nuclear reactor using relatively innocuous thorium fuel.

ACCELERATORS: COMPETITION OR COMMON CAUSE

While the first accelerators were invented by an electrical engineer, many of the subsequent innovative ideas have come from physicists. Of the many landmarks in progress towards larger accelerators, the more successful in terms of economy and rapidity of construction, reliability and performance, have been the product of leaders who have combined the skills of both the applied physicist and the engineer to plan, assemble and control a huge project. The masters of this art have included Fermilab builder R R Wilson on one side of the Atlantic and J B Adams (Sir John) in Europe—two men with very different personalities, the former, an artist, forthright and full of ideas, and the latter, pipe smoking, reflective, a brilliant pathfinder in the highways of reason and the byways of compromise. Each was matched to the style of their technical and political environment. Among others in the same league, but less fortunate in the scale of the project that they commanded, are Maury Tigner and Burton Richter in the USA, Gus Voss and Bjorn Wiik at DESY, and Kjell Johnsen and Giorgo Brianti at CERN.

Engineering must be said to have won the day in the competition between the USA and Europe to build the best accelerators. At the end of the Second World War, the USA was much better placed to provide the engineering needed for both Cosmotron and Bevatron, while Birmingham's graduate physics students served Oliphant's ambitions less well. Europe only found its feet when it centralized its effort at CERN. Even then, although peopled by the best engineers in Europe, it took time to assemble an equivalent team of experienced physicists. Speed of reaction has always been on the side of the USA. Politics in Europe and the way that budgets are granted caused large projects such as the SPS to be delayed by as much as five years. However, builders of European projects, once approved, have usually been fortunate enough to be left to get on with it.

Right from the outset, when the USA and Europe decided to construct the AGS and the PS, specialists were eager to share their ideas. The AGS and PS were neck and neck during their construction but this did not prevent a close exchange of experience, and Hildred Blewett from the AGS was in the PS control room in Geneva when the PS won the race by a short head. In the electron community the CEA and the machines at Cornell led the field, but Europeans were among their teams and, as DESY gave Europe its first large electron machines, CEA staff were there to help.

The USA and Europe parted company in the 1960s to build the Fermilab machine and the ISR at CERN. Here there was less opportunity for pooling knowledge but, as Europe moved to approve the SPS (its own version of Fermilab), there were contributions by CERN staff to help to get Fermilab going and Fermilab people who helped build magnets and commission beams for the SPS. The SPS of course was later than Fermilab but quickly turned the tables by operating as a proton–antiproton collider. The exchange continued as Fermilab built the Tevatron.

For a brief period the SSC and LHC were seen by their builders more as rivals for political and financial support, but now the world community turns to the LHC for its next big machine; its builders enjoy the co-operation of accelerator builders worldwide.

CONCLUSIONS

Almost 70 years have passed since a particle physics experiment triggered the search for an effective machine to accelerate charged particles. Circular accelerators and linacs alike have grown from table-top instruments to huge engineering enterprises rivalling the most ambitious undertakings of mankind in their scale, ingenuity and daring. Nevertheless the vast majority of accelerators are used for applications other than particle physics and

their share of the technology is growing rapidly. One may be sure that colliders for pure physics research will certainly be required in the post-LHC era. However, we are clearly entering the era of the 'world machine' where continents must find rational ways of distributing the large projects required by physics and, despite the huge distances involved, develop practical methods of sharing in their exploitation.

FURTHER READING

Livingood J J 1961 *Principles of Circular Accelerators* (New York: van Nostrand)

ABOUT THE AUTHOR

Born in Liverpool in 1938, E J N ('Ted') Wilson was awarded an Open Exhibition to University College Oxford in 1956, where he graduated in Physics and Theoretical Physics in 1959. Leaving academia to work at what became the Rutherford Appleton Laboratory, he combined high-energy physics with the design of secondary beam components for the Nimrod proton synchrotron. Coming to CERN in 1967, he was one of three 'caretakers' for the 300 GeV project, later to become the SPS. Responding to ideas from the newly created Fermilab, he completely redesigned the ring and became the right-hand man of project director John Adams, helping to resolve many design decisions. A sabbatical year at Fermilab in 1963, where with Rae Stiening he helped coax the sleeping giant into life, taught him the realities of accelerator commissioning and he returned to CERN to play a similar role for the SPS. With the historic decision to convert the SPS into a storage ring, he led the first tests before moving to work with Roy Billinge and Simon van der Meer to make CERN's pioneering Antiproton Accumulator store and deliver antiparticles.

Discovering a talent for teaching, he went on to introduce many of today's accelerator experts to their chosen field. Since 1992 he has headed the CERN Accelerator School, bringing accelerator teaching to all of CERN's Member States as well as India, China and the US.

16 THE TOOLS: DETECTORS

David H Saxon

Editor's Introduction: As well as a source of particles (such as cosmic rays or a beam from an accelerator), a particle physics experiment needs a target for the beam, and a 'detector' for monitoring and interpreting the results of these collisions. Crude electroscopes and the human eye were once all that were available. This chapter looks at the development of detector technology as physics has become more demanding, and as detector specialists have become more ingenious, culminating in the huge and sophisticated modular instrumentation used at today's high-energy colliding-beam machines. The high collision rates provided by these machines also brings special requirements for data handling. As well as keeping pace with the increased demands of physics, detector technology has continually led to spin-off applications, especially for medical imaging.

INTERACTIONS AND EVENTS

Elementary particles, like people, exhibit characteristic behaviour. In both cases, this behaviour is displayed as interactions, or responses to stimulation. However, for particles, these interactions have to be understood in terms of quantum mechanics, which gives the probability that, in a given set-up (e.g. an electron colliding at a high energy with its antimatter partner, the positron). one particular interaction or another will occur. Classifying and understanding behaviour needs repeated observations. To build up this experience we need to undertake repeated trials, and to observe, measure and classify the different kinds of 'event' that can occur.

Some types of event will be common; others will be very rare. The strategy then must be to accumulate a library (e.g. half a million photographs taken in a bubble chamber) of repeated observations of the same prepared set-up, and to exercise scholarship on this library of information to classify what occurs into known categories, and to identify new phenomena. Discoveries can be made on the basis of one spectacular event which stands out from a crowd of many less exotic events (such as the Ω^- discovery: see the chapter by Samios), or by identifying an as yet unseen trend by the accumulation of high statistics.

The keys to experimental progress are therefore to be found at high energies (to open up new reaction possibilities), at high intensities (many events, to access rare processes), in selectivity (to capture the latter unambiguously), in precision on individual event measurement (e.g. to isolate a very short-lived particle), in identification of leptons (electrons, muons or neutrinos as signs of rare decays), in high statistics (to accumulate statistical precision) and in completeness of measurement (to reduce ambiguity by describing each event as completely as possible). One describes a detector as 'hermetic' if no reaction product can escape without being observed.

The Ω^- discovery event illustrates many of these points. A beam of K^- particles enters the bubble chamber from the foot of the picture (see figure 3.6). One of them causes the interesting interaction and makes a set of particles. The electrically charged particles produce 'tracks' of bubbles along their paths through the liquid hydrogen. The whole chamber is in a magnetic field so that particles veer left or right according to their electric charge (positive or negative) and momentum; the paths of fast particles are almost straight, and the paths of slower particles are strongly curved. From accurate measurement of these tracks, physicists can reconstruct the charge and momentum of each particle, and their production and decay vertices, separated in this example by a few centimetres.

Neutral particles are not observed directly; they pass through matter without interaction until suddenly they either decay into charged particles (see the Λ^0 in figure 3.6) or interact; two examples of a photon (γ) interacting on a proton in the medium to make an electron–positron (e^+e^-) pair are seen. This event is unusual in that both these photons convert to e^+e^- pairs. The mean distance before conversion in hydrogen for photons is 11 m, much larger than the practical size of a liquid-hydrogen bubble chamber. Most photons ought therefore to escape unobserved; such devices are not very efficient as photon detectors.

An ideal modern 'hermetic' detector therefore has a layered cylindrical structure surrounding the interaction zone (e.g. at the Stanford Linear Collider (SLC) in figure 16.1). Closest to the interaction zone are fine-grained silicon microstrip detectors that can resolve tracks less than a tenth of a millimetre apart. Further out is a zone for the reconstruction of the curved tracks of charged particles, next perhaps a region dedicated to particle identification, and then two regions of dense material (a 'calorimeter' designed to absorb photons and to measure their energy, followed by a deeper zone to absorb the energy of all hadrons). Muons will penetrate intact through this armoury and can be identified outside, but neutrinos will escape all this without interaction. Their production can be inferred only from the 'missing momentum' in the final state, as something that must be added to bring the initial and final states into balance.

CHARGED PARTICLE DETECTION; CLOUD CHAMBERS AND BUBBLE CHAMBERS

An electrically charged particle with an energy typical of elementary particle interactions (say, 100 MeV) passing through a material medium encounters many atoms. Most of these atoms are unaffected, but in a few cases an electron is ripped from the atom and is then free to move; the atom is said to be 'ionized'. The electrical force from the particle strips the electrons (negatively charged) away from the nucleus (positively charged). With a sufficiently vigorous close encounter, an electron is freed; the moving particle loses about 30 eV of its energy in the process.

To use this phenomenon to detect charged particles needs something to amplify the effect of these few ionized atoms into a visible phenomenon. This problem was solved finally by C T R Wilson in 1911 and subsequently won him

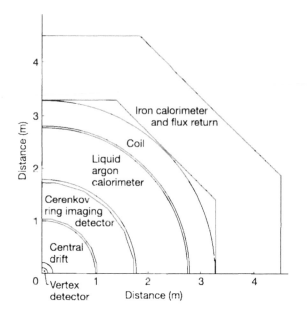

Figure 16.1. The concentric layers of a modern 'hermetic' detector ensure that no reaction product can escape without being observed in some way. This shows a section through one quadrant of the cylindrical detector used to observe e^+e^- annihilation at the SLC (Stanford, CA), viewed at right angles to the line of the colliding e^+ and e^- beams which pass along the axis of the detector.

the Nobel Prize for Physics. His work as a holiday relief in the meteorological observatory at the top of Ben Nevis in Scotland (at 1343 m the highest point in Britain) gave him time to study the formation of clouds at midsummer dawn and he saw how they always started from a very small disturbance.

He invented the cloud chamber in which a supersaturated atmosphere is created in a small container (see figure 16.2(a)). The ionization caused by the passage of α particles from radioactive decay seeds a trail of condensation along their paths, revealing the direction and length of each track. The example shown in figure 16.2(b) has a set of tracks of equal length. This shows that the α particles are emitted all with the same initial energy. They lose 30 eV at each ionization and are brought to rest when their initial kinetic energy is exhausted.

This phenomenon should be less surprising today than

(a)

(b)

Figure 16.2. (a) A cloud chamber with a sensitive volume of about 20 cm³, sold for 15 guineas (about 35 Swiss francs) in 1934. Turning the crank causes a sudden expansion of the chamber, and tracks are formed for a few seconds after expansion. (b) A set of α-particle tracks taken in the same chamber.

it was in 1911. We are all familiar with the vapour trails left by high-flying aircraft. Often the plane is too high for us to see, but it makes a track that can grow in size and which tells us precisely where the plane went.

The bubble chamber uses the same phenomenon in reverse; a superheated liquid is suddenly expanded, just before injecting a pulse of, say, ten particles from an accelerator. For a millisecond or so, boiling will commence along the trail of ionization left by the passage of charged particles. A trio of stereoscopic photographs is taken, and a library of beautiful events is accumulated. Figure 3.6 shows a classic example of the art. The chamber is filled with 1000 l of liquid hydrogen which is both the target material for the interactions to be studied, and the medium for the detection of outgoing particles.

Visual techniques using principally bubble chambers became the detector workhorses for the 1960s. They have many advantages: the clarity with which complex events are laid out to see in detail, the simplicity and homogeneity of the device, the permanent record that can be studied in detail at leisure.

Pictures can be taken quickly at the accelerator: one every few seconds, say 250 000 in a week or two, and transported home for study. However, to analyse and interpret them takes longer. The events of interest in a particular investigation are scattered randomly throughout the film library. Every picture must be scanned individually before finding them, scanned again to reduce and quantify inefficiency and bias, and measured with great accuracy and care to reconstruct the particle directions and momenta.

An army of cheerful and careful scanners was recruited to undertake this labour in shifts. The technology in the early days was primitive; measuring machines converted the information on the photographs into thousands of reels of five-hole paper tape which sliced the hands of the graduate student who rewound them after feeding them into the mainframe at the computer centre. Computer time was a very scarce resource. In the early days, one flew every weekend to work shifts at a remote computer centre. The error rate in hole punching by measuring machines was high. Before a reel of paper tape could be read in completely, every error had to be found by computer, edited by hand using glutinously adhesive red tape, and the tape offered once more to the computer.

In such labour-intensive work, it made sense to share it by collaborating with other institutes. So groups of a few universities with a common interest got together, sharing the batches of film around the collaboration. Each site took responsibility for calibrating the optical distortions of the film assigned to them, and measured on these reels the events of interest not only to themselves, but also to the whole collaboration. This instilled the discipline of cooperation and, after a year or more, the final event sample was ready. The intellectual input of the whole collaboration raised the quality of the analysis. The sharing of data and of methodology made the whole collaboration take responsibility for the validity of a result. The collaboration would act as a stringent internal referee for proposed papers, with parallel analyses doing battle on their strengths and weaknesses. The result was a method of working that has benefited the field. The rule that a paper is submitted for publication in the name of the whole collaboration provides the best guarantee of quality. The competition from rival collaborations provides both a spur and a check.

HIGHER ENERGIES AND HARDER QUESTIONS

Despite their compelling immediacy, visual techniques have disadvantages. First, there is only limited scope for increasing the rate at which interactions are studied; the passage of say 100 or 1000 input charged particles per frame would hopelessly confuse the pictures. With only a few incident particles per picture, it is difficult to study rare processes as the time spent at the accelerator would stretch into years. Secondly, the scale and slowness of the scanning and measuring operation tend to standardize technique and so holds back innovation. Thirdly, to make useful measurements of momenta at higher energies requires an ever larger or more unwieldy chamber. The 2 m long CERN chamber (volume, 1.05 m^3) did excellent physics at the 28 GeV CERN proton synchrotron. For the opening of the Russian 70 GeV accelerator at Serpukhov, the French Saclay laboratory constructed a 4.5 m (volume, 11 m^3) chamber, whose explosive decompression every few seconds was memorably loud. The technique cannot be scaled indefinitely to ever-larger energies. Finally, the inability to discriminate between pions on the one hand and electrons and muons on the other limits the physics that can be studied.

With the first physics from the 500 GeV accelerator at Fermilab, USA, in 1972, interest turned to the production in hadronic reactions of electrons and muons as signatures of new physical phenomena (one looked for new kinds of quark, and for the W and Z carriers of the weak force and their decay to leptons; predictions of the masses of these were not yet available). Reliable identification of one electron or muon in a background of 10 000 pions was needed. The search for weak neutral currents (see the chapter by Sutton) also placed a premium on lepton identification.

Figure 16.3 illustrates one of the earliest experiments (Fermilab E-70, a Columbia–Fermilab collaboration, 1972–74) to have these features. To reach the rarest processes, the full-energy and full-intensity proton beam is extracted from the accelerator and interacts on a thick target. No attempt to identify or reconstruct complete events or to be hermetic is made. Instead a small aperture ('keyhole') spectrometer is used to sample particle production. A strong magnet is used to sweep away low-momentum particles, and to 'trigger' logic selects for recording those events where hits are found simultaneously in several layers of scintillator strips, indicating the passage of a high-momentum charged particle. The magnet and the scintillator strips together form a spectrometer, which can reconstruct the trajectory of the particle, and its momentum from the angle of deflection by the magnet.

Electrons are differentiated from the much more common pions by the segmented lead–glass calorimeter following the spectrometer. An incident electron causes an electromagnetic cascade ('shower') of photons, electrons and positrons in this dense medium. As lead–glass has a high refractive index (up to 1.67), light travels 40% slower in the glass than it does in a vacuum. No particle can travel faster than light in a vacuum but, since light goes more slowly in glass, an electron in glass can exceed the speed of light there. Such a particle emits a shock wave of light, called Čerenkov radiation, rather as an aircraft flying faster than the speed of sound makes a sonic boom.

All the electrons and positrons in the shower do this, and the total amount of light emitted is proportional to the incident electron energy. So an electron is identified by a triple signature: by its calorimeter energy deposit, by a scintillator track pointing at the relevant part of the

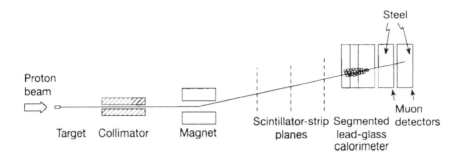

Figure 16.3. Schematic diagram of the E-70 detector used at Fermilab in 1972–74. The small-aperture ('keyhole') collimator enables one to study single particles arising from complex interactions. The diagram is not to scale. The calorimeter is 70 cm deep, and the whole apparatus extends over 60 m.

lead–glass array, and by the agreement in its energy as measured by the lead–glass with its momentum found using the spectrometer. On the other hand, most pions travel through the glass without interaction and deposit only minimal energy. Occasionally one has an interaction that transfers most of its energy to a neutral pion; this in turn decays rapidly to two photons which produce electron-like showers. Careful techniques can suppress this effect to isolate an electron signal at the level of 0.01% of the pion rate.

Muons penetrate through both the lead–glass and a stack of steel plates (which absorb pions) without interaction and emerge as single unaccompanied particles. The E-70 spectrometer made the first measurements of direct production of leptons in proton interactions and was upgraded later to discover the Υ particle which decays to e^+e^- or $\mu^+\mu^-$. However, through delays in the accelerator schedule, it missed an earlier and bigger prize.

THE NOVEMBER REVOLUTION

A Nobel Prize discovery was dramatically captured by two different experiments on 11 November 1974 (see the chapter by Schwitters); experiment one, by the group of Ting at Brookhaven, pursued at lower energies similar techniques to E-70, but the J/ψ particle was discovered simultaneously in a different way. Figure 7.2 in the chapter by Schwitters shows the layout of the Mark I magnetic detector at the SPEAR electron–positron storage ring at Stanford, CA, and figure 7.7 shows a typical event reconstructed within

it. The main tracking detector consists of four cylindrical layers of spark chambers. (In a spark chamber the trail of ions is used to seed an electrical breakdown at the same location after a high-voltage pulse is applied: a variation on the principle of the cloud and bubble chambers.) Each track in figure 7.7 has four spark chamber hits, and an electromagnetic shower counter hit at the outer radius; the two low-momentum (strongly curved) tracks are pions, and the two high-momentum tracks are muons. These pass through the iron magnet flux return and are tagged in the muon spark chambers outside.

This discovery was a revolution not only in our philosophy of physics but also in the that way we do experiments.

Instead of a beam incident on a fixed target (such as the protons in the hydrogen of a bubble chamber), two beams of matter (electrons e^-) and antimatter (positrons e^+) collide head on. Not only does this maximize the energy available, but it also creates a state (after the matter–antimatter annihilation) of pure energy with no bias towards one form of matter or another and no legacy of strong interactions to complicate the final state. The detector approaches the visual clarity and hermetic coverage of the bubble chamber (except for a small hole at each end for the vacuum tube along which the colliding beams pass). Since the colliding particles never enter the sensitive region of the detector, a huge flux can be tolerated, and the reaction rate can be increased to study rare processes.

After pioneer work with the AdA (Annello d'Accumulazione) ring, first at Frascati, near Rome, and then at

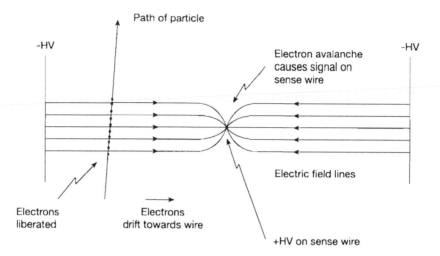

Figure 16.4. Principle of the drift chamber. The passage of a charged particle ionizes the gas (typically an argon-based mix), liberating electrons which drift under the chamber's electric field to the sense wire. High electric fields close to the wire cause avalanche multiplication of the number of electrons, giving an electric signal large enough to be detected. The position at which the particle traversed the chamber can be inferred from the drift time.

Orsay, near Paris, e^+e^- annihilation had been exploited for physics at the ADONE facility at Frascati, ACO at Orsay, VEPP-2 at Novosibirsk in Siberia, and at the CEA (Cambridge Electron Accelerator) at Harvard–MIT. However, the spectacular discoveries at SPEAR brought this technique to centre stage.

The detector is a complex construction with several distinct elements to be built and operated. This is naturally suited to a collaboration where different institutes take responsibility for fabrication of different components and for their operation. The development and construction time for a complex detector can extend over several years.

By contrast the time taken to reconstruct events is much diminished, compared with bubble chamber methods. Once the optical spark chambers have been replaced by drift chambers (see figure 16.4), event reconstruction could become almost in real time. 'Bicycle-on-line' jobs monitor the detector performance ever hour or two to look for malfunctions. Physicists suddenly had to get used to a highly pressured way of life, with 24 hour turnround from data taking to reconstructed and selected events on tape, ready for physics analysis.

GASEOUS IMAGING TECHNIQUES

Georges Charpak won the Nobel Prize for Physics in 1992 for developments begun some twenty years earlier: the invention of gaseous imaging techniques. Figure 16.4 shows the basic idea. A charged particle passes through a gaseous medium (say, argon with 10% carbon dioxide) and knocks electrons, perhaps thirty per centimetre, off atoms in its path. An applied electric field causes these freed electrons to drift along the field direction quite uniformly at about 50 km s^{-1} through the gas. If certain impurities (particularly oxygen) are avoided, the electrons are not recaptured by gas molecules. Eventually they arrive at a thin (0.03 mm diameter, say) wire at a high voltage. Close to the wire, the electric field is very high and gives the drifting electrons enough energy to knock secondary electrons out of the gas atoms. Thus one electron becomes two, then four, then eight and so on as an avalanche of electrons cascades towards the wire. The electric charge in this avalanche is large enough to be detected by sensitive electronics. The timing of the output signal gives the time for which the electrons have drifted through the gas, and hence the position where the incident particle passed through the detector.

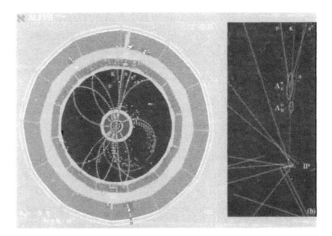

Figure 16.6. An electron–positron annihilation event reconstructed in the ALEPH detector: (a) the event seen in the Time Projection Chamber (TPC), with the electromagnetic calorimeter, hadron calorimeter and muon detectors in successive layers around it; (b) the vertex region, showing the 3 mm flight path between the interaction point (IP) and the decay vertex of a Λ_b^0 (containing a beauty quark) to $\Lambda_c^+ \pi^-$. The Λ_c^+ then decays to $pK^-\pi^+$. The TPC identifies the proton and K^- by their rate of ionization energy loss. The lines show particle trajectories extrapolated back to the production zone, and the ellipses (IP, Λ_b^0 and Λ_c^+) indicate the accuracy with which we can reconstruct the production and decay points. (See colour plate 3.)

Figure 16.5. The ALEPH detector at LEP (with its end cap removed). The elements are the same in sequence as at SPEAR and SLC: close to the axis are tracking regions, then an electromagnetic calorimeter, a magnet coil, iron return yoke instrumented as a hadron calorimeter and finally muon detectors. Prominent ALEPH physicists (right to left) Pierre Lazeyras, Lorenzo Foà, Jack Steinberger and Jacques LeFrançois are dwarfed by the detector. (See colour plate 4.)

With these three elements (electron production, electron transport, and electron amplification) we have what we need. The very act of reading out has cleared the detector volume of electrons, and it is ready for another event. The information is already digitized and on its way to the computer for event reconstruction.

MODERN COLLIDING-BEAM DETECTORS

This style of collider and detector has become the dominant form at all colliders. There is an unbroken evolution from the SPEAR magnetic detector, via the Large Electron–Positron (LEP) collider (1989–99 according to present schedules) to the Large Hadron Collider (LHC) (scheduled to take data in 2005).

Figure 16.5 shows the ALEPH detector at the LEP e^+e^- collider as an example. The collision energy is a factor of forty higher than at SPEAR and there are more, and closer, tracks in each event. Instead of four layers of spark chambers there are 120 layers of Time Projection Chamber (TPC) (a finely divided gaseous imaging detector) giving visual continuity along the tracks that compares well with the bubble chamber.

Figure 16.6(a) shows an event in which e^+e^- annihilate to make a Z^0 boson which decays to a quark–antiquark pair, making two opposite jets of particles. In this particular event a Λ_b^0 baryon is produced. This can be considered as a neutron but with one down quark d replaced with its heavier cousin, the beauty quark b (quark composition, bud). The Λ_b^0 decays after about 3 mm, corresponding to a decay time of 10^{-12} s (see figure 16.6(b)) to $\Lambda_c^+ \pi^-$. The Λ_c^+ then decays to $pK^-\pi^+$. The TPC is used to measure the particle momenta, and to identify the proton and K meson by their rate of ionization energy loss. In the lower part of the event, a muon is identified by its penetration as a lone particle through the calorimeter and by hits in the muon chambers outside.

To provide this visual clarity the detector has tremendous granularity. The 120 TPC layers consist of 45 000 independent parallel readout channels, each with 500 consecutive time bins, giving a total of 20 million independent volume elements. The electromagnetic calorimeter has grown from 45 parallel channels in E-70 to 225 000 channels. Most notably the track reconstruction close to the vertex is dramatically improved by the use of silicon microstrip detectors. Charged particles traversing thin layers of silicon raise electrons from the conduction to the valence band, a process akin to ionization. In 0.3 mm depth a signal of 24 000 electron–hole pairs are produced. The detector is divided into independent strips, each 0.05 mm wide, and knowledge of which strip was hit provides the track position information. The complete vertex detector consists of two cylindrical layers up to 11 cm in radius and 40 cm long, giving 98 000 independent parallel channels. This allows particle trajectories to be reconstructed with the accuracy that enables decay vertices such as that in figure 16.6(b) to be unambiguously reconstructed. Short-lived decays are a signature of heavy quarks. Isolating clean samples of these rare decays opens new areas of study.

A detector of 98 000 parallel strips is not the limit. Using charge-coupled device technology, a silicon layer can be divided into pixels, each a cube of side 0.02 mm. The vertex Stanford Linear Detector (SLD) at the SLC at Stanford, CA, has a total of 307 million pixels in three layers, giving a truly three-dimensional event reconstruction.

THE HIGHEST ENERGIES

There is great excitement in physics over the possibility that we may understand why things have mass through the Higgs mechanism (see the chapter by Veltman), and there is experimental evidence that the solution will be found by raising the accelerator energy to give an effective collision energy of some 14 000 GeV. As the energy increases we run into a number of technical challenges. The reaction rate for processes of interest falls as the energy increases, whilst the total reaction rate stays constant. We therefore have to learn how to work with increasing interaction rates, and with ever-higher background-to-signal ratios obscuring the physics that we wish to study (the same issues as were faced at Fermilab in 1972). This produces problems of radiation damage to components, and of data rate, of selectivity in real time, and in final analysis.

At the same time the events of interest themselves become more complex. Figure 16.7 shows a view of a simulated LHC event seen in part of the ATLAS inner detector; one needs the full repertoire of fine segmentation: lepton identification, hermetic closure and vertex detection. The increasing complexity and precision of the detector in turn places demands on alignment accuracy, on heat loads, and on thickness of material, to avoid unwanted interactions in successive detector layers.

The problem of high rates had been faced already at the HERA electron–proton collider in 1992. Table 16.1 compares various facilities.

Table 16.1. Rates of interaction at various colliders.

Facility	Effective collision energy (GeV)	Year	Crossing interval	Events per crossing
PETRA	12–44	1978	3.6 μs	Rare
LEP	90–190	1989	22 μs	Rare
HERA	300	1992	96 ns	Rare
HERA B	—	1997	96 ns	4
LHC	14 000	2005	25 ns	25

The crossing interval is the time interval between collisions of successive bunches of particles in the vacuum ring of the collider. Interactions occur only at the moment of collision; in between is an interval of quiet. As the energy increases, we must pack bunches closer together to achieve the desired interaction rate, and the crossing interval decreases. Depending on the technology chosen, it takes roughly a millionth of a second to read out the information in a particle detector into a temporary data store.

At LEP and earlier colliders it is practical to take a 'snapshot' approach, in which at every crossing information

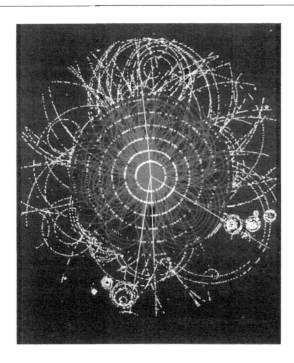

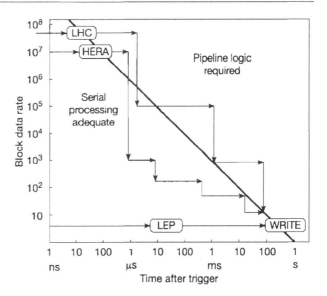

Figure 16.8. The challenge in data transfer that particle physics experiments present. Below the diagonal line, serial processing of single events is adequate. Above the line, at the highest rates, events arrive too fast for this to be possible, and pipelined logic and real-time data compression are essential. At lower rates, but still above the line, parallel processing of many events using computer arrays ('farms') can be used.

Figure 16.7. Preview physics: a simulated event in the central part of the ATLAS inner detector at CERN's LHC. The inner layers are silicon tracking detectors with 0.08 mm granularity. The outer layers are 4 mm diameter proportional drift tubes ('straws'). The event shown is at half the maximum accelerator intensity and includes the production of a B_d^0 meson decaying to $J/\psi K_s^0$, with $J/\psi \rightarrow e^+e^-$, $K_s^0 \rightarrow \pi^+\pi^-$. The LHC is scheduled to run in 2005. (See colour plate 5.)

transmitted into a temporary buffer store. We then enquire rapidly whether there is any sign of interesting activity in that event. If so, new data are blocked from coming in, while the event is transferred on to tape or disc for subsequent processing.

At HERA and at the LHC this approach will not do. The crossing interval is less than the time that it takes to transfer data out of the detector. To avoid wasting valuable interactions provided by the accelerator, every crossing must be interrogated although their readout sequences overlap in time. Further, to avoid drowning rapidly in floods of data we must be highly selective. If crossings occur at a rate of 40 MHz (25 ns interval) and it takes 1 ms to process an

event through a selection algorithm, then events are arriving 40 000 times faster than they can be handled.

The solution to this challenge lies in pipelined logic (see figure 16.8). The decision whether to accept or reject an event is built up through a cascade of hundreds of logical steps. Different events follow one another closely down the cascade, the final decision on each being made as it reaches the foot of the cascade. The LHC event sizes will be such that the front end data flow rate exceeds the total of the information that could be carried by all the television sets in Europe. At the first level, simple selections are made on energy deposit in the calorimeter and the like. If a 'yes' is given after 1 μs, the event is passed out of the pipeline memory into a buffer, where it is compressed (empty channels removed) and moves on to subsequent computation, using a 'farm' of many computers working in parallel. Each computer processes a separate event. New events arriving from the pipeline for computation are passed to any unoccupied computer. At the final level reconstructed events are written out to permanent store at a rate of 10–100 Hz.

Recall that each selection stage implies rejection, and by their very nature we know nothing about the rejected events. It is essential therefore to construct trigger selections which are efficient for the desired processes but, more important, selections whose efficiencies for passing wanted event types and for eliminating unwanted backgrounds are both well known and continuously monitored. One simple rule is that all selections must be made in two different ways, so that they can continuously monitor each other. Secondly, experience teaches us that real problems arise in unexpected ways, so that flexibility of approach is essential in trigger design.

DETECTOR EVOLUTION AND TECHNOLOGICAL SPIN-OFF

Particle physics is sometimes criticized for 'gold-plated' engineering and ever-increasing cost. Let us look at the facts. As the energy increases by orders of magnitude, the accelerators are in fact getting cheaper. At CERN the Super Proton Synchrotron (SPS), a 7 km ring, was built around 1974 on a rising budget worth in annual terms at 1994 prices some 1150 million Swiss Francs (MSFr). The LEP collider, a 27 km ring, was built in the 1980s on a level annual budget of 950 MSFr. The LHC is to be built around 2000 on a falling annual budget of 870 MSFr. World efforts are focused on to a single project, with contributions from outside Europe of some 450 MSFr. The cost per GeV of useful energy has fallen from the SPS to the LHC by a factor of over 500.

Far more advanced accelerators are built at lower costs, partly because of the advance of technology, and partly because each builds on the infrastructure and civil engineering of the earlier projects. If the CERN proton synchrotron was originally gold plated, it has paid for itself many times over. By 2015, serving as an LHC injector, it will have withstood some 55 years of use, setting new standards of machine reliability. Between 1976 and 1996 the number of CERN users increased from 1300 to 5800; so the cost per research head has fallen dramatically. At the LHC, as at the LEP collider, detectors will cost roughly half

as much as the accelerator, but will handle an event rate one billion times higher.

As we saw earlier, the ability to discover and to measure accurately is driven by technological progress, which is in turn driven by scientists' desire to find out more. The last ten years in particular has seen great emphasis on new detector techniques, whether in semiconductor technology, in survey and alignment, in construction in beryllia for precision with lowest material content, in fibre-optic readout, in large-area inexpensive detectors for muons, in massive photomultiplier arrays for neutrino astronomy, in low-noise electronics, in data flow and real-time processing, or in devising the World Wide Web (created by Tim Berners-Lee at CERN in the late 1980s to handle the document flow in collaborations that now can embrace some 150 institutes in 43 countries).

Much of the technical innovation needed arises in laboratories around the world and, as the results come together into a collaborative whole, diplomatic and managerial skills are required. Such technically driven international collaboration provides a highly relevant training at PhD level for the leaders of tomorrow's industry. Young scientists become confident in the skills of international teamwork and leadership.

Since the start of the field, particle physics has driven ever-increasing computer performance, from the mainframes of the 1960s to massively parallel Teraflop theoretical computation in the 1990s. The skills developed for particle physics in the technology of large superconducting magnets will find application elsewhere, possibly first in new accelerators to provide beams for radiotherapy. From the discovery of x-rays a hundred years ago, via nuclear magnetic resonance imaging, medical science has continually made immediate use of the technologies opened up by subatomic physics. Charpak's Nobel Prize for the invention of gaseous imaging techniques was in large part in recognition of the technology developed for experiments such as ALEPH being brought effectively to the benefit of medicine.

Perhaps the application that many people find most surprising is the use of antimatter for medical imaging. The same particles (positrons—the antimatter of the electron) that are used to mimic conditions shortly after the Big Bang in the LEP collider, provide the tools for beautifully accurate imaging of the body's internal organs. The introduction of solid-state technology, now developed for vertex detectors and to be used on a large scale for the first time at the LHC, will in due course open up new avenues of real-time x-ray imaging, and at vastly reduced radiation exposure levels for patients. In terms of spin-off, driven by the technical needs of the field in making precision measurements of very small effects under difficult conditions, particle physics still has much to offer mankind.

FURTHER READING

For the general reader

Riordan M 1987 *The Hunting of the Quark* (New York: Simon and Schuster)

Textbooks

Gilmore R *Single Particle Detection and Measurement* (London: Taylor and Francis)

Fernow R 1986 *Introduction to Experimental Particle Physics* (Cambridge: Cambridge University Press)

Kleinknecht K 1986 *Detectors for Particle Radiation* (Cambridge: Cambridge University Press)

ABOUT THE AUTHOR

David Saxon is the Kelvin Professor of Physics at the University of Glasgow.

He has worked at accelerators in the UK, France, Germany, the USA and at CERN, using bubble chambers, spark chambers and drift chambers. At present he works on the ZEUS experiment studying electron–proton collisions at the HERA collider in DESY, Hamburg and on preparations for the ATLAS experiment at CERN's LHC.

17 PARTICLE PHYSICS AND COSMOLOGY

John Ellis

Editor's Introduction: The 'Big Bang' picture of the origin of the Universe, one of the major scientific contributions of the twentieth century, was slowly pieced together as the interests of astronomers, astrophysicists, cosmologists and particle and nuclear physicists converged.

For centuries, astronomers had to be content with the faint light which manages to pierce the Earth's atmosphere. While huge new terrestrial telescopes continue to be a major weapon in the astronomers' armoury, the technological advances of the twentieth century, particularly following the Second World War, enabled astronomers to view the Universe's 'light' over a wide range of the electromagnetic spectrum, and to mount sensitive detectors aboard satellites which orbit in 'empty' space. The outcome was a radical reappraisal of our understanding of the Universe around us, a reappraisal in which particle physics has played a major role.

Einstein's development of the General Theory of Relativity in 1915 refined the classical ideas of Newtonian gravity, which still remained exactly as Newton had published them in 1686. Einstein's new theory set the stage for a new understanding of the Universe around us and provided a framework for the new science of cosmology—the study of the origin and structure of the large-scale Universe. Very soon afterwards, astronomers led by Edwin Hubble discovered that the Universe is much bigger than had been thought, and that the distant galaxies are receding at a rate proportional to their distance.

These two important developments ushered in a new concept of the origin of the Universe in a massive initial cataclysmic explosion—what is now called the Big Bang. The ingredients of this Big Bang were the same particles and forces which particle physicists study, but under very different conditions.

It took about half a century before particle physics was sufficiently well explored and its mechanisms understood for it to be ready for use as a cosmological tool. Most of the material in this book can be considered as necessary 'input' for modern cosmology. In his book *Before the Beginning* (New York: Simon and Schuster), Cambridge astrophysicist and cosmologist Sir Martin Rees says 'Cosmology's current buoyancy owes a lot to the incursion of particle physicists with (their) robust intellectual confidence'. In the next two chapters John Ellis and Qaisar Shafi describe this 'incursion' and the new insights it has produced.

INTRODUCTION

The Universe is, by definition, the largest possible physics laboratory, able to produce conditions more extreme than those achieved with our accelerators. Almost all conceivable experiments are presumably being carried out somewhere in the Universe, but the problems are to discover where, and to observe and interpret the results. The crucial difference between cosmological and accelerator experiments is that we have direct control over the conditions of the latter, whereas we are passive observers of whatever the Universe chooses to show us.

Before accelerator experiments gained their present sophistication, many particle discoveries were made in cosmic rays. As we shall see in the section on cosmic rays, these are now providing tantalizing hints of possible physics beyond the Standard Model, which may also have profound implications for the conventional Big Bang cosmology, developed in the subsequent two sections. A generation ago, advances in nuclear physics revolutionized our understanding of astrophysics, enabling us to calculate the properties of stars and their evolution. Analogously, particle physics provides the framework in which fundamental issues in cosmology may be resolved, including the origin and dominance of matter in the Universe, discussed in the section on Big Bang baryosynthesis, and possible candidates for dark matter, discussed in the subsequent section. Together with the primordial density perturbations revealed by the COBE satellite, this dark matter provides us with a theory for the formation of structures in the Universe, including galaxies and their clusters, as discussed in the section on cosmological inflation and structure formation. The earliest epoch that we can discuss rationally with our current physical theories is the Planck time, when the Universe was about 10^{-43} s old; typical particle energies would have been of order around 10^{19} GeV, and quantum gravity reigned. This may leave traces such as gravitational waves and superheavy relic particles in the present Universe, as discussed in the penultimate section. The final section reviews some speculations about the symbiosis of particle physics and cosmology in the new millennium.

COSMIC RAYS

These messengers from space played a formative rôle in the early days of particle physics, before accelerators were developed (see the chapter by Lock). They revealed antimatter with the discovery of the positron, mesons with the discovery of the pion, the second generation of fundamental particles with the discovery of the muon, and strange particles. However, during the past 50 years, the development of the Standard Model of particle physics has largely been the achievement of experiments at accelerators. Charm, the third-generation particles (bottom and top), the gluon, the W and the Z were all discovered in accelerator experiments, and the details of the Standard Model could not have been unravelled without the controlled environment which they provide.

However, hints of new physics beyond the Standard Model are now emerging from a new generation of cosmic-ray experiments using neutrinos. Historically, the first of

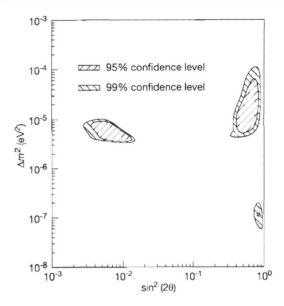

Figure 17.1. Possible scenarios for neutrino transformations ('oscillations') using input data from several experiments. Several allowed regions emerge for the neutrino mass and mixing parameters. The confidence level for the outer regions is 99% and for the inner regions 95%.

these were those looking for neutrinos emitted by nuclear reactions in the Sun. Their observation was an experimental *tour de force*, but the remarkable fact for us is that all experiments find electron neutrino fluxes considerably below the predictions based on the known energy output of the Sun. No explanation based on modified solar physics, such as a different equation of state or lowered central temperature, seems able to explain this deficit. The least unlikely interpretation may now be that neutrinos have properties not predicted by the Standard Model; perhaps the electron neutrinos produced in the solar core change into some other, less detectable species before they reach our detectors? Such transformations would be possible if the neutrinos had masses (figure 17.1).

An analogous phenomenon has recently appeared in neutrinos produced by cosmic-ray collisions in the Earth's atmosphere. In this case, it is the flux of muon neutrinos that appears to be deficient, and transformations between neutrinos of different masses again provide the most plausible interpretation.

Such neutrino masses could have dramatic consequences for cosmology. According to the standard Big Bang theory of cosmology reviewed later, there should be a billion times more neutrinos in the Universe than conventional matter particles. Thus, if individual neutrinos weighed 1 eV or more, collectively they would outweigh all the matter in the Universe, providing a possible explanation for the dark matter long advocated by astrophysicists.

Other types of cosmic-ray experiment also have the potential to reveal new physics beyond the Standard Model. For example, many underground experiments around the world are looking for weakly interacting massive non-relativistic particles (WIMPs) arriving from space. If such particles existed in numbers comparable with protons and neutrons, and with masses around 30–100 times larger, they could provide all the dark matter suggested in current cosmological models based on inflation, as discussed later. However, so far as there is no experimental confirmation of the existence of any such WIMPs.

At the other end of the energy spectrum, there are puzzles concerning ultrahigh-energy cosmic rays with energies around 10^{20} eV (figure 17.2). Clearly, their production requires some very powerful astrophysical accelerators whose identification is still enigmatic. Moreover, these sources cannot be too far away, or they would be absorbed by interactions with the microwave background radiation discussed in the next section. One very speculative possibility is that they might originate from the decays of supermassive relic particles from the very early Universe, as discussed in the final section.

BIG BANG COSMOLOGY

We now pass from the direct observation of particles from space to their indirect manifestations in cosmology. Since the 1920s, it has been known that our Universe is expanding, and the standard model for this is the Big Bang. Historically, the first evidence for this was the recession of the distant galaxies observed by the astronomer Hubble, now known to occur at a rate of about 50–80 km s^{-1} Mparsec^{-1} of distance. Viewed on a sufficiently large scale above about 100 Mparsec, the Universe appears approximately homogeneous and isotropic.

One immediate question is whether this expansion will continue for ever. Certainly the observable matter would not

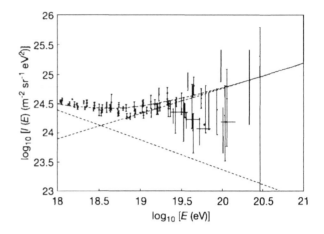

Figure 17.2. The measured spectrum of the highest energy cosmic rays. The upper broken curve is a superposition of the steeply falling power law combined with a flatter distribution at higher energy.

exert sufficient gravitational attraction to halt and reverse the expansion; it has a density which is less than one tenth of the critical density required to cause such a collapse. However, astrophysicists observing the motions of stars and galaxies have long insisted that these are being pulled around by the gravitational attraction of more (invisible) matter out there, the dark matter mentioned earlier. Current determinations of the total density of matter come out at around one third of the critical density. To my mind, it is premature to exclude the possibility that the density may even be critical, in which case the Universe might eventually recollapse.

Setting the cosmological videocassette recorder now on 'fast rewind', we find two crucial additional pieces of evidence for the Big Bang theory of cosmology (figure 17.3). Earlier in the history of the Universe, when it was less than a million years old, its matter would have been much hotter and denser than it is today. When it was more than about 1000 times smaller and hotter, atoms could not have existed, and all the matter in the Universe would have been ionized into nuclei and free electrons. As these electrons cooled and settled into atoms, they would have emitted photons which should be visible today as a diffuse background of microwaves. Predicted by Gamow, this cosmic microwave background was first observed by Penzias and Wilson in 1965. The most complete

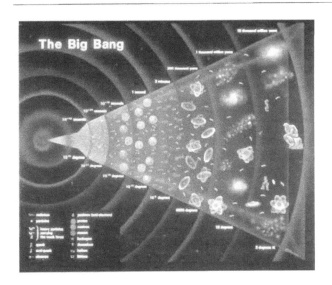

Figure 17.3. Milestones in the evolution of the Universe from the Big Bang. (Photograph CERN.) (See colour plate 2.)

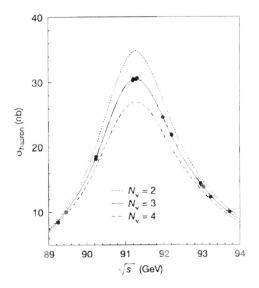

Figure 17.4. The shape of the Z resonance measured at CERN's LEP, compared with predictions using different numbers of light neutrinos. Only the three-neutrino solution (•) corresponds to the observed data.

observations of it have been with the COBE satellite, which demonstrated that it had (to an excellent approximation) a thermal spectrum, and that it was isotropic to one part in 10^5. At this level, minuscule fluctuations appear that may be an exciting window on primordial particle physics, as discussed in the section on cosmological inflation and structure formation.

Rewinding the Universe further, when it was aged less than a minute and about 10^8 times smaller and hotter still, even nuclei would no longer have been sacrosanct, and thermonuclear reactions would have occurred throughout the Universe. Before this epoch, the dominant forms of matter would have been protons, neutrons, electrons, positrons and neutrinos. The Big Bang nuclear interactions would have synthesized deuterium, helium, lithium and other light elements. The efficiencies of these interactions would have depended on the rate of expansion of the Universe, which was controlled by the density of protons and neutrons and the number of different species of neutrinos.

In recent years, there have been extensive calculations of this Big Bang nucleosynthesis and their confrontations with observations of the cosmological abundances of light-element abundances. These are in concordance if the

density of protons and neutrons is about one billionth that of photons, corresponding to a present matter density below one tenth of the critical density, as observed. This concordance also requires that there be no more than a handful of species of light neutrinos. Laboratory experiments at the Large Electron–Positron Collider (LEP) and elsewhere have now determined that there are precisely three neutrino species, leading to successful calculations of the abundances of light elements (figure 17.4).

This agreement constitutes the most successful confirmation of the Big Bang theory; when it was less than a minute old, the Universe was a thousand million times smaller and hotter than it is today, and the fundamental laws of microphysics describe its behaviour. The next sections will extrapolate this success back to earlier times when the Universe was orders of magnitude smaller and hotter, the experimental evidence correspondingly smaller and the theoretical speculation correspondingly hotter.

THE PRE-HISTORY OF THE UNIVERSE

Before the 'written records' provided by the microwave background radiation and the cosmological light elements, the Universe experienced several periods of smooth expansion in a state close to thermodynamic equilibrium, punctuated by the occasional phase transition.

In everyday life, we are familiar with the transitions from the frozen to the liquid to the gaseous state as matter is heated. The ionization of atoms to the plasma phase of nuclei and free electrons is another example that we have already met in the early Universe. The next transition we meet as we go back in time is probably the transition from ordinary matter such as protons and neutrons to an analogous plasma phase of quarks and gluons. Theoretical calculations suggest that this should have occurred when the Universe was a few microseconds old, with a temperature of around 10^{12} K!

This quark–gluon plasma phase transition is currently the object of intense experimental searches at accelerators colliding beams of heavy nuclei. Experiments at CERN have recently found tantalizing evidence for this new phase of matter (figure 17.5). Certain bound states of heavy quarks and antiquarks are produced less copiously than expected naïvely on the basis of individual proton–proton collisions. Is this because the collisions of heavy nuclei produced an extended hot plasma region where these bound states cannot exist? Only future rounds of experiments at Brookhaven and with CERN's Large Hadron Collider (LHC) will be able to answer this question definitively.

The results of these experiments will confirm our understanding of the Universe back to when it was less than a microsecond old and may also be relevant to the interiors of dense astrophysical objects such as neutron stars. However, it is not yet clear what the observational signatures of the cosmological quark–gluon plasma transition might be.

The next step back is to the electroweak transition, where the quarks, leptons, W and Z acquired their masses. The Standard Model says that these were generated by interactions with a universal Higgs scalar field. However, there is no direct evidence for this mechanism; in particular, LEP experiments have searched intensively for a particle associated with this Higgs field and have so far only been able to establish a lower limit that is currently just below 90 GeV.

Going even further back, to higher temperatures above about 10^{15} K, when the Universe was about 10^{-10} s old,

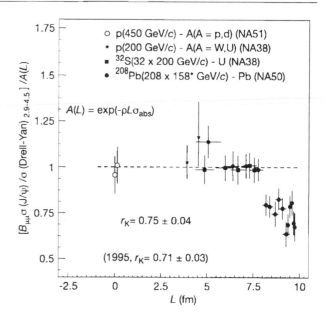

Figure 17.5. The NA50 experiment at CERN studies the production of J/ψ particles in high-energy nucleus–nucleus collisions. Under certain conditions (right), the production of J/ψ particles falls, suggesting that conditions are nearing those required to create the quark–gluon plasma.

thermal fluctuations would have disrupted the normal Higgs mechanism for mass generation, and the quarks, leptons, W and Z would have lost their masses. The LEP lower limit on the mass of the Higgs particle suggests that this transition was probably relatively smooth and uneventful, but this conclusion could be reversed if LEP or another accelerator such as the LHC discovers additional particles beyond the Standard Model, such as those suggested by supersymmetry. An abrupt (first-order) transition could have interesting consequences for the history of the Universe; in particular, it could explain the origin of the matter in the Universe, as discussed in the next section.

Before the electroweak phase transition, the pre-history of the Universe becomes more speculative and less subject to experimental test. Within the framework of conventional Grand Unified Theories (GUTs), the next phase transition would have been associated with the recovery of an underlying symmetry between the strong and electroweak interactions. In simple models, this occurs at an energy of around 10^{15} GeV, corresponding to a temperature of around 10^{28} K. There are two possible observable implications of this conjectured phase transition; one is the synthesis of matter already mentioned, and the other is the suggestion of cosmological inflation discussed later.

If one wishes to speculate even more outrageously, one is restricted by the fact that our conventional notions of space and time probably break down when quantum gravity effects become important and particle energies reach the Planck energy of around 10^{19} GeV, corresponding to an ultimate temperature of around 10^{32} K. Beyond this mark, the notions of distance and time probably break down under the incessant impacts of high-energy particle collisions. Since the concept of time breaks down, it does not make sense to ask: what happened before the Big Bang? In my view, the appropriate framework for discussing this Planck epoch is provided by string (perhaps in its latest incarnation of M theory), which is the only candidate we have for a Theory of Everything (TOE) (see the chapter by Ross). Does this lead to a 'pre-Big Bang' or a phase in which only mathematical topology is meaningful?

Let us try to get our feet back on the ground by discussing some concrete cosmological probes of this pre-historical framework.

BIG BANG BARYOSYNTHESIS

Where does all the matter in the Universe come from, and why does it predominate over antimatter? This is typical of the sort of fundamental issue that would have been left to theologians until the advent of modern particle physics and cosmology. It was the physicist Sakharov in 1967 who set out the necessary and sufficient conditions for generating such an asymmetry between matter and antimatter.

First and foremost, one needs particle interactions that change quarks into leptons and vice versa. These are explicitly present in most GUTs and should also lead to proton decay. Unfortunately, despite intense experimental searches, protons remain stubbornly stable. However, all is not lost, even if one is GUT-less, because it has been realized that even the Standard Model should have hidden interactions that induce quark–lepton transitions, although these are not vulnerable to direct experimental observation.

Secondly, Sakharov pointed out that these interactions should distinguish between matter and antimatter. Such a distinction has been observed in the electroweak interactions of neutral kaons since 1964, although its origins are not yet established. There is now a vigorous campaign of experiments on heavier particles containing bottom quarks to study in more detail this matter–antimatter distinction. The Standard Model makes detailed and specific predictions for what should be observed. Regardless of whether these are confirmed or refuted, the new experimental campaign should provide us with understanding of the matter–antimatter distinction that should be essential for any mechanism for Big Bang baryosynthesis.

The third Sakharov condition is that there should be a breakdown of thermal equilibrium. This is because, in equilibrium, particle and antiparticle densities are fated to be equal following Boltzmann's law. Subsequent annihilation of these equal densities would yield a relic matter density that is at least 10 orders of magnitude smaller than indicated by observation. We conclude that a breakdown of thermal equilibrium is essential, and this could only have occurred at a phase transition that was abrupt (first order) and too rapid for the particle and antiparticle densities to keep together.

Two of the transitions identified in the previous section could provide this opportunity: the electroweak transition and the GUT transition. The electroweak scenario has the merit of being closer to experimental test. According to this idea, prior to the transition, the 'hidden' electroweak interactions would have enforced matter–antimatter equality. Then, if the electroweak transition was sufficiently abrupt, residual electroweak interactions could have favoured matter creation, e.g. in the collisions of particles with nucleating bubbles of the low-temperature phase. This scenario is subject to test in the ongoing matter–antimatter experiments, as well as being severely constrained by LEP searches for the Higgs and other particles.

The GUT scenario invokes the out-of-equilibrium decays of some massive particles X spawned by the GUT phase transition. For example, if GUTs produce equal numbers of X particles and their antiparticles $\overline{X}$, but the probability that X decays into matter is larger than the probability that $\bar{X}$ decays into antimatter, a net matter–antimatter asymmetry will be produced. Examples of such massive X particles could include superheavy neutrino-like particles that are associated with the possible neutrino masses mentioned earlier in connection with the deficits of solar and atmospheric neutrinos. As also mentioned there, massive neutrinos are among the particle candidates for dark matter, but there are other candidates as well, as we now discuss.

PARTICLE CANDIDATES FOR DARK MATTER

Astrophysicists have long argued that the Universe must contain large amounts of invisible dark matter. Could this be ordinary matter that does not shine in stars, for some reason? Big Bang nucleosynthesis tells us that there could not be enough ordinary matter to provide the fraction of about one third of the critical density apparently present in clusters of galaxies. However, ordinary dark matter could be significant locally, e.g. in the halo of our own galaxy. Indeed, observations of the Magellanic Clouds have revealed a few stars whose light is briefly amplified by gravitational lensing owing to the passage of an intermediate dark object with mass a fraction of that of the Sun. Indications are that these massive compact halo objects (MACHOs) could not constitute the whole of our galactic halo, but the jury is still out. Theoretically, MACHOs would not help to explain the formation of galaxies, and clusters surely need some extra matter, as already mentioned.

Neutrinos are good candidates, since they certainly exist, may have mass, as discussed earlier, and are only weakly interacting, and so they would escape collapse into concentrations of ordinary matter such as the discs of galaxies. Unfortunately, the indications are that they would also not populate sufficiently galactic haloes, although they might be a significant source of dark matter in clusters.

For many astrophysicists and cosmologists, the favoured dark matter candidates are WIMPs. These would have been non-relativistic ever since they froze out early in the expansion of the Big Bang. As such, they would easily have gravitated together and catalysed the formation of structures in the Universe. They would fill up galactic haloes as well as clusters. The only snag is, no such particles are known to exist!

Many speculative candidates for WIMPs have been proposed; my own favourite is the lightest supersymmetric particle (LSP). The motivations for a supersymmetric extension of the Standard Model are reviewed elsewhere (see the chapter by Ross). Suffice it to say here that it postulates partners of all the known particles with spins that differ by half a unit, but with the same internal properties such as electric charges, baryon and lepton numbers. If the interactions of these supersymmetric particles conserve these quantum numbers, the LSP would be stable, and an excellent candidate for dark matter, since it would have neither nuclear nor electromagnetic interactions. Calculations indicate that the overall density of LSPs in the Universe might be comparable with those of protons or neutrons, whereas the failures of searches for supersymmetric particles at LEP so far indicate that it must weigh more than about 40 GeV. Hence its relic density today might well approach the critical density.

As already mentioned, many underground experiments are searching for LSPs from the galactic halo scattering off nuclei with the deposit of detectable nuclear recoil energy. At the time of writing, the sensitivity of such experiments is beginning to probe some theoretical predictions. Other experiments are looking for high-energy neutrinos released by the annihilations of LSPs or other WIMPs captured at the centres of the Earth or the Sun. Yet another class of experiments is looking for antiprotons, positrons and γ-rays that may be produced by LSP annihilations in the galactic halo.

Before leaving particle candidates for dark matter, it is worth mentioning the axion, a very light bosonic particle that might be present in the Universe in very large numbers in coherent waves. There are also experiments probing whether our galactic halo may be composed of axions.

There seems every likelihood that the dark-matter particles in our galactic halo may be detected within the next decade. Accelerator experiments will also be looking for supersymmetric particles, in particular. LEP will probe LSP masses up to about 50 GeV, and in the long run the LHC will explore all the likely range of supersymmetric particle masses. It will be interesting to see who wins the race between accelerator and dark-matter experiments; super-optimists do not admit the possibility that supersymmetric particles do not exist!

COSMOLOGICAL INFLATION AND STRUCTURE FORMATION

We commented in the section on Big Bang cosmology that the density of matter in the Universe appears to be less than an order of magnitude below the critical density. *A priori*, this is surprising, because the density of a subcritical Universe tends to fall rapidly far below the critical density as the Hubble expansion proceeds. Related questions are why the Universe is so old and large. Its expansion is controlled by Einstein's equations, in which the only dimensional parameter is the Planck mass, which corresponds to a time scale of order 10^{-43} s and a length scale of order 10^{-33} cm. Clearly our Universe must be a very special solution of Einstein's equations, in order to have survived so long and grown so large.

The solution of these conundra may also explain why the Universe is so homogeneous and isotropic. Cosmic microwave background photons, from opposite directions in the sky, when they were emitted during the combination of nuclei to form atoms, started from points that were separated by about a hundred times further than the maximum distance that a light wave could have travelled since the beginning of the Universe. Yet these opposite regions of the sky look identical to one part in 10^5. How did they succeed in coordinating their behaviour?

The answer proposed by cosmological inflation is that the very early Universe underwent a phase of exponential expansion, perhaps driven by energy stored in a Higgs-like field associated with some GUT (figure 17.6). This mechanism would suggest that the density of the Universe be very close to the critical density and explain its age and size by this enormous exponential kick. The isotropy of the cosmic microwave background radiation would be made possible by coordinating the behaviours of different regions of the Universe before they were blown apart.

One of the beauties of cosmological inflation is that it not only predicts the homogeneity and isotropy of the Universe but also their breakdown. During the phase of exponential expansion, quantum fluctuations in the driving field are imprinted on the Universe as variations in its density, and their magnitude is related to the density of energy during inflation. The COBE satellite detected for the first time fluctuations in the cosmic microwave background radiation that could be due to this inflationary mechanism.

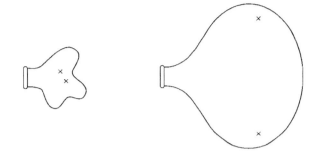

Figure 17.6. As a balloon is inflated, points on its two-dimensional surface appear to move apart. Analogously points in a higher dimensional Universe appear to move away from each other: the 'Hubble expansion'.

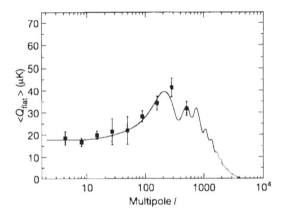

Figure 17.7. Data from a number of experiments measuring the cosmic background radiation, showing the variation in temperature seen over various angles (higher multipoles imply smaller angles).

If so, their magnitude of about one part in 10^5 corresponds naïvely to an energy density around the GUT energy scale. Even if this particular identification eventually fails, there is little doubt that fluctuations in the cosmic microwave background probe particle physics at an energy scale far beyond those attainable directly with accelerators.

Many experiments have by now confirmed the COBE discovery and mapped out the microwave background fluctuations at a wide range of angular scales (figure 17.7). They will be able to measure in detail the spectrum of perturbations and may tell us how they are composed into simple density perturbations and gravitational waves. Within the inflationary paradigm, these measurements should enable us to map out a significant part of the potential energy of the field driving inflation and test different GUT and other models.

The COBE discovery has also caused effervescence in attempts to understand the formation of structures such as galaxies and clusters. Armed with the COBE input and large N-body simulation codes, theoretical astrophysicists have tested many models of structure formation. The most successful has combined inflationary perturbations with some species of massive dark-matter particle such as the LSP or some other WIMP. Not all the detailed comparisons work perfectly, and it may be that this basic recipe requires some spice, perhaps in the form of massive neutrinos, a modification of the fluctuation spectrum, or some energy density in the present-day vacuum—a cosmological constant.

With all the cosmological measurements to be made over the next decade, particularly by the MAP and *Planck Surveyor* satellites, there is good reason to expect soon to have a detailed and quantitative theory of structure formation. This theory will rely upon inputs from particle physics, notably the theory of inflation and the experimental observation of some species of dark-matter particles, either at an accelerator and/or in the galactic halo.

GRAVITATIONAL RELICS FROM THE BIG BANG?

So far we have emphasized the links between cosmology and the non-gravitational interactions of elementary particles. Are there any hopes for observing direct consequences of their gravitational interactions?

One strong hope is that of observing directly gravitational waves, which have already been observed indirectly via their effects on the orbits of binary pulsars. Large laser interferometer devices are now being built that have a fair chance of observing gravitational waves emitted by coalescing binary systems, star capture by black holes and perhaps supernova explosions. The next step would be a laser interferometer array in space. This might even be able to observe gravitational waves emanating from the primordial Universe, as occurs in some 'pre-Big Bang' models.

The evidence for astrophysical black holes is also growing more impressive, with claims being made for hints of a surrounding event horizon, and distortions of the surrounding space–time that are characteristic of a strong field in general relativity. However, direct observation of a quantum particle phenomenon such as Hawking radiation does not seem within reach at present. A recurrent theme is the possible existence of primordial black holes left over from the Big Bang. Suggestions that they might be made during the quark–hadron phase transition are not convincing, and it is difficult to imagine other sources subsequent to cosmological inflation (if this occurred). On the other hand, inflation would have diluted to undetectability any primordial black holes made previously.

The question then arises whether black holes or any other massive relics might have been fabricated during inflation. In the context of current inflationary models, this is unlikely unless their mass is of order 10^{13} GeV. If such relics carry quantum numbers that prevent or inhibit their decays, they could be around today as superheavy relics from the Big Bang. Decay products of such supermassive relics could conceivably be at the origin of the mysterious ultrahigh-energy cosmic rays mentioned earlier.

THE SYMBIOSIS OF PARTICLE PHYSICS AND COSMOLOGY

Microphysics and macrophysics have become so intertwined that it is now impossible to distinguish them. Particle physics provides the building blocks with which cosmologists must work. Cosmology constrains the properties of particles, e.g. the masses of neutrinos or the abundance of WIMPs, that are inaccessible to accelerator experiments. Researchers working in the fields of particle physics and cosmology use arguments from each other's fields as naturally as they draw breath. Experiments at LEP and the LHC provide key inputs to cosmological theories, and our greatest challenge may be the identification of dark matter. The present period of particle cosmology will, in retrospect, surely be considered as seminal as was nuclear astrophysics in a previous generation. Astronomy has already been extended from the electromagnetic spectrum to neutrinos, and perhaps soon to gravitational waves. Many of the most fundamental problems in cosmology may soon find solutions based on particle physics, whereas many particle theories may only be tested by cosmological observations. Within the contexts of the discussions of the previous sections of this chapter, which are the major developments that may be expected in the foreseeable future?

- We shall soon know from the Superkamiokande and other experiments whether neutrinos mix, as expected if they have non-zero masses.
- Underground or other non-accelerator experiments are likely to identify any WIMP component of the galactic halo, if LEP, the LHC and other accelerator experiments do not discover them first.
- We shall soon know to high precision the basic cosmological parameters such as the Hubble expansion rate and the present densities of baryons, other matter and the vacuum, in particular from the data of the MAP and *Planck Surveyor* satellites.
- Laboratory experiments will identify the quark–gluon plasma and will predict the nature of the electroweak phase transition.
- Laboratory experiments will also identify the origin of the matter–antimatter distinction observed in the weak interactions. Together with the electroweak phase transition measurements, these may provide a calculable theory of the matter–antimatter symmetry in the Universe.
- Accelerator experiments at LEP or the LHC will discover supersymmetry, if dark-matter experiments do not beat them to it.
- Observations by MAP and *Planck Surveyor*, together with sky surveys and the discovery of dark matter will provide a tested theory of the formation of structure in the Universe.
- Gravitational waves will be detected.

This list is impressive, but I prefer to end on a more humble note. The most exciting developments will surely be those that I have failed to foresee or predicted wrongly. The reader is invited to laugh at my hubris and myopia when the Universe outsmarts me.

FURTHER READING

Fraser G, Lillestøl E and Sellvåg I 1998 *The Search for Infinity: Solving the Mysteries of the Universe* (London: Philips)

Kolb E W and Turner M S 1990 *The Early Universe* (Redwood City, CA: Addison-Wesley)

Lederman L and Schramm D N 1989 *From Quarks to the Cosmos: Tools of Discovery* (New York: Scientific American)

Rees M 1997 *Before the Beginning* (New York: Simon and Schuster)

Riordan M and Schramm D N 1991 *The Shadows of Creation: Dark Matter and the Structure of the Universe* (New York: Freeman)

Weinberg S 1983 *The First Three Minutes: a Modern View of the Origin of the Universe* (London: Fontana)

ABOUT THE AUTHOR

Educated in England, John Ellis obtained his BA in 1967 and his PhD in 1971, both from the University of Cambridge. After a couple of years at the Stanford Linear Accelerator Center and the California Institute of Technology, he settled at CERN in 1973, where he was leader of the Division of Theoretical Physics for six years and is currently a senior staff member. He is the author of over 500 scientific articles, mainly in particle physics and some related areas of astrophysics and cosmology. Most of his research work has been on the possible experimental consequences and tests of new theoretical ideas such as gauge theories of the strong and electroweak interactions, grand unified theories, supersymmetry and string theory. He was awarded the Maxwell Medal of the Institute of Physics in 1983 and was elected a Fellow of the Royal Society in 1985. The University of Southampton awarded him an Honorary Doctorate in 1994, and he has held visiting appointments at Berkeley, Cambridge, Oxford, Melbourne and Stanford.

18 BIG BANG DYNAMICS

Qaisar Shafi

Editor's Introduction: Echoing the Big Bang message of the previous chapter, Qaisar Shafi sets the scene by recalling Einstein's General Theory of Relativity. He then changes focus to explain the evolution of the early Big Bang universe in terms of particle physics mechanisms, suggesting how the present distribution of matter could have been moulded. Reconciling increasingly detailed cosmological observations with the parameters of these primordial mechanisms remains a major ongoing objective of this work.

Modern cosmology can trace its foundation to the development of general relativity (GR) by Einstein in 1915. It had taken Einstein almost ten years to arrive at this remarkable new theory, which generalized his 1905 theory of special relativity, to include the effects of gravity. The general theory provided a profoundly new way to think about gravity. Space, and more generally space and time, in general relativity are ascribed dynamical characteristics which manifest themselves through 'gravitational' effects especially when matter is included.

Newton's theory of gravitation had reigned supreme for more than 200 years but several pressing issues had raised doubts at the turn of the century about the extent of its validity. First, the theory could not fully accommodate the observed perihelion shift of the planet Mercury as it rotates around the Sun. Second, the notion put forward by Newton that the universe consists of 'stars' spaced uniformly apart and held together by gravitation in a spatially infinite and static universe was shown to be gravitationally unstable, unless one made the highly improbable assumption that the 'uniformity' extended to an arbitrary degree of accuracy. Finally, the theory suffered from the well known Olbers paradox. If the universe is indeed infinite in extent and uniformly filled with stars, it was easy to argue that night should give way to a blazing daylight (this assumes that stars are eternal!). An easy way to see this is to imagine 'surrounding' an observer with spherical shells spaced equally apart, and then to note that the light emitted from any given shell is independent of its distance d from the observer. This follows since the $1/d^2$ dependence of

luminosity is compensated by a corresponding increase proportional to d^2 in the volume.

In formulating the special theory of relativity Einstein had fully incorporated the experimental, but by no means intuitive, observation that the speed of light is always c for all observers in uniform relative motion with respect to one another. This theory leads to some remarkable predictions which become increasingly significant when velocities approach the speed of light. These predictions are routinely tested in high-energy particle accelerators. One of the most fascinating predictions has to do with the lifetime of unstable 'elementary' particles. According to special relativity, the lifetime measured by an observer at rest with respect to a particle moving at relativistic speeds can be significantly greater than what he or she would measure if the particle had been at rest! This fact may one day be exploited to yield a new breed of high-energy accelerators involving beams of 'unstable' particles, such as muons, with far-reaching implications for both particle physics and cosmology.

In order to incorporate accelerated motion, especially gravity, Einstein in 1909 introduced the equivalence principle. This remarkable step was based on the equality between inertial mass (recall Newton's famous second law of motion that force = mass × acceleration) and the gravitational mass (force of gravity = $m \times g$), which had been demonstrated by Galileo, Newton, Bessel and others. Einstein argued that this 'equivalence' between the inertial and gravitational force can be exploited to set up, at every space–time point in an arbitrary gravitational field, a 'locally' inertial coordinate system such that the fundamental laws of physics can be formulated within the framework of special relativity.

In the absence of gravity, the 'local' inertial coordinate system can be extended to 'cover' the whole space–time. Among other things this tells us that, in the absence of gravity, the four-dimensional space–time is 'flat'. The three-dimensional 'space-like' coordinates satisfy the laws of Euclidean geometry (e.g. Pythagoras' theorem holds, the sum of angles of a triangle equals 180°, etc). However, with gravity present, this extension from a 'local' to a 'global' coordinate system becomes non-trivial. The resulting four-dimensional space–time is 'patched' together from a collection of distinct locally inertial coordinate systems, and the final geometric structure that emerges is intimately linked with the underlying gravitational field.

Einstein insisted that the fundamental laws of physics, including gravity, should take a form that is independent of the observer (or the coordinate system). With gravity switched off, they should take a form consistent with the requirements of special relativity. This, as well as arguments related to the equivalence principle, led Einstein to formulate his general relativity using the language of Riemannian (or non-Euclidean) geometry, which had been developed by Gauss, Riemann and others around the middle of the nineteenth century. The two-dimensional surface of a sphere provides an example of a non-Euclidean space. The sum of angles of a triangle drawn on this surface is greater than 180°. Similarly, say on a saddle-like surface the sum of the three angles of a triangle is less than 180°.

It was argued by Gauss, and later generalized to higher-dimensional space, that on such a curved surface one can construct, in the neighbourhood of any point, a local coordinate system in which the laws of Euclidean geometry are valid. The close analogy with Einstein's formulation of the equivalence principle is striking and indeed forms the basis of GR.

In Newtonian gravity, the gravitational field is characterized in terms of a gravitational potential function obeying a very well known differential equation associated with Poisson. The right-hand side of the equation is determined by the matter distribution, and the equation can then be solved (at least for simple enough cases!) to yield information about the gravitational potential and

consequently the gravitational field. Things in GR are much more involved in that the single potential of Newton's theory is replaced in GR by a geometrical quantity with six independent components, which carries information about the gravitational field as well as the ambient space–time.

Some people naturally wondered whether Einstein had made things unnecessarily complicated, but GR was here to stay and the reason was the experimental confirmation of the predictions by Einstein and others.

- The measured value of the perihelion shift of Mercury could not be fully accounted for by Newton's theory. In GR there are additional effects, caused by the 'warping' of space through which Mercury moves, which accounts for the missing contribution. GR fully accounts for the measured value.

- GR predicts the 'bending' of light by massive bodies such as the sun, Jupiter, etc. This was first confirmed in a 1919 solar eclipse expedition, by observing starlight grazing the Sun's edge. Numerous subsequent observations have confirmed the correctness of this prediction.

- Gravitational fields, according to GR, affect the running of clocks. In particular, clocks run slower as the field gets stronger, and this has been tested with high-precision clocks flown on planes and satellites at various heights.

- The equality of gravitational and inertial mass, on which the equivalence principle is based, has now been tested to an accuracy of better than 1 part in 10^{11}!

Finally there are some remarkable predictions of GR which are much more difficult to confirm.

(a) The existence of solutions of Einstein's equations known as black holes, which would correspond in nature to collapsed objects of extremely high density from which hardly anything escapes, but which can suck matter in from their surroundings; evidence for their existence has steadily been mounting during the last decade.

(b) The existence of 'gravity' waves, in analogy with the electromagnetic waves predicted by Maxwell's theory of electricity and magnetism.

In summary, GR provides a more complete and accurate description of gravitational phenomena than Newton's theory, which turns out to be a limiting case of GR under certain special conditions.

We now turn to discuss the main topic of this chapter, the cosmology of the Big Bang. With GR in hand Einstein focused attention on its implications for the universe as a whole. Like most other scientists of his generation, as well as generations preceding him, he believed in a static universe. Much to his dismay Einstein discovered that the GR equations which he had written down did not admit a static solution! The space preferred, instead, either to expand or to contract with time. He came up with a brilliant fix for this apparent conundrum. He added a new term, now referred to as the cosmological constant term, to the GR equations, whose sign and magnitude was suitably adjusted to 'neutralize' this expansion (or contraction).

Just a few years later, however, astronomers led by Edwin Hubble made the remarkable discovery that distant galaxies are receding from us at a speed that is directly proportional to their distance. (The proportionality constant is referred to as the Hubble constant and is one of the key parameters of modern cosmology.) This discovery laid the foundation for Big Bang cosmology. Einstein blamed himself for missing a great opportunity to predict the Big Bang on the basis of general relativity, referring to this episode at the 'greatest scientific blunder' of his life.

How are we to envisage the Big Bang? The enormity of the universe suggests that it may be very old, having evolved from a much smaller beginning. The measurements made by Hubble suggested a universe on the order of 1–2 billion years old, while recent measurements give a value of 10–14 billion years. This means that at the beginning the farthest galaxy must have been on top of us! Does this put us at the centre of the bang?

According to the cosmological principle, the universe, on average, is homogeneous and isotropic. This means that each 'typical' observer finds himself or herself at the centre of the Big Bang in the beginning. Clearly, the Big Bang is quite unlike, say, an explosion in which things fly out from a common centre. A closer, albeit two dimensional, analogy is to think of an expanding balloon with galaxies glued at fixed points on its surface.

What determines the expansion rate of the universe? The answer, according to GR, depends on the equation of state of the universe, and there are two especially important cases to consider as follows.

(i) Matter-dominated (present) universe

This describes, as many cosmologists would agree, the present state of the universe. Indeed, we may be tempted to say that the bulk of the universe's mass resides in ordinary matter that 'glows' and makes up the galaxies. It turns out that this is not the case! In order to explain the motion of stars in galaxies, especially stars in the outer fringes of a galaxy, one needs to invoke the existence of a new type of matter, referred to as 'dark matter'. The 'dark' matter is argued to exist in haloes around the galaxies, does not shine (hence 'dark') and makes itself felt only via its gravitational attraction. The dark matter plays a fundamental role in cosmological theories of large-scale structure formation, and we shall have more to say about it shortly.

In a matter-dominated universe, with pressure essentially zero, the separation between two distant galaxies increases as the $\frac{2}{3}$ power of the elapsed (cosmic) time.

(ii) Radiation-dominated (early) universe

It is widely accepted that, before the 'matter-dominated' era, the 'early' universe was dominated by radiation. In order to see how this comes about, we need to review briefly the foundations of particle physics.

As presently understood, these 'building blocks' consist of quarks and leptons, that come in three families. The (first) family, certainly the most important for our own existence, consists of the electron and its neutrino (leptons), the up and down (flavour) quarks carrying a colour charge (analogous to and yet in some ways crucially different from electric charge), as well as their antiparticles. The interactions between these particles are described through the exchange of a variety of 'particles' known as gauge bosons. Let us recall the four fundamental interactions listed according to decreasing strength, with the strength of the 'strong' (colour) force conveniently normalized to unity (table 18.1).

Some noteworthy points are as follows.

- Quarks and colour carrying gluons combine to form colour neutral 'atoms' (hadrons), the best known

Table 18.1. Four fundamental interactions.

	Strength	Number/name of gauge boson(s)	Range	'Matter' participants
Strong	1	8 gluons	$\simeq 10^{-13}$ cm	Quarks
Electromagnetic	10^{-2}	1 photon	Infinite	Quarks and charged leptons
Weak	10^{-5}	2 $W^{\pm}$ bosons 1 Z^0 boson	$\simeq 10^{-16}$ cm	Quarks and leptons
Gravity	10^{-38}	1 graviton	Infinite	All matter (and more!)

being the proton and the neutron which make the nuclei of ordinary atoms. Free quarks and gluons are not found in nature, and the theory that attempts to describe all this and much more is called quantum chromodynamics (see the chapter by Altarelli).

- The gauge boson of electromagnetic interactions is the particle associated with the electromagnetic waves that were discovered by Hertz, shortly after their existence was predicted in 1867 by Maxwell.

- The weak interactions are responsible for radioactivity and also play an important role in reactions occurring at the centre of the sun. The gauge bosons $W^{\pm}$ and Z^0 were discovered in 1983 at CERN (see the chapter by Rubbia). This discovery was crucial in confirming the so-called 'electroweak model' which unifies, at least partially, the weak and electromagnetic interactions despite their apparent differences (as seen in table 18.1 above). By combining this with the theory of strong interactions, we obtain what is commonly referred to as the Standard Model of particle physics.

- The three components of the standard model can be elegantly unified within a larger framework known as 'Grand Unified Theories' (or GUTs). In GUTs the apparent distinctions between quarks and leptons, as well as between the gauge bosons (and consequently the three forces), becomes blurred. Indeed, at energies that are well beyond our reach for the forseeable future, GUTs predict a complete unification of the three (strong, electromagnetic, weak) forces (see the chapter by Ross). These theories, as well as the Standard Model, belong to a class referred to as spontaneously broken gauge theories. A higher

symmetry is spontaneously broken by the ground state of the system on which the theory is built, and so the underlying symmetry is 'hidden' at 'low' energies. The full symmetry manifests itself at sufficiently high energies. For instance, in terms of strengths, the distinction between the weak and electromagnetic interactions disappears above the characteristic scale $M_W \simeq 100$ GeV (note that the proton mass is 0.938 GeV). The characteristic scale of grand unification is about 10^{16} GeV!

The reader may wonder about the relevance of particle physics for cosmology, especially the Big Bang scenario. The answer is very simple. Combined with Einstein's gravity, these theories enable us to explore the universe when it was very young indeed, say, as early as 10^{-36} s after the Big Bang! This comes about because spontaneously broken gauge theories respond to changes in temperature analogously, say, to a ferromagnetic substance. When the latter is heated above some critical temperature, it loses its spontaneous magnetization and the full rotational symmetry is recovered. Superconductors provide another example of this type of behaviour.

In other words, as we explore an increasingly younger and consequently hotter universe (assuming that $RT =$ constant, where R measures the 'size' of some region in the universe and T denotes the temperature: this is justified if one makes the assumption that the expansion is adiabatic), the underlying spontaneously broken gauge theory can undergo a series of phase transitions signalling the onset of symmetry restorations. The universe turns out to be the ultimate high-energy accelerator!

Let us first consider the early (radiation-dominated) universe, whose physics is determined by the Standard Model in a hot expanding environment. The characteristic energy scale of the electroweak interactions is about 100 GeV, which means that, for temperatures greater than this, the electroweak gauge symmetry is restored (in natural units 1 GeV $(= 10^9$ eV$) = 10^{13}$ K). The universe consists of a hot, weakly interacting 'gas' of quarks, leptons and gauge bosons of the Standard Model. Note that even the 'strong' interactions mediated by the 'coloured' gauge bosons are 'not so strong' at these energies, and the quarks are unconfined! The number density of each of these particles in this 'almost ideal' gas is determined by the standard thermodynamic arguments, taking account of their quantum-mechanical properties. As the universe expands and cools below a 100 GeV, the electroweak phase transition occurs and weak interactions are 'born'! The $W^{\pm}$ and Z^0 bosons become massive, while the photon remains massless. The universe is less than a billionth of a second old at this point in time.

Further expansion cools the 'gas' and, once the temperature falls below 1 GeV or so, the quark–gluon plasma undergoes a phase transition to the 'confined' (strongly interacting) phase. Almost all the quarks and antiquarks annihilate at this stage. However, in order to explain the presence of matter (including us!) in the universe, we require a tiny excess (1 part in 10^9) of quarks over antiquarks. (As we shall see later, GUTs can explain the origin of this excess.) These quarks are pulled together by the strong force to make protons and neutrons (baryons) and so 'nuclear physics' is born. The universe at this stage is barely 10^{-4} s old!

The universe now enters an ever more interesting environment in that it consists of 'matter' (baryons), submerged in a gas of radiation at a temperature of some 100 MeV, made up of electrons, neutrinos and photons which still dominate the energy density of the universe. At a temperature of a few MeV to 10^{-1} MeV, corresponding to a cosmic time of seconds to 100 s, the 'light' nuclei (D, ^{3}He, ^{4}He, ^{7}Li) are created. This is known as nucleosynthesis and is one of the success stories of the hot Big Bang cosmology. It is worth noting that the relative abundance of the light elements vary over a wide range, from about 25% for ^{4}He to 10^{-10} for ^{7}Li, with D and ^{3}He around 10^{-5} or so, and that

the Big Bang cosmology provides a consistent explanation of these numbers. It is believed that the 'heavier' nuclei are processed through nuclear reactions in stars.

The relative abundance of the light elements depends in an essential way on the fraction of critical density contributed by baryons, estimated to be around 3–10%. Among other things, this implies that the bulk of baryons are 'dark'! Moreover, their gravitational pull is not strong enough to ultimately halt the Hubble expansion. Such a Universe expands for ever and is not gravitationally 'closed'.

The expanding universe cools further and a particularly important era begins at a temperature of a few eV, when the cosmic time is about 10^{12} s. It is now 'cool enough' for the electrons to combine with the protons and neutrons to form atoms (often referred to as 'recombination'), thereby giving rise to atomic physics. With the charged particles bound up in neutral atoms, the photon gas is 'liberated' from its opaque environment and expands outwards, following an expansion law which makes its energy density decrease faster than that of matter. We have now entered the early stages of a 'matter-dominated' universe.

The discovery of this 'freed-up' photon radiation is one of the greatest triumphs of the hot Big Bang cosmology. At decoupling, the radiation temperature was a few eV (10^4 K). This temperature falls with the expansion of the universe (the photon wavelengths become stretched by the expansion which lowers the temperature), and today we expect to be 'bathed' in a sea of cosmic background photons with a temperature of just a few kelvins. In 1965 two physicists named Penzias and Wilson accidentally discovered this radiation in the radio-frequency range. By 1994 data from the COBE satellite enabled astronomers to verify that the cosmic background radiation (CBR) has a thermal origin, and to provide the best value for its temperature. It turns out to be 2.726 ± 0.005 K.

The CBR carries far more information than just news about its fiery past. For instance, it turns out that the CBR is isotropic to a high degree of accuracy. In other words, no matter which direction of the sky one studies, one measures 'essentially' the same temperature. Could it be that the early universe was somehow 'tuned' to be completely isotropic?

The answer must be an emphatic 'no' within the Big Bang framework. In order to explain the observed large scale galactic structure, we must invoke the presence of 'primaeval' density fluctuations in the early universe which, through gravitational instability, in an expanding universe, can eventually grow to become galaxies, clusters, etc. How large were these primordial perturbations?

Because matter and radiation were coupled together in the early universe, the perturbations must be sufficiently small in order to be consistent with the measured isotropy of the CBR. However, perturbations involving ordinary (baryonic) matter can grow, at a rate prescribed by GR, only after decoupling from radiation has taken place which, as stated earlier, happens during the era of atomic physics. It turns out that these two constraints are sufficient to rule out the simplest structure formation models with just baryonic matter.

A crucial piece of evidence for this last statement comes from the first CBR anisotropy detection provided by the COBE satellite in 1992. It detected primordial anisotropies in CBR at the level 10^{-5} (or so), thereby providing striking agreement with models of structure formation in which the matter content of the universe is dominated by two kinds of dark matter, referred to as 'cold' and 'hot' dark matter (CHDM), with baryons composing just 5–15% of the total matter content. Roughly speaking, one can think of 'cold' dark matter as particles that move sufficiently slowly to be captured in galaxies and ever-smaller-scale structures. 'Hot' dark-matter particles, on the other hand, move fast enough to escape this pull. They yield only to the pull of much larger structures which require a greater 'escape' velocity.

Primordial anisotropies on the order of 10^{-5} are much too small to yield the observed structure in a baryon-dominated universe. In the CHDM model, on the contrary, the primordial fluctuations start to grow well before the epoch of 'recombination', thereby providing more time to produce the observed large-scale structure. At recombination, the baryons decouple from radiation and 'fall' into the gravitational potential wells created by the CHDM.

Let us briefly review the hallmarks of the hot Big Bang cosmology.

(i) It accommodates the expansion of the universe observed by Hubble and others, according to which the nearby galaxies are receding from one another at a speed proportional to their separation.

(ii) It predicts the existence of CBR, a remnant of the primaeval fireball.

(iii) It explains the observed cosmological abundance of the light elements (D, ^{3}He, ^{4}He and ^{7}Li) through nucleosynthesis in an expanding radiation-dominated universe.

(iv) It can explain the observed large-scale structure provided that one assumes the presence of 'small' primordial density fluctuations which, through gravitational instability, eventually grow and turn into galaxies, superclusters, etc.

Despite these remarkable achievements the 'standard' Big Bang scenario, based on Einstein's GR and supplemented by the standard model of particle physics, fails to provide answers to a number of profound questions.

- The ratio n_b/n_γ, where n_b and n_γ denote the mean cosmological baryon and photon number densities respectively, is found to be about 10^{-9}–10^{-11}. How does this ratio arise? Recall that the baryons play a critical role in nucleosynthesis.

- The highly isotropic nature of CBR is not understood within the Big Bang framework. According to this model, regions in the sky separated by an angular separation of 2° or more could, in principle, display large variations in the background temperature, in stark contradiction to the uniform temperature (2.726 K) that is measured.

- The model fails to 'explain' how the mean density of the universe today lies so close to the critical density which separates a closed universe (which eventually collapses under gravity) from an open universe (which continues to expand for ever). Indeed, it can be shown that the initial density must have been very carefully adjusted to lie incredibly close to the critical value; otherwise the universe would not resemble what it is today. It would have collapsed a long time ago if the initial density was reasonably greater than the critical value. Similarly, for an initial value slightly larger than the critical value, it would reach a state today with density far below the critical value.

- There is no explanation in the model for the origin of the 'tiny' primordial density fluctuations that are needed to seed the formation of galaxies and other large-scale structure in the universe.
- The model fails to provide any explanation of the origin or cause of the Big Bang.
- The Standard Model of particle physics fails to provide any viable dark matter candidate without which structure formation is not viable. The simplest model of structure formation that is consistent with a large amount of diverse data, varying over length scales of around a megaparsec up to the present horizon size of several thousand megaparsecs, requires, as mentioned earlier, the existence of two types of (cold plus hot) dark matter in the universe.

These and other challenges to the standard Big Bang cosmology have aroused a huge amount of interest amongst both particle physicists as well as cosmologists during the last two decades. It is commonly believed that some of the deepest problems can be resolved only through the discovery of a new theoretical framework which provides a satisfactory merger of Einstein's gravity (or some suitable generalization thereof) with quantum mechanics. Superstring theories (see the chapter by Ross) provide the most ambitious attempt along these lines and may one day be able to answer the above questions. Here we content ourselves by exploiting a more modest approach inspired by GUTs which, nonetheless, yield a number of exciting and testable results.

As mentioned earlier, GUTs provide a 'natural' extension of the Standard Model by bringing together the strong and electroweak interactions within a single unified framework. In GUTs, the 'apparent' distinction at present energies between these forces becomes gradually blurred as we approach the GUT symmetry-breaking scale which typically is about 10^{16} GeV. These theories provide a number of experimentally testable predictions.

Quantization of electric charge

In contrast with the Standard Model, GUTs provide a nice explanation of why the electric charge in nature is apparently quantized in units of $e/3$, where e denotes the electron charge.

Baryon number violation

Many GUTs give rise to baryon number violation at a rate which can be exploited to explain the origin of the famous cosmological quantity $n_b/n_\gamma \simeq 10^{-10}$.

Lifetime of the proton

One of the most exciting predictions has to do with the lifetime of the proton. In many examples, one finds that the latter is unstable and decays at a rate which may be experimentally accessible at the Superkamiokande (largest undergound) detector in Japan. The predicted lifetime is about 10^{30}–10^{35} years. Models in which the predicted lifetime is less than 10^{32} (or so) years are already excluded.

Magnetic monopole

The GUT prediction of electric charge quantization is accompanied by the appearance of a new kind of 'extended' supermassive (GUT scale) particle known as a magnetic monopole. Dirac in 1931 had argued that the existence of a magnetic monopole, carrying magnetic charge, would provide a way to understand the observed quantization of the electric charge. Note that, analogous to electric charge, the magnetic charge must be conserved as a consequence of gauge invariance.

Because the monopoles are superheavy and therefore hopelessly beyond the reach of any forseeable accelerator, researchers turned their attention to the 'very' early (and presumably very hot) universe to try to see whether it could provide an arena for 'manufacturing' some. It was soon realized that the combination of GUTs and standard Big Bang cosmology (extrapolated to a GUT scale) was far too efficient (!) in producing 'primordial' monopoles during the phase transition from GUT to the standard model. Because of their huge mass (about $10M_{GUT} \simeq 10^{-6}$ g) and the relatively high predicted abundance, the primordial monopoles would quickly dominate the energy density of the universe, in gross contradiction with observations!

The question that one may ask is: should one give up GUTs or the 'naive' extrapolation of standard cosmology to the GUT scale? The latter alternative has turned out to be a far more productive line of research and, in the process, yielded the so-called inflationary cosmology which, in addition to resolving the monopole problem, has also shown how some of the conundrums of Big Bang cosmology listed earlier can be overcome.

The basic idea behind one inflationary scenario can be stated as follows. Let us suppose that the very early universe undergoes a phase transition from a grand unified symmetry to the Standard Model gauge symmetry which is strongly first order. In other words, it takes place after a certain amount of supercooling. In this case, it follows from Einstein's equations supplemented by the equation of state $p = -\rho$, where p denotes the pressure and ρ the 'nearly constant' energy density of the universe, that the universe undergoes a brief period of extremely rapid (exponential) expansion. A tiny (say atom size) speck of the universe can expand by thirty or more orders of magnitude during this exponential burst and exceed the size of the present universe! After this brief spell of inflation is over, the 'latent' heat is released and the universe reverts to the 'standard' radiation-dominated phase.

The exponential burst lasts for a tiny fraction of a second, say about 10^{-34} s, but the implications for cosmology are truly remarkable.

- All length scales, in particular the radius of curvature, are stretched by an enormous factor. This leads to the important prediction that the universe is spatially 'flat', with a mean density very near the critical density. Note that this 'explains' why the universe is so old! By being near the critical density the universe can 'live' for ever!

- The isotropy of the CBR, a major puzzle for Big Bang cosmology, is explained by the fact that the observed universe evolved from a tiny speck, in thermal equilibrium or in causal contact, that experienced exponential expansion.

- The monopole problem of GUTs is resolved in essentially the same way. Monopoles are 'topological defects' and inflation 'smooths out' such regions in the universe. Whether some survive inflation is a major topic of investigation.

- Last but not least, tiny quantum fluctuations in a certain field that 'drives' inflation can evolve through the inflationary phase to turn into primordial density fluctuations that subsequently lead to the formation of galaxies and other large-scale structures in the universe.

How do we test inflationary cosmology in the same way that standard Big Bang cosmology has been experi-

mentally tested through the CBR and nucleosynthesis? The situation can be compared with the example of 'gauge invariance'. The principle of gauge invariance remains an 'attractive' idea unless it is implemented within the framework of some models which can be tested. The most well known examples are given by the electroweak model and quantum chromodynamics. To test inflation we need to put forward realistic inflationary models with testable predictions. Challenges include the following.

- What is the mean density of the universe? In the simplest models it is predicted to be the critical density, but some very recent measurements suggest a value considerably below that. If this is confirmed, a major revision of the 'standard' inflationary cosmology would be required.

- What is the dark matter content of the universe? Is there more than one type of dark matter? What are the best 'cold' and 'hot' dark matter candidates?

- What are the characteristic features of the primordial density fluctuations? How do they impact the CBR anisotropies in terms of magnitude and angular dependence? Two recently approved satellite-based experiments known as MAP and PLANCK will test this prediction of inflation models.

- What is the present value of the cosmological constant? This turns out to be a particularly noteworthy opponent in that inflation so far has failed to answer the fundamental question: is 'empty' space truly weightless? By 'empty' space we mean the quantum ground state on which rests the entire framework of the underlying physical theory.

- Finally, in inflation and beyond we would like to address such fundamental issues as the origin of or what preceded the Big Bang, etc. We shall defer discussion of such topics to the new millennium!

ABOUT THE AUTHOR

Qaisar Shafi is a Professor at the University of Delaware's Bartol Research Institute. His principal research interests lie in the areas of high-energy physics and cosmology and he is one of the directors of the Annual Summer School covering these subjects at the International Centre for Theoretical Physics, Trieste.

19 RECENT ADVANCES IN ASTRONOMY

A C Fabian

Editor's Introduction: Although astronomy is one of the oldest of sciences, this century has seen the opening of new frontiers in space exploration. Special missions have produced dramatic images of distant parts of the solar system, huge telescopes have peered into the depths of the Galaxy and beyond, showing the distant Universe as it has never been seen before, and ultra-sensitive instruments have picked up faint signals from the dawn of time.

At the same time, the increasing interaction between physicists and astronomers has made the two disciplines more conscious of each other's objectives. Physics and astronomy are now working in tandem to answer the ultimate questions about the origin and structure of our Universe and producing convincing answers. In this chapter, Cambridge astronomer Andy Fabian looks at the current status of astronomy from the viewpoint of physics.

[Most of the figures which accompany this contribution are included in the colour plate section.]

INTRODUCTION

This chapter is a broad overview of some recent advances in astronomy. Particular attention is paid to those areas which have the most direct overlap with physics. Of course, one of the reasons for doing astronomy is to explore those parts of physics which are inaccessible in the laboratory. One good example is the study of gravity which is the dominant large-scale force throughout the Universe. Astronomers explore mass and distance scales which are impossible in the laboratory and, at the extremes, densities that range from the highly tenuous gas in intergalactic space to the nuclear matter density of neutron stars, temperatures of billions of degrees and magnetic fields exceeding 10^8 T. Relativistic motions of bulk matter are commonplace in some situations, with Lorentz factors of tens to hundreds. The highest-energy cosmic rays exceed 10^{20} eV.

Astronomy has generally been driven by observations. There are of course stranger things out in the Universe than we could ever imagine. Many of the extremes in the Universe are not readily observable by ground-based optical telescopes but have required the development of radio and then space-based techniques. Only through space-based observation can the ultraviolet, the x-ray and gamma-ray bands and much of the infrared be explored owing to absorption of the radiation by our atmosphere. As amply demonstrated by the Hubble Space Telescope, observations in the traditional optical bands are also transformed from space, where the seeing effects of our atmosphere are eliminated and the sky is much darker.

The Solar System

All the planets have now been explored thoroughly by Earth-based instruments and most have been visited by at least passing space missions. The early heyday of this work, in which large probes were sent to many of the planets, has now been replaced to some extent by NASA's new approach of 'better, faster, cheaper'. The planetary bodies provide us with conditions different from those on the Earth, not so different from those that could obtained in the laboratory, but on far greater scales in terms of the physical size and lengths of time involved. (See plates 6–8.)

Our knowledge of the Solar System has expanded in the last decade through the detection of objects in the Kuiper belt which lies just beyond the orbit of Neptune. The objects are of the order of hundreds of kilometres in size and may number more than a million. They demonstrate that the (smoothed-out) mass density of the solar system does not stop at the orbit of Neptune but continues to much larger radii. Much further out is the Oort cloud where comets originate.

An exciting recent development has been the discovery of other planetary systems. The first of these was found to orbit a pulsar and has three planets with masses and orbital radii in similar ratios to those of Mercury, Venus and Earth about the Sun. A pulsar, which is the remnant of a supernova explosion, is an unusual host to a planetary system. The existence of at least one such system does, however, emphasize the ease with which planets form, in this case presumably from some cooled debris of the explosion.

Planets around solar-type stars have also been found. Surprisingly they tend to be of Jovian mass (about 1000 times the mass of the Earth and a thousandth of the mass of the Sun), and yet with orbital periods of only a few weeks around the companion star (the orbital period of Jupiter about the Sun is 12 years). Also found are more massive dark objects orbiting around some stars in highly elliptical orbits. These objects are 'seen' via their effects on the orbital period and the radial velocity of the companion star, determined from the Doppler effect on the spectral lines or pulses of the central object produced in response to the orbital motion of the planets. Large radial velocities (large here means walking speed for pulsars and racing car speed for stars) and short orbital periods are the most readily evident. With time the parameter space searched will expand to include Earth-like masses. None of the planets has yet been directly observed and it will be some years before they can be.

THE SUN AND STARS

The detection of solar neutrinos over the last thirty years, from underground particle detectors, confirms that the Sun is powered by nuclear fusion. There remains a problem however. The detected solar neutrino fluxes from all the experiments (Homestake, Sage, Gallex and Kamiokande) are a factor of two or so smaller than the predictions of the standard solar model. This model is based upon the laws of stellar structure and through numerical computation takes into account the distribution of pressure and density in a self-gravitating star, the generation of nuclear fusion and the diffusion of the radiation out through the star, either by radiation or by convection, and, of course, the changes in the molecular weight that occur as the star evolves. The origin of the solar neutrino discrepancy may lie in neutrino oscillations, which could be an example of astrophysics guiding particle physics.

The interior structure of the Sun has been probed by helioseismology, which is the study of the propagation of acoustic oscillations in the Sun observed through perturbations of the photosphere. Results now give the distribution of helium, the distribution of sound speed and the rotation speed throughout the Sun. The core of the Sun rotates relatively slowly. As for the surface of the Sun, we know that it is magnetically active. A small fraction of the solar power emerges in solar flares and also powers the corona. The precise mechanism by which solar flares convert magnetic energy into particle energy with resultant emission of x-rays, gamma-rays and radio waves is still problematic, but observations with satellites such as the Solar Maximum Mission and Yohkoh have helped the understanding of this process of magnetic reconnection.

There is a small wind from the Sun, of the order of 10^{-14} $M_{\odot}$ year^{-1} ($M_{\odot}$ = solar mass) which washes out over the planets to the heliosphere where it interacts with the interstellar medium. In the lifetime of our Sun, in its present hydrogen-burning phase, the solar wind will take away very little of its mass, but in more evolved, or more massive, stars a stellar wind takes away a very significant fraction of the mass of the star. It is understood in terms of radiation and thermal pressure and its overall effect is uncertain, although it can be very significant.

The evolution of individual stars is fairly well understood apart from the uncertainties of stellar mass loss. A star undergoes hydrogen burning when it has exhausted the hydrogen in its core, which is the innermost 10%, the core then collapses and the outer part of the star swells up to become a red giant. Eventually, either the core of the star collapses to form a white dwarf or there is a supernova explosion and the core forms a neutron star or black hole. (See plates 9 and 10.)

NEUTRON STARS

There is considerable interest in stellar remnants. Neutron stars, in particular, are wonderful laboratories for physicists.

Much is known about neutron stars because they spin rapidly and have strong magnetic fields which cause them to emit pulses. The precise origin of the pulses, particularly in the radio band, is not thoroughly understood, but the clock-like behaviour of pulsars allows very precise measurements to be made. Timing observations are of the greatest importance throughout astronomy and many of the most precise measurements derive from time measurements. Some pulsars are much more precise than single atomic clocks in the laboratory. Pulsars spin down steadily owing to emission of electromagnetic dipole radiation apart from some abrupt changes or 'glitches', which are probably changes due to the superfluid vortices within the core of a neutron star interacting with its crust as the star tries to readjust to a reduced spin speed. Much of the radiation seen from pulsars is due to electron–positron pairs created by large voltages occurring over the polar caps of the stars (e.g. 10^{12} V). In the radio band the radiation must be extremely coherent.

Many precise observations of stars and stellar remnants, such as masses and radii, derive from observations of binary stars, particularly those which eclipse. Roughly one half of all stars in the sky are members of binary systems, a third are members of triple systems and so on. Binary systems appear to be formed with a variety of separations, perhaps a uniform distribution, and it is those which form close binaries that are often of most interest. In that case as the more massive star, which is the one to evolve first, expands to form a red giant, it can fill the equipotential surface which is common to both stars and so cause matter to spill onto its companion.

This process of mass transfer can lead to many interesting phenomena, particularly if one of the stars has evolved to a compact remnant: a neutron star, white dwarf or a black hole. Of course, the first star to evolve into a remnant will have been the more massive in the original system but after mass transfer it is the less massive. When matter is transferred onto a compact star the gravitational potential well down which it falls is very deep, which means that an enormous amount of gravitational energy is released. This accretion power can make the system highly luminous and is responsible for some of the most luminous objects in our galaxy. It is also the ultimate source of the power of quasars. Throughout the galaxy there are many x-ray binaries where

mass transfer occurs from a companion star onto a neutron star or black hole. Thermodynamic considerations require that the high luminosities seen from regions as small as the surface of a neutron star are predominantly emitted as x-rays.

Mass transfer in a binary system means that the accreting matter has significant angular momentum which causes it to orbit the compact star. Friction between matter at different radii then causes most of the matter to spiral inwards and the angular momentum to be transported outwards, thereby forming a disc. The accretion of matter and angular momentum from the accretion disc by the neutron star at the centre of the disc causes it to spin up.

In some cases a neutron star which has spun down as a radio pulsar has then been spun back up by accretion as an x-ray source. Provided that the surface magnetic field has decayed, it can spin up until its period is a few milliseconds. Such an object is of about a solar mass, with approximately 10^{57} neutrons together with a smaller number of protons and electrons, and a diameter of about 20 km (the size of a city), all spinning at about 1000 revolutions per second. Studies of the properties of neutron stars, particularly when the accreting matter undergoes nuclear flashes, enable the surface radius and other properties of a neutron star to be determined. With measurements of neutron star masses, the mass–radius relationship for neutron stars is obtained, which is a handle on the equation of state of nuclear matter at densities above that of an isolated atomic nucleus.

Some close binaries consist of two neutron stars orbiting each other and are observed if at least one is a pulsar. One such system was discovered early in the 1970s and is known as the binary pulsar. The companion is also a neutron star (but not seen as a pulsar). The importance of the system lies in the pulsar behaving as an accurate clock orbiting in a deep potential well. The orbital period is about eight hours and the orbit eccentric. This allows precession of the orbit to be observed. A general relativistic effect, it is about 48 arcsec per century for the orbit of Mercury about the Sun, and $4.2°$ per year for the binary pulsar. More importantly the period of the binary system is seen to decrease slowly, meaning that the two objects are spiralling closer together and so moving faster and faster. The rate of decrease is directly in agreement with that predicted by the emission of gravitational radiation and so verifies its existence. Over the next decade or more it should be directly measured, not necessarily from this object but from other binary systems and also from stellar collapses during supernovae.

The fate of a system such as the binary pulsar, when the orbit of the two neutron stars is finally ground down over a hundred million years until the two neutron stars touch, is a gigantic explosion. About 10^{51} erg is released in a fireball, some of which may be detectable over cosmological distances. There is now good evidence that such fireballs are the origin of gamma-ray bursts, first discovered about 30 years ago. (See figure 19.1.)

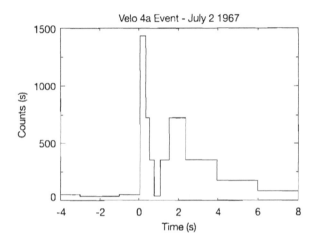

Figure 19.1. The first gamma-ray burst detected, by the Vela 4a satellite launched to monitor terrestrial nuclear tests. The gamma-ray flux arriving at the Earth during the brightest bursts can briefly rival the optical flux from the brightest stars. Following recent studies of the bursts by the BATSE detectors aboard the Compton Gamma-Ray Observatory, and with bursts found by the SAX satellite optically identified by the ground-based William Herschel and Keck telescopes, it seems that these bursts originate at cosmological distances, far outside our Galaxy. This burst may have occurred over 5 billion years ago. If the emission in some extreme bursts is isotropic, the total energy release can exceed 10^{53} erg.

SUPERNOVAE

Neutron stars and black holes form in supernovae. We were particularly lucky that in 1987 a supernova occurred very nearby, in our satellite galaxy, the Large Magellanic Cloud. Supernova 1987A was the first for which the neutrino flash was detected, by large underground detectors originally looking for proton decay. This demonstrated that the collapse energy was mostly emitted as neutrinos, about 10^{53} erg in total, and was probably due to the formation of a neutron star. Now, as the remnant has cooled, we do not see any evidence of that neutron star as a pulsar, or through any additional energy that is being given out. It may have since collapsed into a black hole. The gaseous remnant carries about 1% of the energy of the explosion and is seen to expand outwards with velocities of $(5-10) \times 10^3$ km s^{-1}. Most is now about half-way out to an enigmatic surrounding ring. (A small fraction is already interacting with it.) The ring may be some remnant of the early evolution of that star or may be because it had been a member of a binary system. (See plate 11.)

Supernovae are predicted to occur in our own Galaxy about once every thirty years. Most, however, occur deep in the plane of the Galaxy where they are hidden from direct view by dust clouds and so have not been detected. As space instruments become more sophisticated and monitor the Galaxy continuously, then galactic supernovae may be seen directly through gamma-rays and other radiation. In the case of supernova 1987A the decay of the observed brightness for the first 1.5 years followed precisely what was expected for the radioactive decay of ^{56}Co (half-life of 77 days) made by the explosion. This was supported by direct observation of cobalt emission lines with gamma-ray detectors. As a last comment on supernova 1987A, the arrival of all the neutrinos within 20 s provided a useful limit on their mass.

Another class of supernova occurs because a white dwarf has been pushed over the Chandrasekhar limit in a close binary system through mass transfer. These type Ia supernova explosions do not leave a compact remnant but the binding energy of the equivalent of a neutron star is emitted as neutrinos and kinetic energy. As the physics of the process implies that all such explosions are very similar, it is hoped that they can be used in cosmology as 'standard candles'.

GALAXIES, CLUSTERS AND DARK MATTER

Studies of individual galaxies have demonstrated the presence of much dark matter around them. The rotation velocity of gas and stars within many galaxies, measured from Doppler shifts of spectral lines, is roughly constant beyond the galaxy core out to large radii, unlike that predicted if mass followed light within a galaxy; the mass of a galaxy must be much more extensive than the light.

The existence of dark matter has been suspected for over 60 years in the context of clusters of galaxies. A few per cent of galaxies occur in large clusters which may have several thousand members. If the mass from all the individual stars in a cluster (deduced from the total light) is compared with the mass derived from application of the virial theorem to the motions of the galaxies, then a discrepancy of about a factor of 10 is found. There is 10 times more mass in the cluster than can be seen in the individual stars in the member galaxies. Some of this dark matter is in a hot intra-cluster medium lying between the galaxies, seen by its x-ray emission. It accounts for about 20% of the total mass of a cluster so that with the stars the fraction observed is about 25–30%. The remainder appears not to radiate in any form and is probably not composed of baryons. (See plate 12.)

Much of it is probably the same dark matter which constitutes the bulk of the mass of our Universe. If one makes a similar approach to the mass of the total Universe, where the virial theorem is replaced here by the equations of a Friedmann–Robertson–Walker Universe, the discrepancy between the mass observed in galaxies and the critical density required if the Universe is just bound, i.e. stops expanding at infinite time, is about a factor of 100.

The nature of dark matter remains a dominant puzzle in astrophysics and cosmology. It is most evident on the largest scales and yet probably has an explanation on the smallest, in exotic non-baryonic matter such as theorized by particle physicists and formed in the earliest phases of the Big Bang.

GRAVITATIONAL LENSING

One prediction of general relativity is the bending of light. This was confirmed in 1919 when the positions of stars close to the Sun observed during a total solar eclipse were compared with their relative positions 6 months earlier. This lensing effect of gravity can also be seen in distant objects which lie behind nearby objects in the Universe.

For example, some quasars are seen to be double because of lensing by an intervening galaxy.

Spectacular examples have been found over the last 10 years of gravitational lensing by clusters of galaxies. (See plate 13.) The large mass of dark matter in a rich cluster of galaxies can often achieve the conditions necessary to lense background galaxies. One of the most distant known objects is a galaxy with a red shift of 4.92 which has been made visible as a consequence of being lensed by a foreground cluster.

Of course individual stars can produce lensing but on a very small scale. Such microlensing has been observed within and around our own Galaxy. It is seen in the dense star fields near the Galactic Centre when one out of millions of stars briefly brightens in a characteristic manner. Very large numbers of stars must be monitored for an event to be seen but, with digitized imaging on wide fields and powerful computers to scan the images, a hundred or more such events have now been detected. An important set of microlensing events has been seen in stars of the Large Magellanic Cloud, our nearby satellite galaxy, which indicates that some of the dark matter in our galactic halo is due to objects of mass 10–50% that of the Sun. These are called massive compact halo objects (MACHOs). Within observational uncertainties, the fraction of the mass of the halo mass of our Galaxy in MACHOs is consistent with them being baryonic. Exactly what they are is debated.

BLACK HOLES AND QUASARS

Of much current interest is the discovery of massive black holes in the nuclei of many galaxies. These were certainly expected because extremely luminous sources of radiation, namely quasars, have been known for the past 30 years. The era of quasars was when the Universe was maybe 3–5 billion years old; it is now about 13 billion years old. The luminosity of a quasar outshines the 10^{11} stars of the host galaxy by factors of tens to hundreds. The central engine must be very massive, $(10^7–10^{10})$ $M_{\odot}$, in order that gravity can stop it from being blown apart by the enormous pressure of radiation. Variations in the luminous output of quasars are seen on time scales of days to weeks, indicating that the emitting object is very compact for such high masses. The observed broad-band spectrum, extending from the radio through the infrared, optical, ultraviolet to x-rays and in some cases to gamma-rays, indicates that non-stellar processes are involved. The simplest way to power such objects is for interstellar matter to fall into a massive black hole in the centre of the galaxy. Accretion like this onto a black hole can liberate 10% or more of the rest-mass energy of the matter as radiation. (See plates 14 and 15.)

Dead quasars are a significant observational issue. The large numbers of quasars in the past suggest that their remnants should be relatively common in galaxies today. Our own Galactic Centre has a black hole of about $2.5 \times 10^6 \ M_\odot$, measured from Doppler measurements of gas and stars and recently from studies of the proper motions of stars very close to the Galactic Centre. (The proper motion of a star is the motion tangential to the line of sight and is often measured from images spaced apart in time.) The motion of the stars in the Galactic Centre is sufficiently large that complete orbits of some of the stars may be seen over the next 20 years. The Galactic Centre is not luminous in the quasar sense but is highly quiescent, despite the large amounts of gas, dust and other material around it. (It may appear quiescent because the radiative efficiency of a weak accretion flow can be so low that the black hole swallows both the matter and the gravitational energy.) Other galaxies, e.g. the nearby Andromeda Galaxy, have more massive black holes of about $10^8 \ M_\odot$ and the nearest giant elliptical galaxy, M87, has one of $3 \times 10^9 \ M_\odot$. They are inferred to be present in about one third of all galaxies.

It is difficult to see and make measurements of stellar motions around luminous quasar-like objects because the quasar light completely swamps that from the stars. Progress has been made in two ways. One is by observations of water maser emission from a disc of material orbiting at, say, 100 000 gravitational radii ($10^5 \ GM/c^2$), provoked by the high radiation field of the central engine. This approach gives the most precise measurements of the mass of the central objects but is rare (there must be a disc and it must be seen almost edge on). The second uses x-ray spectrometers, such as on ASCA, which detect emission from matter orbiting very close to the central black hole.

The material accreting onto the black hole is likely to have some angular momentum and thus to be in a disc. Most of the energy is released close to the black hole, at

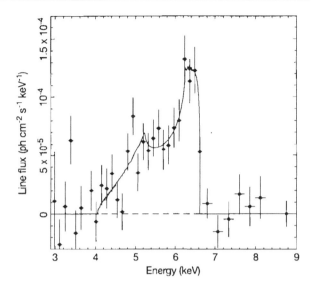

Figure 19.2. ASCA satellite x-ray spectrum of fluorescent iron line emission from the nucleus of the active galaxy MCG-6-30-15. A power-law continuum has been subtracted. The emission line occurs at 6.4 keV in the laboratory and should appear at 6.35 keV when the small cosmological red shift of the galaxy is taken into account. The spectral resolution of the charged-coupled device used to make the spectrum is about the width of the bars shown. The emission drops sharply at about 6.5 keV and a long tail continues to about 4 keV. The curve indicates the expected emission from matter in a disc extending down to six gravitational radii about a black hole. Most of the emission is from within 20 such radii. The line profile clearly shows the effects of relativistic beaming and gravitational red shift due to the matter orbiting so close to a black hole.

10–20 gravitational radii. The enormous release of energy from the innermost radii of a luminous accretion disc again implies, from thermodynamic considerations, that much of the radiation is in the far ultraviolet or in the x-ray band. (See figure 19.2.) In practice a significant fraction of the emission comes from flare-like activity above the accretion disc. The x-rays irradiate the disc, causing fluorescence of iron in the disc. It is this x-ray fluorescence line which is a diagnostic of the near environment and strong gravity of the black hole. The profile of that line indicates the presence of large gravitational red shifts and transverse Doppler shifts. There is tentative evidence from such observations that at least one of the black holes is very rapidly spinning, i.e. is a Kerr black hole. This work is only in its infancy now but is turning black holes and strong gravity into an observational science. Over the next few decades much more precise measurements of spectral variability and other properties will enable the mass of the black hole and its geometry to be determined and theories of strong gravity to be tested.

Jets

A major puzzle concerns the jets seen in about 10% of quasars and lower-power active galaxies. Throughout the Universe, a fraction of most classes of accreting objects appear to be able to produce jets of matter readily. (See plate 16.) These have been known for decades and yet our understanding of what powers them, what collimates them and, in the case of quasar jets, whether they are composed of electrons and protons or electrons and positrons, is still rather slim. Observationally they are well studied. Two jets of matter emerge from either side of the compact object. The matter is highly collimated and moves relatively rapidly. In many cases, the motion is relativistic and provides beautiful macroscopic demonstrations of special relativistic effects. The jet side moving towards us is brighter because the radiation emitted by the jet is beamed in our direction and is blue shifted. The jet on the other side is much fainter, often undetectable, because its radiation is beamed away from us. Even within our own Galaxy there are some jet-like systems which are small versions of the large ones in quasars. In the case of SS433 the jet direction moves in a cyclical way, making plain a wide variety of special relativistic effects.

One particular relativistic effect occurs if there is a disturbance in a jet pointing close to our line of sight. This is the appearance of superluminal motion; the disturbance appears to move with a velocity exceeding that of light. It is of course a relativistic illusion. There is no relativity problem if all components of the motion are taken into account. The effect was actually first observed in 1901 when a nova explosion occurred and light was seen to be reflected off a nearby interstellar cloud but was not understood at the time. (See plate 17.)

Jets are also seen from young stars or protostars. Star formation is a complex process because it involves turbulence, magnetic fields and accretion from giant magnetized molecular clouds. It is likely that much of a young protostar is accreted from a surrounding disc and in this case the jets are not relativistic. As a rough guide the jet velocity is of the order of the escape velocity from the central object. In the case of jets in active galaxies much of the energy is dumped into particle acceleration at the ends of the jets producing the radio lobes. They constitute the radio galaxies which have long been studied by radio astronomers.

LARGE-SCALE STRUCTURE

The past 15 years have seen the exploration of the distribution of galaxies closest to us. Observers, using ground-based techniques, map the positions and red shifts of individual galaxies brighter than some flux limit and construct three-dimensional maps of our surroundings. These now extend out beyond red shifts of 0.1 and are expected to extend to red shifts of 1 or so over the next ten years. They show that the distribution of galaxies is not uniform but is frothy. It consists of walls, voids and filaments. A cluster of galaxies is often found where walls and filaments intersect. (See figure 19.3.)

Much of the frothiness of the galaxy distribution is understood in terms of the gravitational instability of cold dark matter. The length and mass scales of the unstable regions have been steadily increasing with time. The normal baryonic matter was coupled to the radiation (now seen as the microwave background) for the first million years and could only fall into the dark-matter clumps thereafter. The galaxies and clusters that we see today are the result of this process which has happened in a hierarchical manner on galaxy scales since the Universe was about a billion years old (i.e. a red shift of about 5). The peak of galaxy formation took place at a red shift of about 2, at a similar time to the peak era of quasars.

It is unlikely that any large galaxy formed in one go; it probably took many billions of years. The gravitational energy released when gas collapsed in protogalaxies must have heated much of that gas. The visible stellar mass of galaxies is then that gas which has been able to cool.

A very striking example of distant galaxies can be seen in the Hubble Deep Field. (See plate 18.) The Hubble Space

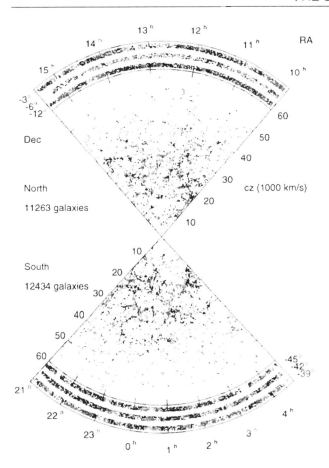

Figure 19.3. Galactic froth: the positions of 26 418 bright galaxies measured from the Las Campanas Observatory in Chile. The galaxies lie in narrow strips of the sky (seen top and bottom). The new measurements give the red shift of each galaxy so that they can be plotted in space. Occasional radial streaks arise because several galaxies are in a cluster, where motions within the cluster far exceed the local (Hubble) flow. The most distant galaxies shown have a red shift of 0.3 and are at about 3 billion light-years.

Telescope has observed that region very deeply through several filters. This enables the galaxies to be classified and, where they are very distant and absorption by intergalactic clouds takes place, red shifts and thus ages to be estimated. The mean star formation rate in the Universe as a function of time has been deduced using these and much other data.

THE EXPANSION RATE AND AGE OF THE UNIVERSE

The scale of the Universe is determined by the Hubble constant H_0 defined by $v = H_0 d$, where v is the cosmological recession velocity of a galaxy (i.e. the velocity deduced from the red shift) at distance d. The inverse of the Hubble constant is (to within a factor close to unity which depends on the cosmology) the age of the Universe. Obtaining the Hubble constant is difficult. Measuring the red shift of a galaxy is straightforward, but finding some other way to measure its distance is complicated. A complete distance ladder has been built up where we make use of measurements (e.g. radar) within the Solar System, parallax measurements to nearby stars, statistical methods extending to more distant stars and then, having found some so-called 'standard candles' amongst distant stars, the ladder can be extended to other galaxies. A major problem is that an error at one point means that everything beyond is wrong.

Perhaps the most observationally powerful standard candles are Cepheid variable stars. These vary regularly in brightness over days to weeks. The period of the variation is directly related to their intrinsic brightness. What is needed is an accurate distance to a few of them in order to calibrate the scale for all. Space observations with the Hubble Space Telescope now resolve Cepheid variables in the nearest cluster to us and the Hipparcos astrometric satellite has enabled an accurate calibration of the Cepheid distance scale in our Galaxy to be obtained. Together this has led to a several-fold increase in the accuracy of measurement of the Hubble constant. This and many other methods (e.g. using supernovae or brightest cluster galaxies as standard candles) now indicate that

$$H_0 = 60 \pm 10 \text{ km s}^{-1} \text{ Mpc}^{-1}.$$

Until very recently there was a possible problem with the age of the Universe since the age of stars in galactic globular clusters (tight clusters of about 100 000 coeval stars which can be dated accurately and are amongst the oldest known) appeared to exceed that of the Universe for most preferred cosmological models. A similar problem has occurred before in the history of the Hubble constant and was resolved by a change in the distance scale of

Cepheids. The Hipparcos measurements show that the globular clusters are slightly further away, which make their stars more luminous. Because more luminous stars burn hydrogen faster, they must be younger. Younger globular clusters and a larger distance scale (and thus smaller Hubble constant) make the Universe older, so resolving the problem.

THE MICROWAVE BACKGROUND

On the largest scales, much of the current Hot Big Bang Universe model is supported by observations of the microwave background. This is the radiation from the fireball of the Big Bang which has been expanding with us and has now cooled to a temperature of 3 K. The COBE satellite launched in 1990 has provided the most accurate measurements to date of the spectrum of the microwave background and of its smoothness over the Universe. The spectrum is a precise fit to a black body of 2.73 K. It is slightly hotter over one half of the sky compared with the other half of the sky by a fraction of a per cent owing to aberration and frequency shifts induced by our motion through the microwave background. This motion is the vector sum of our orbit round the Sun, of the Sun around the centre of our Galaxy, of the Galaxy in the Local Group, and so on. Most is due to gravitational interactions between local large-scale mass enhancements (e.g. superclusters). (See plate 19.)

At the level of 10^{-5}, the whole microwave background sky shows significant variations. These fluctuations are related to the origin of large-scale structure. They appear to fit a scale-free fluctuation spectrum and to support an inflationary Universe. This explains away many of the early problems in cosmology to do with why the Universe is so flat, and why different parts of the microwave background sky which have been out of causal contact since the Big Bang are so similar. COBE measured the fluctuations in the microwave background on scales down to about $10°$. Ground-based experiments are now measuring them on smaller scales yet, particularly trying to get down to the scales within which the individual parts of the microwave background as we see them were in causal contact and there are fluctuations due to acoustic oscillations of the gas and trapped microwave background photons. The angular

scales of the peak amplitude of these fluctuations will enable many cosmological parameters (e.g. H_0, Ω_0 and Ω_b) to be measured with fair precision, particularly with the next generation of space-based detectors. (See plate 20.)

THE DENSITY OF THE UNIVERSE AND OTHER PROBLEMS

Support for the standard Hot Big Bang theory is also obtained from the relative abundances of the light elements. In the first few minutes of the Big Bang, helium and traces of lithium and deuterium were formed from hydrogen. All the other elements have been formed in stars. The rate of expansion, and density, of the Universe at the time of helium formation dictate the relative abundance of these light elements. Comparison of the observed relative abundances with predictions from cosmic nucleosynthesis thus pins down some of the properties of the Universe at that time. In particular it indicates the fraction of the closure density of the Universe that can now be in baryons: $0.05 \lesssim \Omega_b \lesssim 0.1$.

A simple comparison of that ratio with the fraction of the mass of a cluster which is baryonic, assuming that the baryonic-to-dark-matter ratio within an object as large as a cluster is similar to that of the whole Universe, suggests that the total fraction of the closure density which is in any sort of matter is $\Omega_0 \approx 0.3$. The Universe is therefore open and will continue to expand for ever.

We have, however, ignored here a contribution due to neutrinos, which may have a small but finite mass, and which may not cluster exactly as assumed. They are, however, unlikely to make $\Omega_0 = 1$.

There are both physical and philosophical reasons for supposing that our Universe is a critical one, such as inflation, and why Ω is so close to, if not precisely equal to, unity. In that case, either one of the above suppositions is incorrect or there is a cosmological constant, $\Lambda = 3H_0^2\Omega_\Lambda$, so that $\Omega_0 + \Omega_\Lambda = 1$. Such a term in the cosmological equations implies that the vacuum has an energy density.

EPILOGUE

The night sky is the largest physics laboratory known or indeed possible. The view from the ground is often cloudy and murky and at many wavelengths opaque. Much can be done from the right site, but it improves considerably from space. Athough we cannot actually carry out experiments on most of the objects to be seen, there is so much variety that often something somewhere is behaving in a useful manner.

Most of what is out there is of unknown form, dark matter, the nature of which will continue to be next century's goal. Vacuum energy, which appears through a cosmological constant, may further indicate our ignorance and open up new areas of physics. Neutrons stars and black holes, the near effects of which are largely unobservable in the optical band, allow us to test our knowledge and understanding of the properties of dense matter and strong gravity.

Such a giant laboratory waits to be explored by astronomers, by physicists and, perhaps soon, by biologists.

SOURCES

The figures from this contribution which appear in the colour section are from the following sources.

Plates 6 and 7: NASA/JPL/CALTECH.

Plate 8: Sojourner™, Mars Rover™ and spacecraft design and image © 1996–97 California Institute of Technology. All rights reserved.

Plates 9, 10, 11, 13, 14, 15, 17, 18, 19 and 20: AURA/STScI.

Plate 16a: WSRT.

Plate 16b: NRAO.

FURTHER READING

Rees M 1997 *Before the Beginning—Our Universe and Others* (New York: Simon and Schuster)

Begelman M C and Rees M *Gravity's Fatal Attraction: Black Holes in the Universe* (New York: Freeman)

Shu F 1992 *The Physical Universe* (University Science Books)

Weinberg S 1983 *The First Three Minutes: a Modern View of the Origin of the Universe* (London: Fontana)

Thorne K 1994 *Black Holes and Timewarps* (Papermac)

Crosswell K *Planet Quest* (Oxford: Oxford University Press)

Shapiro S L and Teukolsky S A 1983 *Black Holes, White Dwarfs and Neutron Stars: The Physics of Compact Objects* (New York: Wiley–Interscience) (textbook)

Luminet J-P 1992 *Black Holes* (Cambridge: Cambridge University Press)

ABOUT THE AUTHOR

Andy Fabian has been a Royal Society Research Professor at the Institute of Astronomy in the University of Cambridge since 1982. He obtained his PhD in 1972 for rocket studies of the cosmic x-ray background. His present research interests lie mostly in extragalactic astronomy, particularly in active galactic nuclei and clusters of galaxies which he studies in the x-ray and other wavebands. He has worked on data from most x-ray astronomy satellites and combines those observations with theory and interpretation. Recent discoveries on which he has collaborated include the very broad iron emission line seen from matter orbiting close to black holes in active galaxies and massive cooling flows in clusters of galaxies. He was elected a Fellow of the Royal Society in 1996.

SUBJECT INDEX

Name Index